THE MICRO-HYDRO PELTON TURBINE MANUAL

The Micro-hydro Pelton Turbine Manual

Design, Manufacture and Installation for Small-scale Hydropower

JEREMY THAKE

Practical Action Publishing Ltd
27a Albert Street, Rugby, CV21 2SG, Warwickshire, UK
www.practicalactionpublishing.org

First published 2000
Reprinted 2007, 2009, 2010, 2011, 2014

ISBN 13 Paperback: 978 1 85339 460 7
ISBN Library Ebook: 978 1 78044 551 9
Book DOI: http://dx.doi.org/10.3362/9781780445519

A catalogue record for this book is available from the British Library.

Since 1974, Practical Action Publishing has published and disseminated books and information in support of international development work throughout the world. Practical Action Publishing is a trading name of Practical Action Publishing Ltd (Company Reg. No. 1159018), the wholly owned publishing company of Practical Action. Practical Action Publishing trades only in support of its parent charity objectives and any profits are covenanted back to Practical Action (Charity Reg. No. 247257, Group VAT Registration No. 880 9924 76).

Typeset by Dorwyn Ltd, Rowlands Castle, Hants

CONTENTS

PREFACE

The aim of this book is to give readers with a general workshop or engineering background, but with no experience of water turbines, sufficient information to allow them to build a small Pelton turbine. It is the book I needed when I started working in micro-hydro some years ago, but which I could not find.

Information on Pelton turbines is somewhat scarce. Pelton technology is well-developed, but the knowledge is mostly within large commercial companies, who like to keep it in-house. The information that is published seems to fall into three categories: academic articles, descriptions of individual installations, and textbooks on hydropower. Academic articles detail advances in theory, materials, and construction. This is interesting, but not much help if you have never built a turbine before. Articles describing individual sites, be they do-it-yourself micro-hydro schemes, or a power-station of many hundreds of megawatts, may be useful, but they are of limited help if your site is different. Textbooks on hydropower do have sections on Pelton turbines, but these books are rather too general. Many standard books are also old and out-of-date. Old turbine designs will often have quite different features to modern ones, and it pays, when you find an author making recommendations or saying such things as 'The optimum dimensions are . . .', to check the publication date.

The term 'micro-hydro' is used in the title of this manual because the book is primarily about small turbines. Definitions vary, but 'micro-hydro' typically refers to plants with powers of up to 100kW. Pelton turbines can be made very small, down to outputs of 100W or less. Some writers classify the smallest turbines as 'pico-hydro' or even 'nano-hydro', but all such units are adequately covered by the material here.

Micro-hydro often, though not always, implies a degree of simplicity too. Large Pelton turbines have to be fully engineered, but small units can work well even if they are quite basic. The emphasis here is on simple technology, so that the turbines can be made in small workshops, or in countries where only basic engineering facilities are available. It can easily be adapted for more advanced manufacturing technology; if better methods are available, do use them.

Despite the emphasis being on smaller turbines, large parts of the book are relevant to any size of Pelton turbine. The basic theory is independent of the size, and many of the design features can be used for larger turbines. The only major omission for larger schemes is that the book assumes that the flow is adjusted manually, and does not cover the design of spear valves or deflectors for use with a governor. Some practices that are only used in large hydro-schemes are mentioned for interest, and also because they show the way micro-hydro can be developed if the technology is available.

I have tried to make this book as comprehensive as possible. While my primary aim has been to cover the practical aspects of the selection, design and manufacture of Pelton turbines, I have also included a detailed discussion of their theory, plus a good deal of relevant engineering reference information. Since I personally find it frustrating to be presented with formulae without knowing where they come from, I have included the derivations of most of the equations used, but have relegated mathematics of this sort to the appendices. The main text does still have a good number of equations in it, but in many ways this is unavoidable. All engineering involves a certain amount of calculation. However, some sections are more important than others. For example, though a procedure is given for the calculation of pipework losses in Section 10.7, it is not usually necessary to go through this fully for each installation. The idea of the section is to illustrate the effect of various features, so that the reader can see why certain designs are better than others. In such cases, where it is not necessary to do all the calculations listed, I have tried to make this clear in the introductory remarks.

While every effort has been made to ensure the accuracy of this book, and most designs contained within it have been tested in the field, the author and publishers cannot accept any liability for damage, injury, breakdown or poor performance arising from the application of designs or data contained in it. The reader is responsible to check designs and conduct such tests as may be necessary to ensure that any components suggested by this publication are suitable for the use to which they are put.

ACKNOWLEDGEMENTS

This book has been written while I have been working at DCS, an appropriate technology development organization in Nepal that has had over 20 years of experience in micro-hydro, and installed nearly 300 schemes. The plentiful water tumbling down the hills gives Nepal great potential for hydropower, but the lack of roads and infrastructure make it a challenge to build hydropower plants there. My colleagues have worked for years in difficult, remote locations, often spending months at a time away from home, and working with very rudimentary equipment, to install mills and electrification schemes for villages in the hills. This book is dedicated to them, and for the willing help they have given in installing, testing, and proving many of the ideas presented here.

Special thanks go to Mark Waltham, who used to work for ITDG both in Nepal and UK. He has had many years of experience with micro-hydro Peltons. The easy-to-manufacture Pelton bucket used in this book was originally designed by him (though his design is slightly modified here), and many of the detail design features derive from Mark's work. He has freely provided much of the information.

Thanks also go to a number of people who have supplied me with advice and relevant information: John Burton and Simon Lucas of Reading University in England, who gave me a stack of information on Pelton turbines, in sundry languages; Alex Arter of SKAT in Switzerland for academic articles; John Williamson who hunted through libraries in UK for relevant information when I was in Nepal; Vijay Shresta of Shresta Industries, Bairahawa, Nepal, for making many batches of Pelton bucket castings and going through the processes involved in detail; Svein Aspesletten, who has worked with large, commercial Pelton turbines for many years, for information on the more high-technology way of doing things; Glenn Creelman of DCS for digging out information I had left in Nepal; Nicholas Talbot of Terrill Bros. Founders, Hayle, England for information on casting. Richard Goss produced a model of the bucket on CAD. My father and sister helped in a variety of ways to print out the text of the book. Various individuals and organizations have kindly given permission to reproduce extracts of published materials, photographs, or illustrations: BSI, ITDG, Kvaerner Energy, Rupert Evans, D.M. Miller, Newmills Hydro, Ringspann, Nigel Smith, and Sulzer Hydro Ltd.

Nearly all the design features in this book have been used in micro-hydro installations, and thanks go to the staff of Nepal Hydro and Electric, Butwal, for building most of these units. The Intermediate Technology Development Group, through its offices in Nepal and UK, has worked with DCS in the development of micro-hydro, and this long term partnership has been very fruitful. I would like to thank them for their assistance.

Two more personal thanks are due. Firstly to Tear Fund, the Christian relief and development organization in the UK who supported our family while we were in Nepal, and who continued that support in the UK so that I could finish this book. We have been glad to be part of Tear Fund, and are grateful to the many staff, individuals and churches behind it who make its work possible. Finally, thanks go to my wife and children. Writing a book of this sort is a time-consuming business, and they have been a help through the whole process, from my frequent absences as I travelled to sites around Nepal, to long hours writing and drawing.

DCS is an appropriate technology research organization within Nepal. It develops small-scale technologies that are appropriate for the conditions within Nepal and other developing countries, working with local communities to help them improve their standard of living. It has a particular emphasis on micro-hydro, having installed hundreds of water-powered mills and village-scale hydroelectric schemes throughout Nepal. DCS also works on rural electrification, building technologies and materials, agricultural and crop-processing machinery, mechanical tools, food-processing equipment and water systems, and is active in a wide range of local training and consultancy.

DCS, PO Box 126, Kathmandu, Nepal.
Fax: +977 71 4131
Email: dcs@umn.org.np

INTRODUCTION

The origins of this book

This book arises out of the micro-hydro work of DCS, which is an appropriate technology development centre in Butwal, on the lowland plains of Nepal. DCS was established in 1972 by a Christian mission, the United Mission to Nepal. Micro-hydro work began in Butwal in 1975 when a simple prototype crossflow turbine was made and successfully tested. This turbine was installed in a crop-processing mill. A type of vertical mill (called a *ghatta* in Nepal), which uses wooden paddles to drive a grindstone, was common within the country, so water-power technology was already familiar. However, the modern turbines allowed a range of other crop processing operations to be done, not just the flour grinding of the traditional mills. The new mills were simple, rugged, and easy to maintain. They were also relatively cheap, and close co-operation between DCS and a government bank, the Agricultural Development Bank of Nepal, gave villagers throughout the country access to the finance needed to construct them. The programme was very successful, and DCS built more than two hundred crossflow mills in the succeeding years. Other local workshops began to make and install turbines, so that, to date, there are nearly a thousand water-powered mills in Nepal.

In the 1980s it was realized that mill turbines could be used in the evenings to drive generators, and supply electricity to the surrounding villages. DCS embarked on a programme of bazaar lighting. Reliable and inexpensive electronic load controllers, ELCs, were just becoming available, and these were often used as governors for the systems. DCS also developed a number of simple control circuits for these stand-alone hydro-electic plants, and a low-wattage cooker that could make use of the electricity. As with the mills, the electrical technology had to be cheap, rugged and simple.

Large areas of Nepal are not served by the national electricity grid, and even today, only about 10% of Nepal's population has electricity. It soon became clear that there was a demand for micro-hydro plants just for electricity, without the milling. Some sites could still use crossflow turbines, but often the steep, mountainous terrain of Nepal gave very high heads and small flows. This was ideal for Pelton turbines. Over the last few years, DCS has been working to produce a Pelton turbine design that can be made using the basic engineering facilities available in Nepal. Pelton turbines have been installed in sizes ranging from 100mm to 400mm PCD (Pitch Circle Diameter – see Section 1.2), with turbine outputs of 2kW to 200kW. At the time of writing, DCS has built near seventy micro-hydro electrification schemes, of which fifteen are for electricity only.

Fig. 0–1: Cooking on electricity generated by a 50kW micro-hydro power station (Ghandruk, Nepal)

1
THE PELTON TURBINE

1.1 Why Pelton Turbines?

Any water turbine is a means of extracting energy from falling water. Different types of turbines prefer different operating conditions. Propeller turbines work best at low head and high flow. Crossflow turbines require medium heads and flows. Pelton turbines work best under conditions of high head and low flow. For many micro-hydro sites, a Pelton turbine is the only option.

Pelton turbines have other advantages too. Firstly, because they can utilize high heads, they can produce

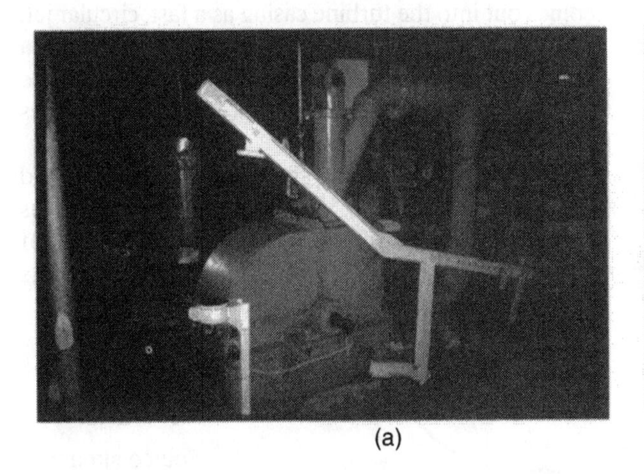

(a)

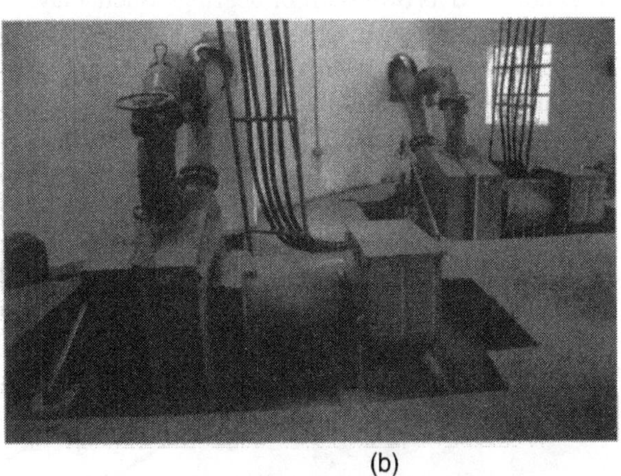

(b)

(c)

(d)

Fig. 1–1: Various Pelton turbine installations: (a) 2-jet, 25 kW at Muktinath, Nepal (b) 3 × 2-jet, 180 kW at Jhankre, Nepal (c) 1-jet, 50 kW at Ghandruk, Nepal (d) 1-jet Peltric set, 4 kW at Mango, Pakistan (Mango photo – Nigel Smith)

a lot of power from a small unit. Secondly, they are reasonably easy to make. Thirdly, a given turbine can be used for a range of heads and flows; unlike propeller or Francis turbines, the runner does not have to be designed for specific flow conditions, so Pelton runners can be made and kept in stock. Fourthly, the runner has space around it, making it much easier to inspect and work on than most other types of turbine.

Peltons are particularly useful for driving small electrical generators. Even at modest heads, small Pelton wheels can be made to run at high speeds, which allows them to be matched to generators without the need for a belt drive or gearbox to change the speed. In many cases, the Pelton runner can be mounted on the generator shaft, so that the turbine does not need its own shaft or bearings. Such a layout greatly simplifies the installation, and is cheap. A very popular use for Pelton wheels has been for small, self-contained generating units of a few kilowatts, often called *Peltric sets* (Figure 1–1). The Pelton runner is fixed directly to the shaft of an induction motor, which is run as a generator to produce electrical power.

Electricity generation is probably the most common use for Pelton turbines, but the mechanical power they produce can be used to drive any rotating machinery: crop-processing machinery, sawmills and woodworking machinery, irrigation pumps, power looms, potteries, forges, mechanical workshop machinery, food production machinery etc.

1.2 General description of a Pelton turbine

Pelton turbines are generally used for medium to high head sites. For small, 'micro-hydro' turbines (those with powers of up to about 100 kW), Pelton turbines will typically run on heads from 25m to 200m. Water is brought down the penstock pipe to a nozzle, and it comes out into the turbine casing as a fast, circular jet. The jet is directed at a wheel, or *runner*, which has a number of *buckets* around its edge (Figure 1–3). The force of the jet on this wheel makes it turn, and gives the output power.

The buckets have the shape of two cups joined together, with a sharp ridge between them. There is a notch cut out of the bucket at the outside end of the ridge. The jet hits the *splitter* ridge and is

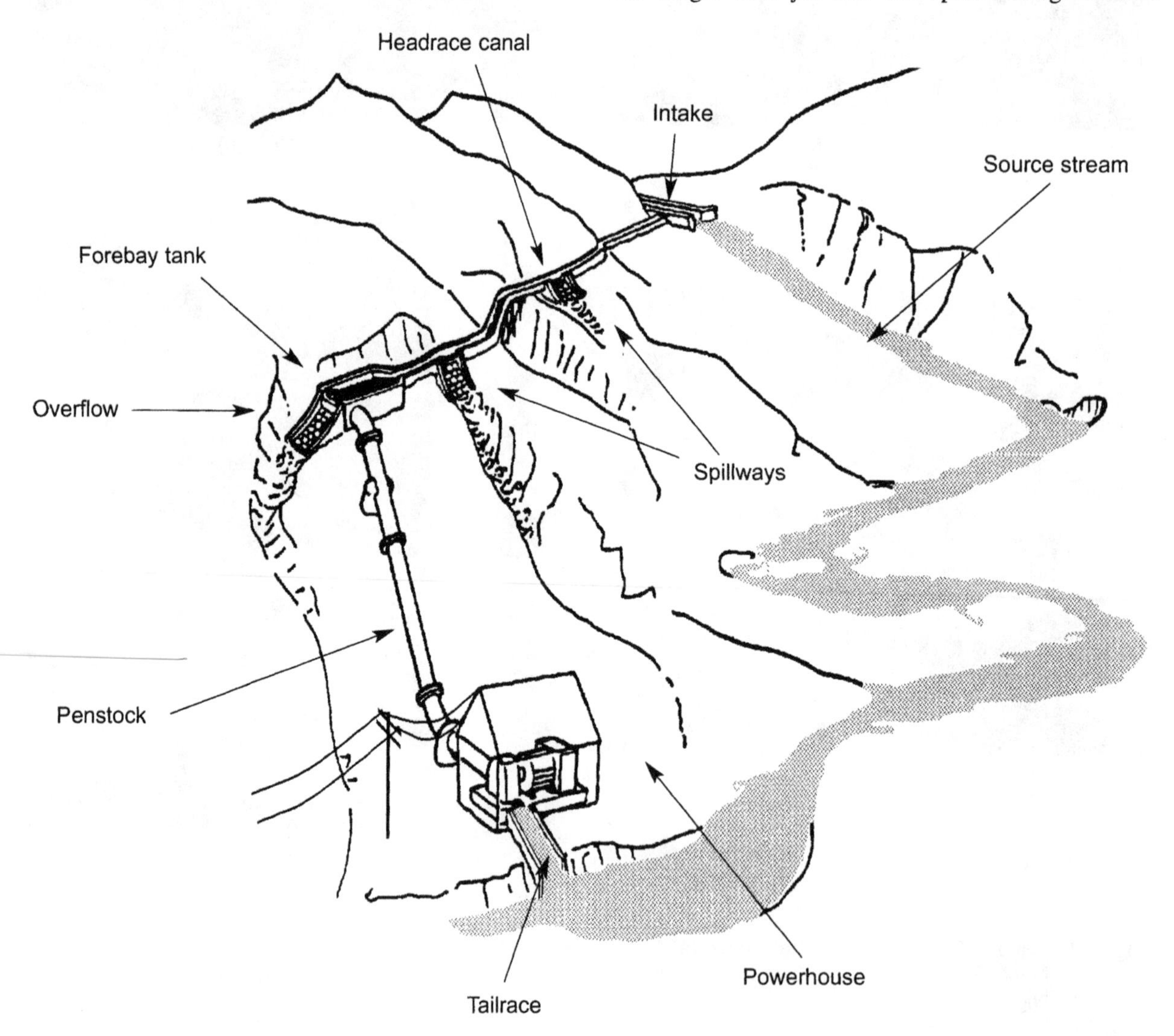

Fig. 1–2: General components of a Pelton turbine hydro-scheme

Fig. 1–3: A Pelton turbine runner (courtesy of Kvaerner Energy)

divided in two (Figure 1–4). The water flows around the insides of the cups, and exits from the sides of the bucket. The notch allows the jet to pass through the outside edge of the preceding bucket so that the first place it touches each bucket is at the end of the splitter (see Figure 1–5). Note the jet passing through the notch of bucket *a* before hitting the splitter ridge on bucket *b*.

The pressure of the water in the penstock produces the speed of the jet, but the water in the jet itself is at atmospheric pressure. The turbine casing is not full of water but has air in it; it is connected to the outside world through the tailrace channel, which is only partly full of water. The power developed in a Pelton wheel is completely due to the momentum change of the jet in the buckets. There is no change in water pressure across the runner. Pelton turbines are specified by their Pitch Circle Diameter, PCD. This is the tangential diameter at which the jet centreline passes the wheel, as shown in Figure 1–5.

Jets are normally controlled by *spear valves* or *injectors*, as shown in Figure 1–6. A spear valve consists of a nozzle with a spear-shaped needle in the centre of it. The spear is moved in and out to vary the flow, and some position of the spear will be

Fig. 1–4: A high speed photograph of a jet entering a Pelton turbine runner (courtesy of Sulzer Hydro Ltd., Zurich)

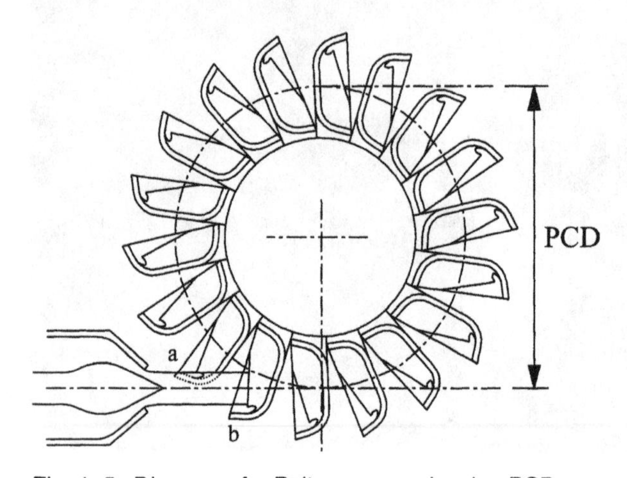

Fig. 1–5: Diagram of a Pelton runner showing PCD

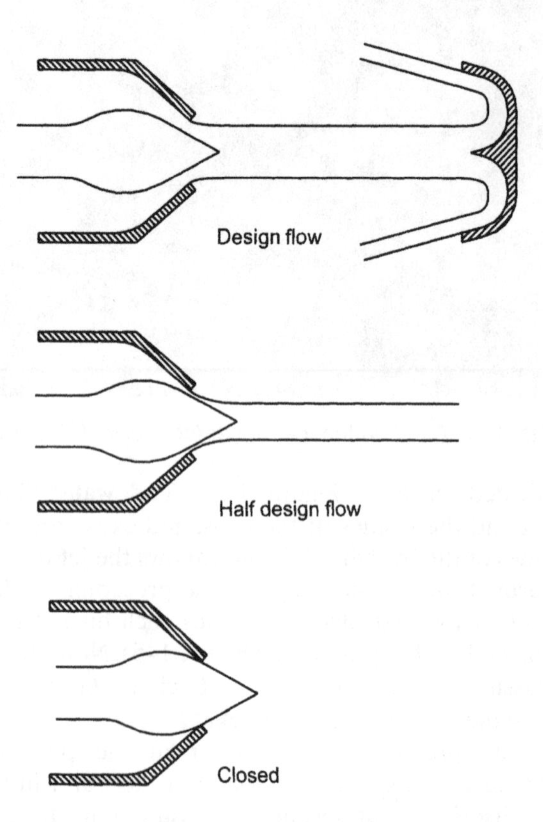

Fig. 1–6: A spear valve controlling the flow into a bucket

the 'design' position, which gives the design flow. In Figure 1–6, the top figure shows the design flow, the middle figure half the design flow. Spear valves have good efficiency over a wide range of flows. Small turbines can have plain nozzles, with valves in the manifold upstream of them to close off the flow.

The Pelton turbine has a wide application range. Pelton wheels as small as 100mm across are used for outputs of only a few hundred watts, on heads down to 15m. In large power-stations heads of over 1800m are used, and single Pelton runners can give outputs of 400MW or more, with runner weights of

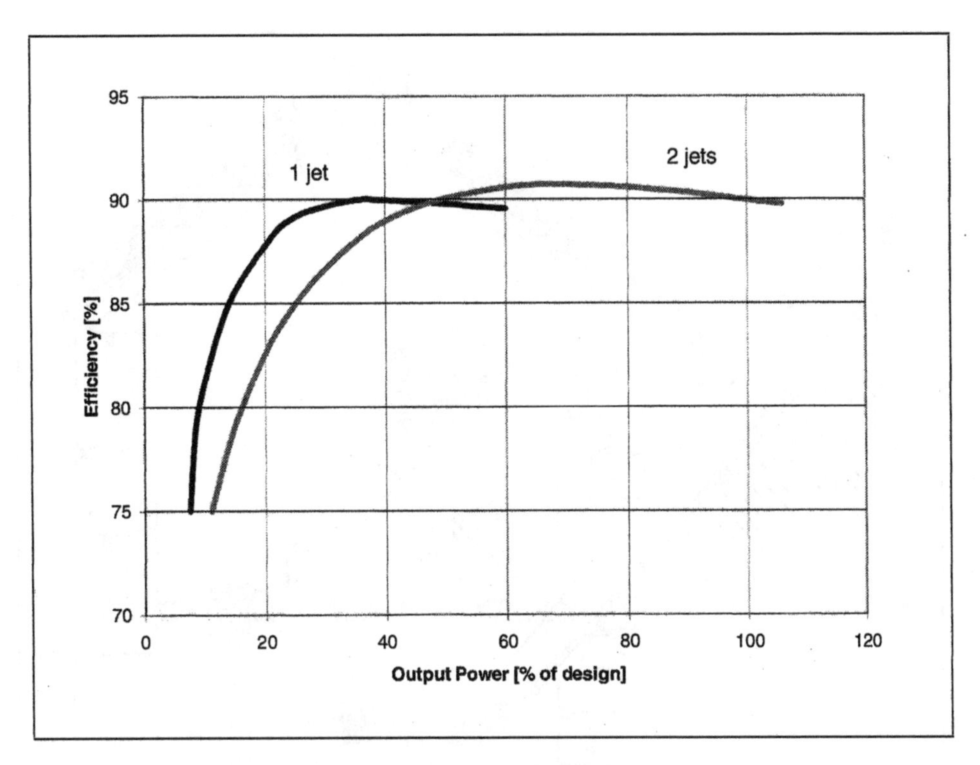

Fig. 1–7: Typical efficiency curves for a commercial 2-jet Pelton turbine

around 60 tonnes, and pitch circle diameters of over 8m. They can operate under higher heads than any other type of turbine, and hold up their efficiency at partial flow better than most other designs (Figure 1–7). The chief advantage of the Pelton design for micro-hydro is that it is relatively easy to make. Provided casting facilities are available, Peltons can be made with simple workshop machinery, and still give acceptably high efficiencies.

1.3 The origins of the Pelton wheel

1.3.1 Early water mills

Turbines have a long history. It is probable that the first water mills were *vertical mills* (Figure 1–8). These worked by dropping water down a hollowed-out log on to flat, wooden paddles on a wheel. This sort of turbine, also known variously as the *Greek mill*, the *Norse wheel*, and by other names in other places, has ancient origins, and is known to have been used for many centuries in China, Greece and Northern Europe. The wheel turned grinding stones to make flour. This mill design spread over the centuries throughout the world, and almost identical designs were found from South America to Asia. There are still thousands in use today in Nepal. Like the Pelton turbine, vertical mills use the impact of the water on the wheel to turn the shaft.

The vertical mills typically work off heads of 3–4m and flows of 25–35litres per second. Centuries ago, when more power was required for industrial uses, alternative designs were tried. References to the other sort of ancient turbine, the water-wheel, are found from the first century BC. The first known literary reference to a water wheel is a poem by Antipater of Thessalonia, written, according to Moritz (1958) in 85BC, though given in Hill (1984) as around 30BC.

> Cease your work, ye maidens, ye who laboured at the mill,
> Sleep now and let the birds sing to the ruddy morning;
> For Ceres has commanded the waternymphs to perform your task;
> And these, obedient to her call, throw themselves upon the wheel,
> Force round the axle tree and so the heavy mill.

It is possible that the water-wheel (Figure 1–9) predates the vertical mill, but the history is unclear. The simplest water wheel is the stream wheel, where the paddles are turned by the water in a stream, but other wheels use both the impetus and the weight of the water to turn the wheels. Many different layouts were tried: undershot, overshot, breastshot, pitchback etc. It seems logical that there should have been a progression from the vertical mill, through the stream wheel to the more complex wheels, but it now seems that they developed independently, in different areas, according to regional needs and traditions (Smith, 1984).

The Poncelot wheel developed in the 1820's by the French engineer General J.V. Poncelot, shown in Figure 1–9(e), is interesting in the context of Pelton development because it includes an impulse element, with the water being jetted on to the paddles. Poncelot also understood that for efficient

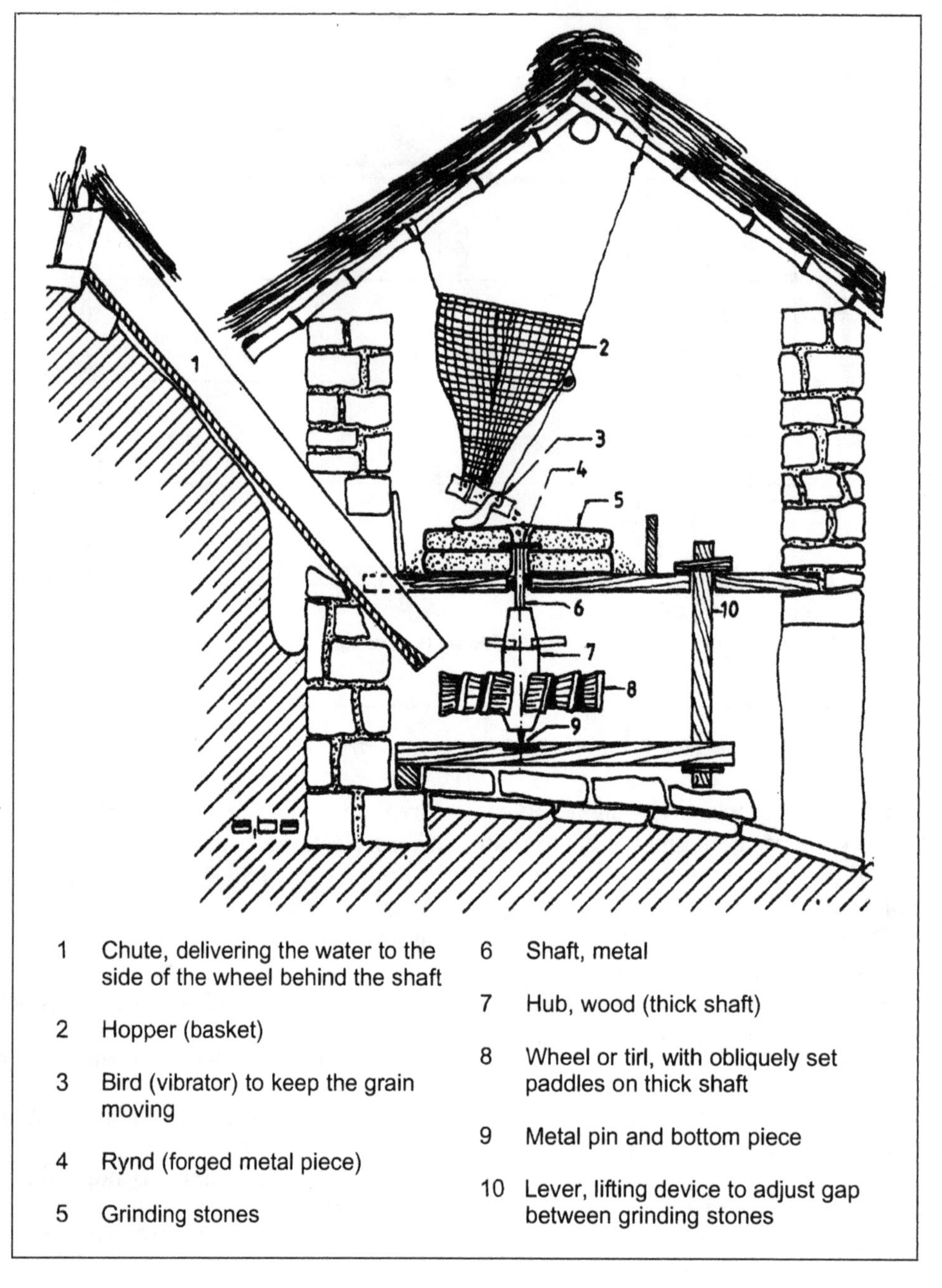

Fig. 1–8: A traditional vertical mill (Saubolle & Bachmann, 1978)

operation the water needed to enter and leave the paddles smoothly. In the eighteenth and nineteenth centuries some of the water-wheels constructed were enormous and gave outputs of up to 100kW, but they were unwieldy and not very efficient; modern turbines giving the same power are a fraction of the size.

1.3.2 Tangential wheels

The first known reference in literature to a wheel driven by a tangential jet is by Branca in 1629, where a steam-driven impulse turbine is illustrated driving a hammer-mill, in Loretto, Italy (Doble, 1899). The jet of steam hits the flat blades of the wheel, causing it to turn. Through the eighteenth and nineteenth centuries, many people experimented with different forms of turbines, including famous names from the history of hydraulics such as Euler in 1754, Borda in 1767, Navier in 1819, and Poncelot in 1827 (Doble, 1899). It was realized that flat vanes were not efficient, and a variety of curved vanes were used. As early as 1767, Borda wrote of tangential turbines: 'To produce its total mechanical effects, the water serving as the motive power must be brought onto the wheel with impulse, and quit it without velocity' – which remains the basic principle of Pelton turbines. A number of turbines were developed using semicircular buckets, such as the Girard waterwheel of Madame de Girard (1843), (Doble, 1899).

When the Gold Rush started the mining industry in California in 1849, simple tangential wheels were used. Most of the available sites had high heads, and water was brought down in pipes and squirted

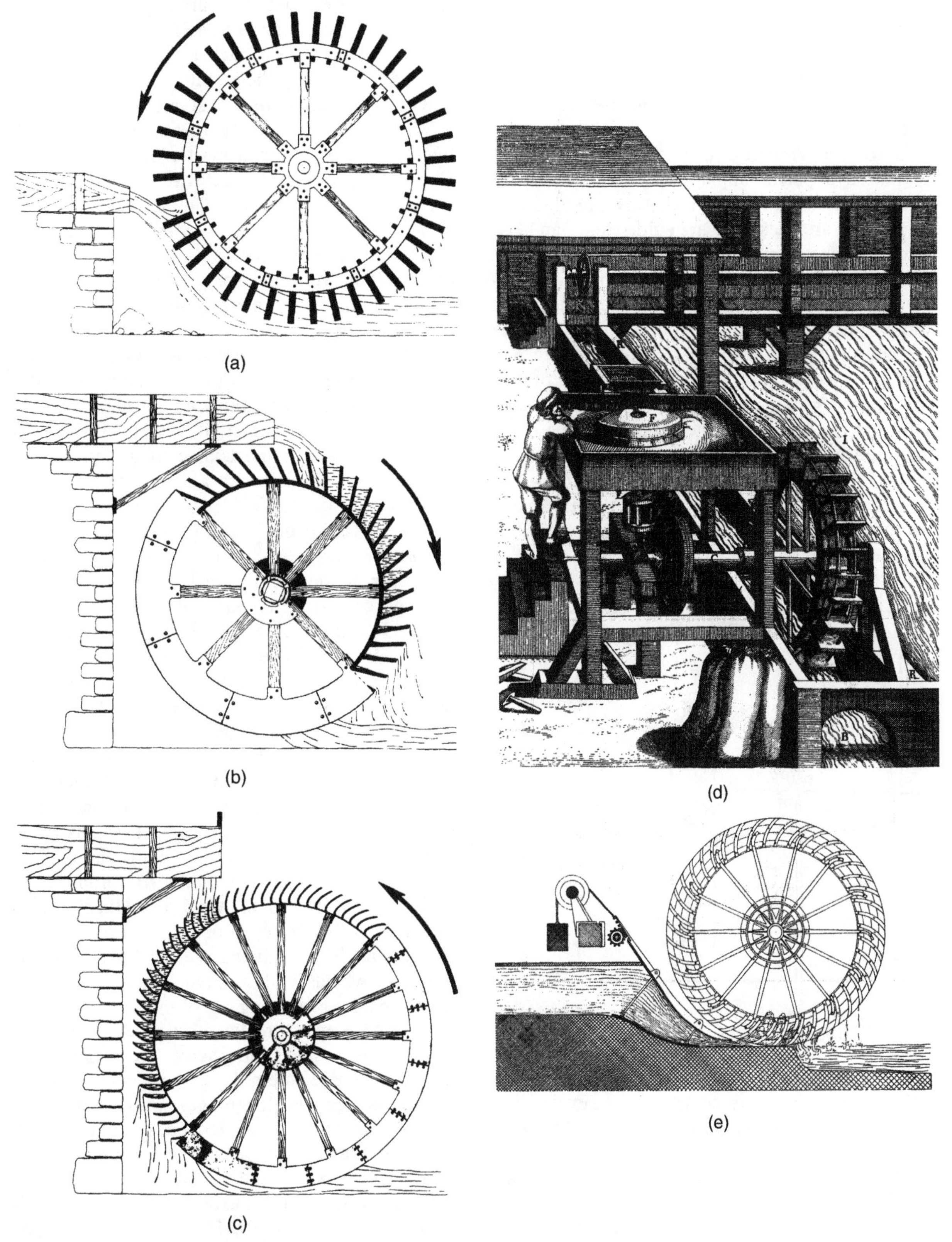

Fig. 1–9: Various water wheels: (a) undershot (b) overshot (c) pitchback or backshot (d) a stream wheel driving a corn mill (e) Poncelot (a, b, c and e, © Quarry Bank Mill Trust; d, Science Museum, London, Science & Society Picture Library)

through nozzles onto vertical wheels. The development of these wheels in California followed the route pursued earlier by the European inventors. Initially the wheels had crude, flat wooden paddles, four inches thick, fixed on to a hub to look like the teeth on a circular saw. The jets were controlled by valves upstream of the nozzles that choked back the flow. These 'Hurdy-Gurdy' wheels (named after a mechanical musical instrument like a violin in which a wheel takes the place of the bow) had efficiencies of 40% at best. Later, the paddles were replaced with hemispherical cups, with the jets were directed into the centre of the cups. This improved the efficiency to around 65%, and by using large heads and flows outputs of nearly 1MW were achieved.

In 1853, an American engineer, Jearum Atkins, submitted a patent application for a rather different form of tangential wheel, shown in Figure 1–10. Parts D are buckets, F are guide vanes, and E is a scroll casing carrying the water into the turbine. In this patent, Atkins gave the basic principles behind the design of a tangential turbine: that the jet should hit the wheel tangentially, that the blade velocity should be roughly half the jet velocity, and that the wheel should reverse the direction of flow so that the water should leave without velocity. This was the first clear statement of the theory of impulse wheels. Atkins' patent was not granted until 1875. There was little interest in the design in America, and he sold the rights to the European manufacturer Escher, Wyss and Co. in 1882 (Doble, 1899).

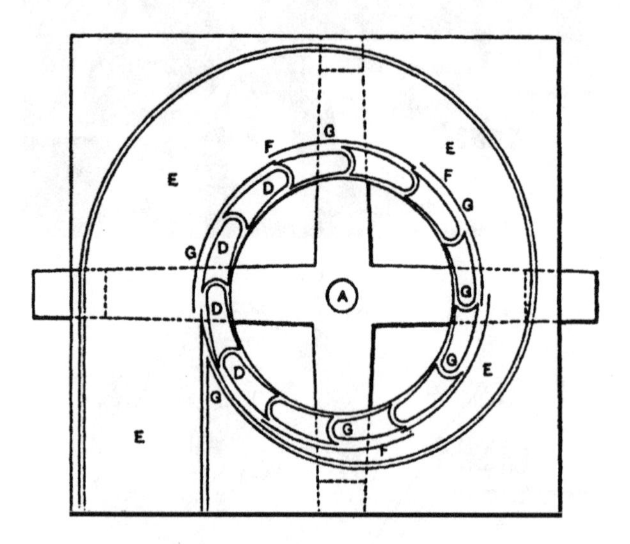

Fig. 1–10: Jearum Atkins tangential wheel (Doble, 1899)

Other manufacturers continued to develop the more standard type of tangential wheel, with one or two nozzles directing water on to them. The hemispherical buckets were replaced by more complicated shapes. The water was generally caught by the outside lip of the buckets, and then discharged at the sides or in towards the centre of the wheel. All these wheels had single cups on them, and did not have a splitter ridge to divide the flow. This was the next development.

The story is told (Daugherty, 1920) of a man watching a turbine one day when the wheel slipped along its shaft so that the water entered the cups on one side, ran around inside them, and came out on the other side. He was surprised to see that the speed increased and the output power went up. This man was Lester Pelton (1829–1908), and he went on to produce a bucket which was split into two halves, dividing the jet into two parts, and patented this as the Pelton wheel in 1880.

Whether Pelton was the first to invent the split bucket was, however, a matter of some dispute. The first patent for a bucket with a splitting wedge was granted to Nicholas J. Coleman of California in 1873. It appears that no commercial turbines were ever made to Coleman's design, but the patent was held to cover some of the principles of Pelton's bucket, so that a royalty had to be paid of $1 for every foot of diameter of every Pelton wheel made while the patent was valid. Another designer, Joseph Moore of Risdon Iron and Locomotive Works, San Francisco, produced a different design of tangential wheel with a split bucket in 1874. In the course of designing this, Moore corresponded with a certain F. G. Hesse, Professor of Mechanical Engineering at the University of California, and in the debate in later years as to the priority of invention, both claimed credit for the design. Lester Pelton started a programme of testing a range of bucket shapes in 1878, patenting the design that came to be known as the Pelton wheel two years later. Pelton claimed not to have known of either Coleman's design, or of Moore and Hesse's wheel, when he did his work. In summary, Coleman appears to have been first with the idea of a splitting wedge, Moore and Hesse invented the first version of the bucket that became the accepted form later, and Pelton patented a similar design. It seems to be primarily Pelton's commercial skills that led to this type of turbine being named after him.

During the period in which the Hurdy-Gurdy wheels were being developed in America, similar 'spoon wheels' were being used in Europe. These began as wheels with a large number of spoon-shaped blades projecting out of the rim of the hub. The orifices were rectangular. Regulation was achieved by a plate sliding across the orifice, or by the whole nozzle assembly being swung towards or away from the wheel. Strength considerations led to a stiffening rib being extended out from the hub along the centre of the spoons, becoming a sort of splitter ridge. So as the Pelton wheel was introduced in America, the European design was heading in the same direction. However, between 1900 and 1910, the Pelton wheel either replaced or assimilated all the other designs. The spear valve was superior to any of the alternatives, and it quickly became the standard.

1.3.3 The Pelton wheel

Lester Pelton's patent detailed a bucket made up of two symmetrical cups on either side of a central splitter ridge. This bucket did not have the notch of modern Pelton buckets, but had an angled lip on the outside edge of the bucket whose function was to catch the water without deflecting it. Pelton described how the buckets could be attached to the hub, and how it could be used with one or more jets. These buckets had efficiencies of just around 82%. One of Pelton's turbines is illustrated in Figure 1–11.

Fig. 1–11: Lester Pelton's turbine (Bodmer, 1900)

The modern shape of the Pelton bucket was actually invented by W. Abna Doble in America in 1899. He introduced the smooth, ellipsoidal bowls, and the notch. Doble also invented the spear valve. In many ways, it would seem more just to speak of the Doble turbine, but tradition names it after Pelton. Subsequent developments have led to the refinement of the bucket shape and improved designs for both the spear valves and the turbine casings. Early machines nearly all had horizontal shafts (as in Figure 1–11), but from the 1960s it became common to use vertical-axis machines, particularly when using more than two jets (see Figure 1–14). Some innovations have gone in and out of fashion. For example, regulation valve spears used to be made with elongated tips, as shown in Figure 1–13, but these were later

Fig. 1–12: A 2-jet, 120kW horizontal axis Pelton turbine (Siklis, Nepal)

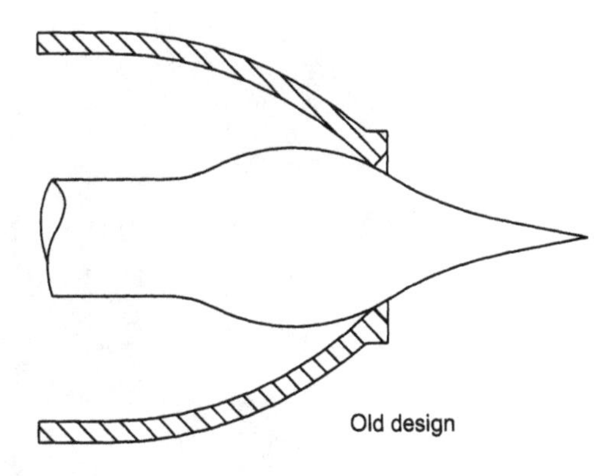

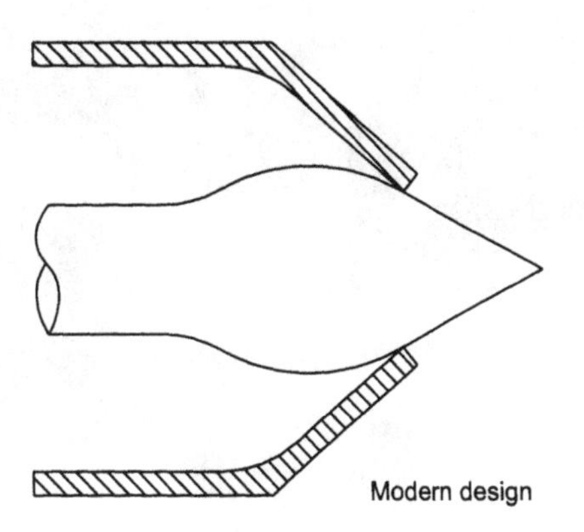

Fig. 1–13: Old and modern spear designs

discarded as they were found to erode much more quickly than straight spears.

Improvement in materials and design have meant that runners can be made smaller for the same size buckets. Turbines can be run at higher speeds, which makes associated electrical generators cheaper. While buckets were bolted to the runner in early designs, commercial runners are now nearly always cast in one piece, with integral buckets. For the largest runners, foundries may cast segments of the runner, each comprising a number of buckets, and weld these together. A new process for large runners involves forging a disk with the bucket roots on it, and then depositing weld on to the roots to build up the shape of the bucket (Brekke, 1994). Single-piece runners are quite common even in microhydro, but welded or bolted runners are used when local casting technology is not adequate for single-piece designs.

The most recent developments in turbines have been in improving the efficiency of multijet Peltons. Careful study of the interaction between jets, and of the shape of the casing, have allowed larger numbers of jets, up to six, to be used satisfactorily. Coupled with the trend to use ever higher heads, this gives very high power from individual units. At the beginning of the twentieth century, the world's most powerful Pelton turbine was less than 50MW. Sizes have increased steadily since then. The

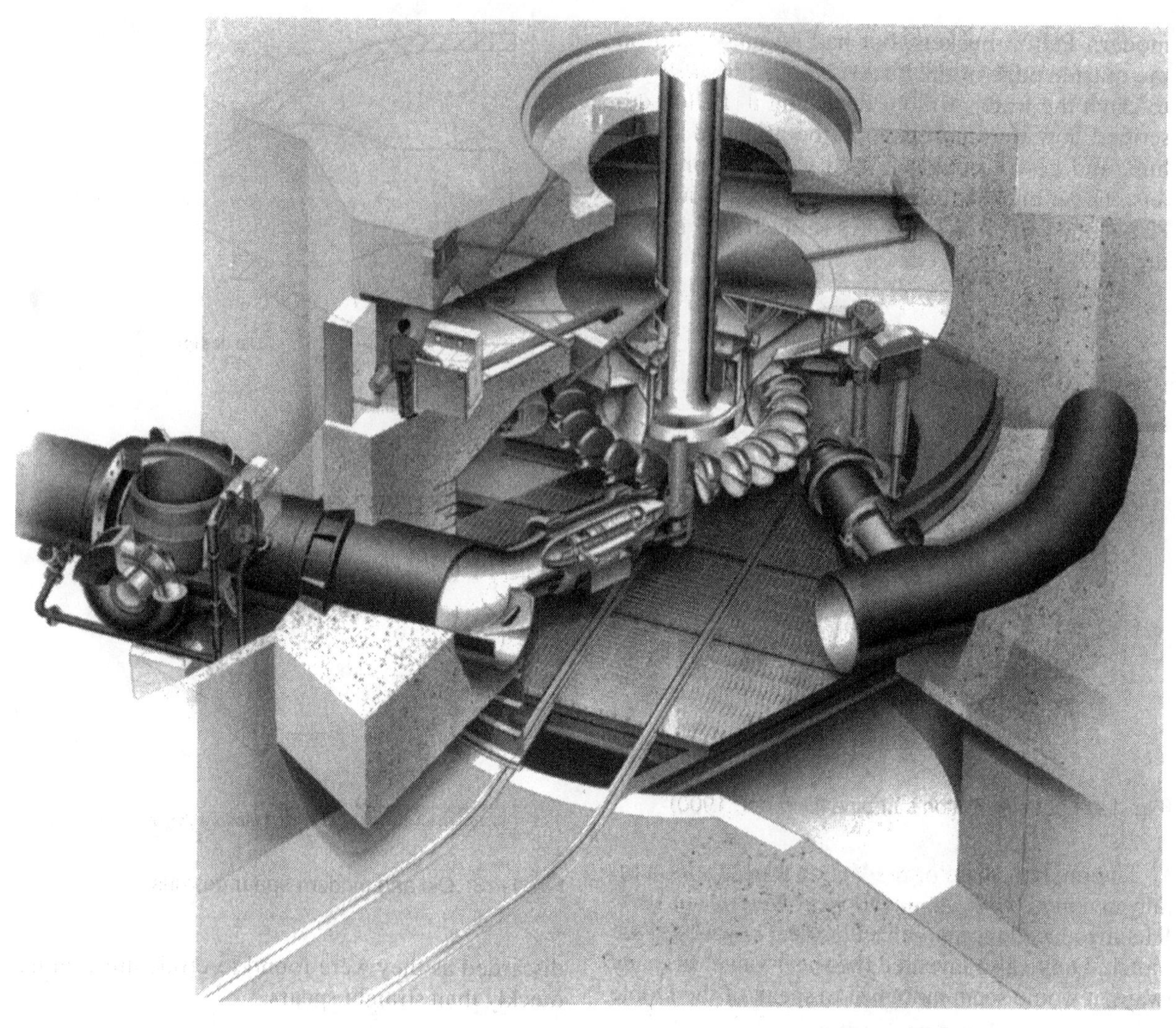

Fig. 1–14: A large power-station multijet, vertical axis Pelton turbine (courtesy of Kvaerner Energy)

recently completed turbine at Cleuson Dixence hydroelectric power station in Switzerland runs on a head of 1869m, and has five-jet Pelton turbines each giving 423MW. The water-jet speed is 690km/h, the same as a jet airliner! (Dansie & Bonifay, 1996). A head of around 2000m seems to be the limit using current technology, as beyond this the life of the runner is unacceptably reduced by fatigue and erosion. Even so, at 2000m, a six-jet unit producing 820MW is thought possible (Brekke, 1994).

Refinement of the Pelton design continues, but there is now considerable similarity between the designs of all the major manufacturers. Efficiencies for large runners are usually over 90%, and under carefully controlled conditions, runners can achieve efficiencies of 92%. Micro-hydro runners, with their rather simpler designs and technology, should nevertheless be able to manage 75–85% efficiency. The efficiency of the runner design given in this book has been measured at around 82%.

2
BASIC THEORY

All water turbines convert the power of falling water into mechanical energy. Water running downhill in a stream dissipates the energy as noise, which you can hear, and as heat. The heating effect is small, though; water dropping 100m only heats up by just over 0.2°C. A hydropower station slightly cools the stream it works on, but the temperature drop is hardly noticeable.

2.1 Impulse and reaction turbines

Pelton turbines are often called impulse turbines. The runners of impulse turbines are surrounded by air. Their power is produced by jets of water hitting the runner, and this power comes entirely from the momentum of the water. The jets are at atmospheric pressure, and the pressure energy in the penstock is converted into kinetic energy in the jet. The Pelton turbine is by far the most widely used and important impulse turbine, but there are other impulse turbines, such as the *turgo* turbines.

Reaction turbines are the other type, and here the runners are submerged in water. The penstock pressure is dissipated through the turbine, as the water flows through rotating vanes. The power produced still comes largely from momentum. Francis propeller, and Kaplan turbines are the most important reaction turbines.

2.2 Water jets

In Pelton turbines, the water interacting with the runner is at atmospheric pressure. Except for friction losses, the potential energy represented by the pressure in the penstock is fully converted to kinetic energy in the jet. This happens in the nozzle or spear valve. For ease of analysis, a simple nozzle is considered first.

2.2.1 Nozzles

A typical water jet emerging from a nozzle is shown in Figure 2–1. Note that the drawing is not to scale; both the contraction and the spreading of the jet are exaggerated for clarity. The water converges within the nozzle, and continues to converge outside the nozzle until the jet reaches a minimum diameter, which is called the *vena contracta*. At the vena contracta the lines of flow within the jet are parallel. The pressure in the jet does not settle down to atmospheric until the vena contracta, and the velocity of the jet actually increases slightly between the nozzle and the minimum diameter. After the vena contracta, the jet slowly diverges, because of friction between the air and the jet.

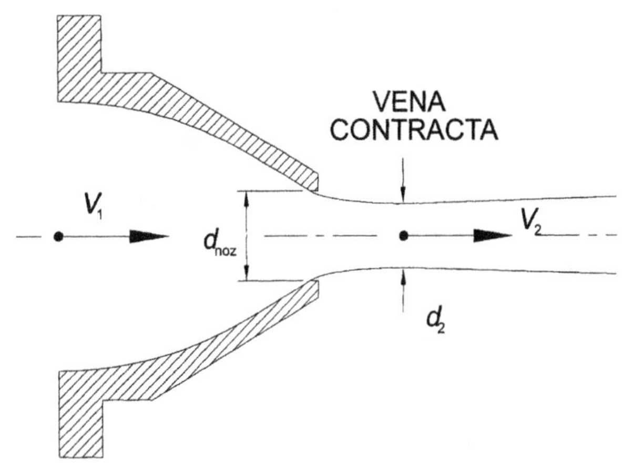

Fig. 2–1: Jet emerging from a simple nozzle (not to scale)

The velocity of the jet at the vena contracta is:

$$v_2 = C_v \cdot \sqrt{2g \cdot H_n} \qquad \textit{Eq. 2–1}$$

v_2 – jet velocity at the vena contracta (m/s)
C_v – coefficient of velocity
H_n – nett head at the entrance to the nozzle (m)
g – acceleration due to gravity (m/s^2)

This is derived in Section 10.1. C_v is usually between 0.95–0.99 for turbine nozzles. A well-made micro-hydro nozzle should achieve 0.97.

The contraction of the jet down to the vena contracta is described by a contraction coefficient C_c such that:

$$A_2 = C_c \cdot A_{noz} \qquad \textit{Eq. 2–2}$$

A_2 – jet section area at the vena contracta (m^2)
A_{noz} – area of the nozzle opening (m^2)

The discharge coefficient C_D gives the actual flow from a nozzle:

$$Q = C_D \cdot A_{noz} \cdot \sqrt{2g \cdot H_n} \qquad \textit{Eq. 2–3}$$

Q – flow through nozzle (m^3/s)

Note that:

$$C_D = C_c \times C_v \qquad \textit{Eq. 2–4}$$

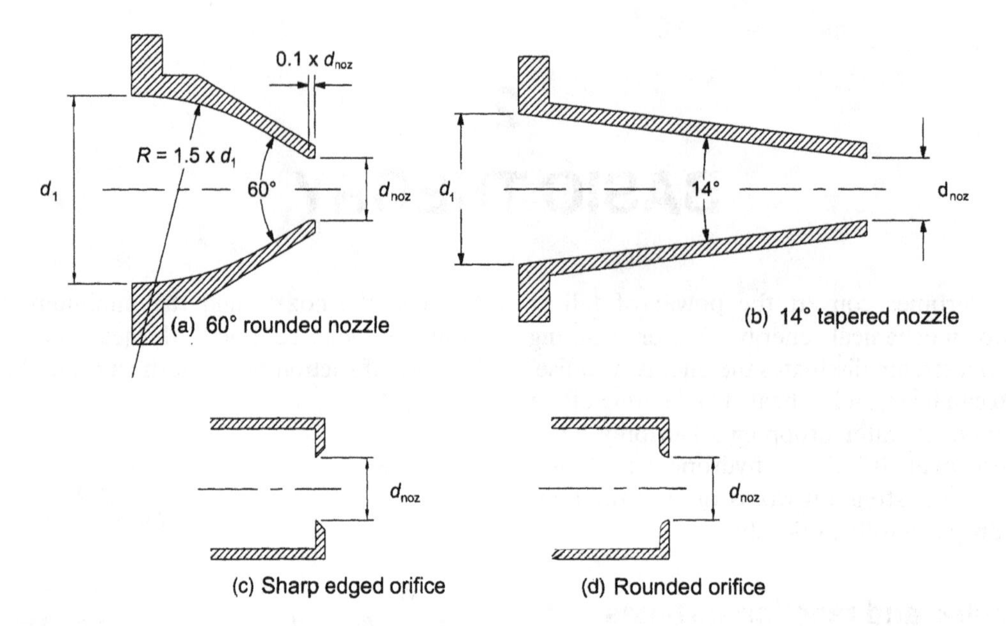

Fig. 2–2: Various nozzles

Various nozzles are shown in Figure 2–2. The characteristics of these nozzles are as follows.

(a) 60° rounded nozzle: C_v = 0.97, C_D varies linearly between 0.805–0.825 for A_1/A_2 = 0.1–0.2 (Miller, 1990), corresponding to C_c of 0.83–0.85. This is a well-designed short nozzle.
(b) 14° tapered nozzle: C_v = 0.98, C_c = 0.98, C_D = 0.96 (Harvey et al, 1993). This nozzle is thought to have the best discharge coefficient possible. It is a compromise between keeping the nozzle short to reduce friction losses, yet having a smooth reduction. Unfortunately it is often too long to fit in a turbine housing.
(c) Sharp-edged orifice: C_v = 0.98, $C_c = \pi/(2+\pi)$ = 0.611 theoretically, actually 0.61–0.69, C_D = 0.60–0.68. Simple to make and easy to change. Prone to wear at the sharp edge, when the efficiency drops off dramatically (to C_v = 0.8 or less) and the clean jet can break up into a spray.
(d) Rounded orifice; C_v = 0.97, C_c = 1.00, C_D = 0.97 (Harvey et al, 1993).

2.2.2 Spear valves

Spear valves are extremely efficient regulating devices. A good spear valve will have a C_v of 0.97–0.98, corresponding to an efficiency of 94–96%, from full flow down to a quarter flow or less. The excellent performance characteristics of spear valves are one reason why Pelton turbines are so good over a wide range of flow.

Spear valves need to be designed to give minimum losses, and to produce a good, coherent jet. Early designs of spear valves had long, pointed spears and shallow-angle nozzles (Figure 1–13). These were thought to give the smoothest transition, the greatest efficiency, and the smallest jet divergence. Their problem is that the flow is not directed strongly enough into the centre to make the water fill the hole left by the spear. This leads to a low-pressure area at the centre of the jet, and severe cavitation on the spear itself. Modern valves generally use greater angles. This manual uses spear/nozzle angles of 55°/80°.

Figure 2–3 shows how spear valve flow changes with spear valve movement (assuming the net head is constant. When the spear is sitting in the nozzle (i.e. the spear travel is zero), there is no flow. As the spear is moved, the flow increases. There will be some nominal maximum amount of travel that gives the design flow through the nozzle; in Figure 2–3 this is taken as 42% of the nozzle diameter, so $s_{max} = 0.42 \times d_{noz}$. The spear can be moved further than this, but the control over the flow becomes less linear, and the losses can increase.

The diameter of the jet that emerges from a spear valve can be calculated from the following equation, which is derived in Section 10.2:

$$d_{jet} = \sqrt{2.\sin\frac{\alpha}{2}.\left(2s.d_{noz} - s^2.\sin\alpha\right)} \qquad Eq.\ 2\text{–}5$$

α – spear angle [°] – here 55°
s – spear travel [same units as d_{jet}]

2.2.3 Jets

Some jets diverge very little. This would be true, for example, for a jet from the 14° nozzle in Figure 2–2 at low heads. Most jets, however, do diverge.

The vena contracta generally occurs about one jet diameter from the nozzle, and the jet divergence starts about half a diameter further on. An accurately-made, finely-machined nozzle can produce a jet whose surface, at the vena contracta, is as smooth as glass. Usually, though, jets will usually have lines on their surfaces, and spray mixed with them.

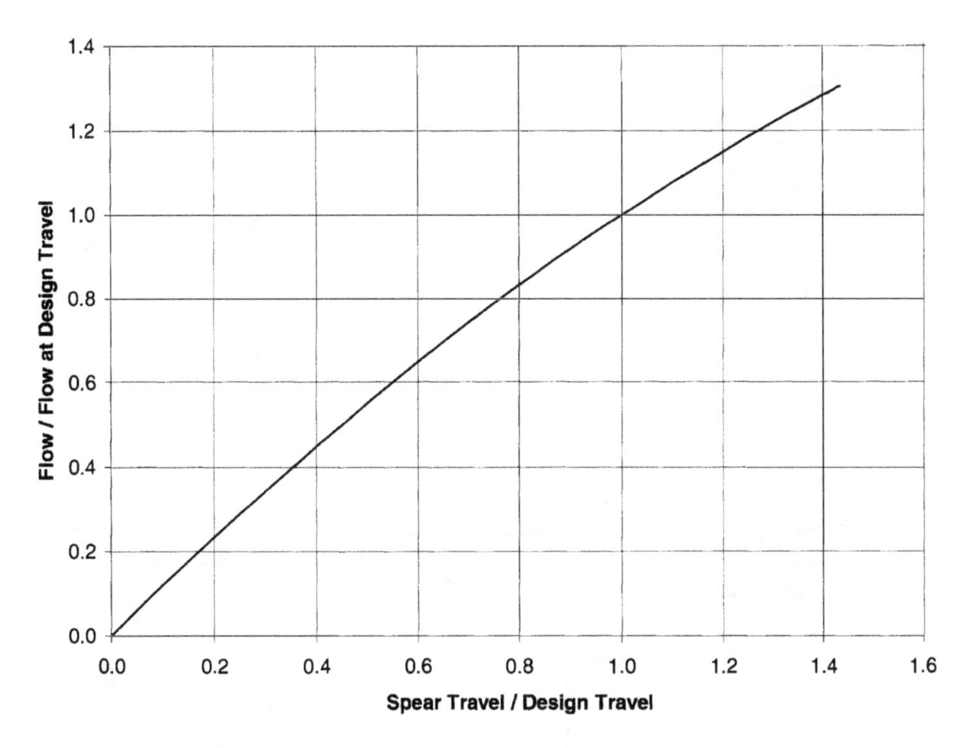

Fig. 2–3: Variation of jet flow with spear valve opening for a 55°/80° valve

Away from the vena contracta, the surface of any jet starts to roughen and break up. It is hard to see what is happening with the naked eye, but using high-speed photography it can be seen that the surface develops long, wavy, unstable furrows, which become deeper and deeper until the ridges between them start to shed droplets. The diameter of the jet is very hard to define as the surface is constantly breaking up and reforming. A true average can only be estimated by comparing a number of high-speed photographs.

It is important to keep the nozzle as close to the runner as possible, for two reasons. Firstly, the jet diverges, but the bucket will have a maximum diameter of jet that it can handle. If the nozzle is too far away, the jet may be too large by the time it reaches the buckets, and some water may miss the buckets or not travel around the cups correctly. Secondly, the energy of the jet is dissipated by the air friction, and the shorter the distance travelled the less the energy loss.

The jet diverges because of turbulence inside it, and friction against the air at its surface. It is hard to predict accurately the divergence because the turbulence inside the jet is affected by so many factors. A general illustration of the velocity distribution across a jet as it diverges is shown in Figure 2–4. At the vena contracta, the velocity is fairly uniform across the jet. There is slight retardation at the outside surface caused by wall friction in the nozzle. Notice too that for a spear valve there is a small area of low velocity in the centre caused by friction against the spear. Further along, at d_3, air friction has slowed the outside of the jet considerably.

The angle of divergence is influenced by:

- The speed of the jet, which is related to the head: the higher the head, the greater the divergence.
- The geometry of the nozzle or spear valve: sharply contracting nozzles give jets which diverge faster.
- The surface finish of the nozzle and the spear, and especially the condition of the orifice edge.
- The position of the spear within a spear valve: divergence is greater at small openings than at a full flow.
- The upstream flow: if the flow upstream of the nozzle is regular and linear the divergence will be limited, but if the flow is turbulent or rotating the jet can degenerate into a spray almost immediately.
- Interference from other jets: just a few droplets of water landing on the vena contracta can make the jet break up. If a small part of the water from another nozzle hits the jet, it can completely disperse it, so that the jet does not put any power into the runner. Many turbines have covers over the jets to prevent this happening.

The exact angle of divergence is difficult to predict, but an indication of the values that can be expected is given in Figure 2–5. The graph is plotted for a jet diameter of 11% of the PCD at the vena contracta, a distance of 50% PCD away from the vena contracta. The continuous line follows data in (Vivier, 1966), the dashed line is extrapolated, and should give reasonable values for well-designed spear-valves or nozzles, with minimal swirl at the inlet, within the range of jet sizes and heads used in micro-hydro. At low heads the divergence can be negligible, but it becomes increasingly important as the head increases.

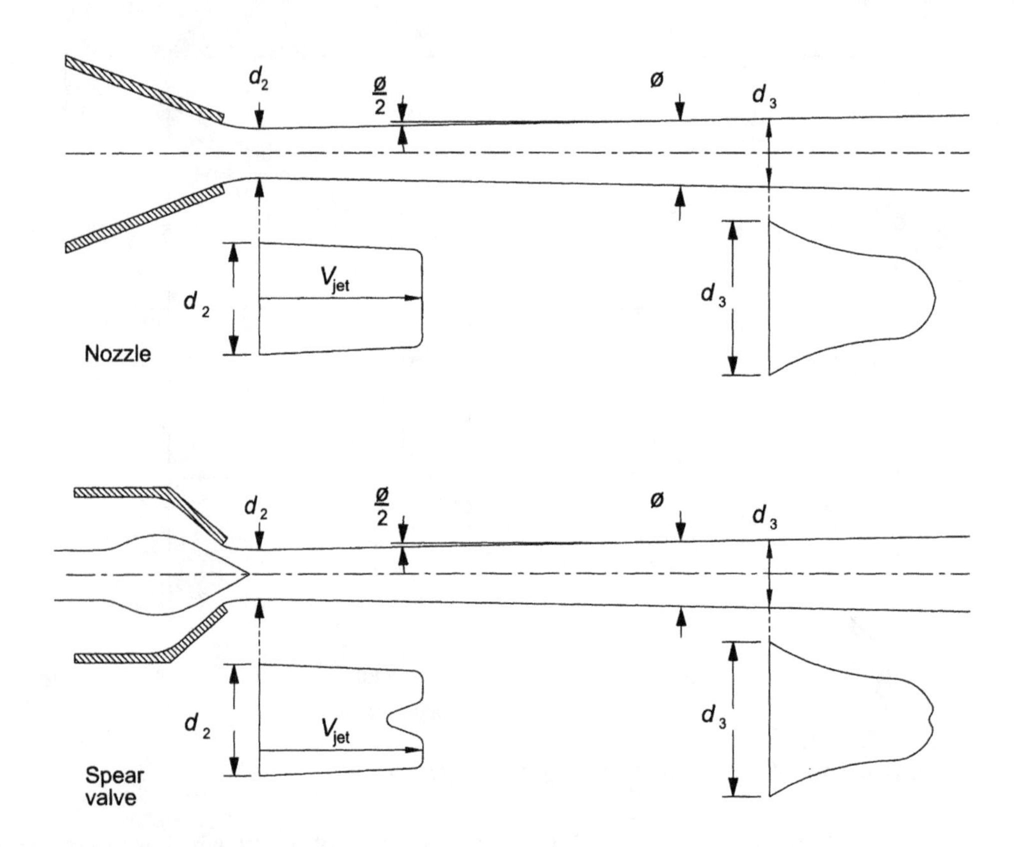

Fig. 2–4: Velocity distribution across a diverging jet. Drawn for $\frac{ø}{2} = 1°$

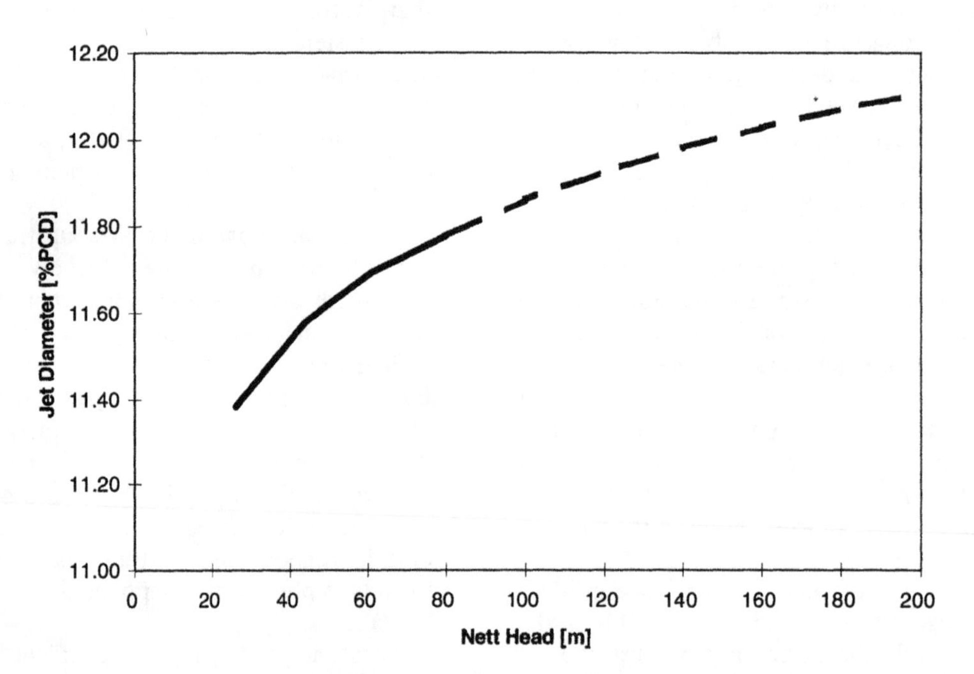

Fig. 2–5: Approximate diversion of an 11% PCD jet – see text for details

For turbines using the runner described in this manual, the nozzle is generally from 0.5 × PCD to 0.6 × PCD away from the mid-point of the jet's action on the runner. The jet does not start expanding until about 1–1.5 × d_{jet} away from the nozzle, and the design jet diameter is 0.11 × PCD. So the length over which the jet diverges is approximately 0.35–0.50 × PCD. The bucket design described in this book will take a maximum jet diameter of 12% PCD. The graph indicates that for a 150m head, a jet which is 11% PCD at the vena contracta will be about 12% in diameter as it acts on the runner. It should be emphasized that there is no accepted formula to calculate jet divergence, and these figures are indicative only.

Jet divergence was used in an interesting way by one turbine manufacturer, English Electric, to stop the turbine. Vanes were placed inside the spear valve, normally parallel to the flow. In an emergency, the vanes were turned slightly so that they imparted swirl to the flow before the nozzle. The effect of this was to turn the jet into a wide, hollow cone of spray, which then did not have the power to turn the runner (Guthrie Brown, 1984).

2.3 Buckets and the runner

The sections above have described the first part of a Pelton turbine, the nozzles or spear valves which produce the water jets. These convert the pressure energy of the water in the penstock into the kinetic energy of the jet simply and efficiently. The function of the runner is to remove as much as possible of that kinetic energy from the water, and to deliver it as rotational energy to the shaft. Modern runners are very good at doing this, giving out over 90% of the power in the jet. A micro-hydro turbine should manage 75–85%. The following sections deal with that conversion.

In order to avoid confusion, the following convention will be used when referring to Pelton buckets. The *front* of a bucket is the concave side, the side that the jet hits. The *back* of the bucket is the convex side, which does not normally come into contact with the jet. The *outside* of the bucket is part furthest away from the axis of rotation, and is the edge with the notch cut into it. The *inside* of the bucket is, conversely, the edge of the bucket that attaches to the hub. The *sides* of the bucket are the edges that run parallel to the central splitter ridge. These terms will be used whatever the orientation of the runner – for horizontal or vertical axis Peltons – and wherever the jets are positioned.

All the theories dealt with below are covered in much greater detail in Section 10.3.

2.3.1 Simple theory

Figure 2–6 illustrates how the buckets remove the power from the jet. The runner is rotating, and the jet is directed at it, on the PCD. The speed of the jet, v_j, is generally about twice the speed of the bucket, v_b. Because the runner is moving away from it, the speed at which the jet enters the bucket is $(v_j - v_b)$. The jet is split by the ridge, and flows round the two cups to emerge at the sides of the bucket. The inside of the bucket is smooth and evenly curved, so the water does not loose much speed as it goes round. If there were no losses, it would leave the bucket at a velocity of $(v_j - v_b)$ too.

If the jet speed v_j is twice the bucket speed v_b, then the water, relative to the bucket, leaves at a speed of $(v_j - v_b) = (2v_b - v_b) = v_b$. So the bucket is moving at a velocity of v_b, and the water is coming out of the bucket at a speed of v_b, in almost the opposite direction. This means the speed of the water coming out relative to the housing is nearly zero; its energy has been removed by the bucket.

The best imaginable runner would remove all the kinetic energy from the jet, leaving the output water stationary in the air. The water would then fall under its own weight into the tailrace. In practice, this is not achievable. The water leaving the buckets has to have some speed to take it out of the way of the runner, otherwise each bucket would hit the water left by the one in front of it. The angle γ, usually about 15°, gives the water leaving the runner a small sideways velocity to keep it away from the next bucket. Note that the speed of the water leaving represents wasted energy, and needs to be kept to a minimum. The energy in the exit water is termed the *discharge loss*.

Actually there are some losses as the water flows around the inside of the bucket: friction between the water and the bucket, turbulence in the flow, and friction between the water and the air. So the exit speed of the water, v_2, relative to the bucket, is somewhat lower than the entry speed.

$$v_2 = \zeta.(v_j - v_b)$$

ζ (zeta) is the efficiency of the flow in the bucket, and can be as high as 0.98, for a good commercial bucket, though it can be much lower for a poorly finished or badly eroded bucket. For micro-hydro, 0.85 would be more typical, and it can be 0.65 or worse for a very bad bucket (Daugherty, 1920).

The forces, power, and efficiency of a runner can be calculated using this simple theory. If the ratio of the speed of the bucket, v_b, to the speed of the jet, v_j, is called x, this simple theory shows that the runner is most efficient when $x = 0.5$, or when the bucket speed is half the jet speed, regardless of the values of the friction ζ or the angle γ.

2.3.2 Flow into the buckets

Unfortunately, a real runner is not so easily analysed. The simple theory discussed in Section 2.3.1 is two-dimensional, and neglects the fact that the runner is turning, that the buckets move in and out of the jet, and that they rotate while in the jet.

Consider Figure 2–7, which shows three buckets as they interact with a jet. It is drawn for the bucket pattern described later in this manual. At (a) the jet is striking bucket 1, while the end of the splitter ridge on bucket 2 is just about to enter the jet at point X. As the runner continues to rotate, bucket 2 starts to cut off the flow to bucket 1. Diagram (b) shows the position at which the end of the splitter ridge of bucket 2 is just at the bottom of the jet, at point Y. The flow to bucket 1 is completely cut off at (b), but there is still a portion of water, disconnected from the main jet, continuing on towards bucket 1. Remember that the jet is travelling faster than the buckets. Point X' shows where the particle of water at X in (a) has arrived by time (b). The runner continues to rotate, and bucket 3 enters

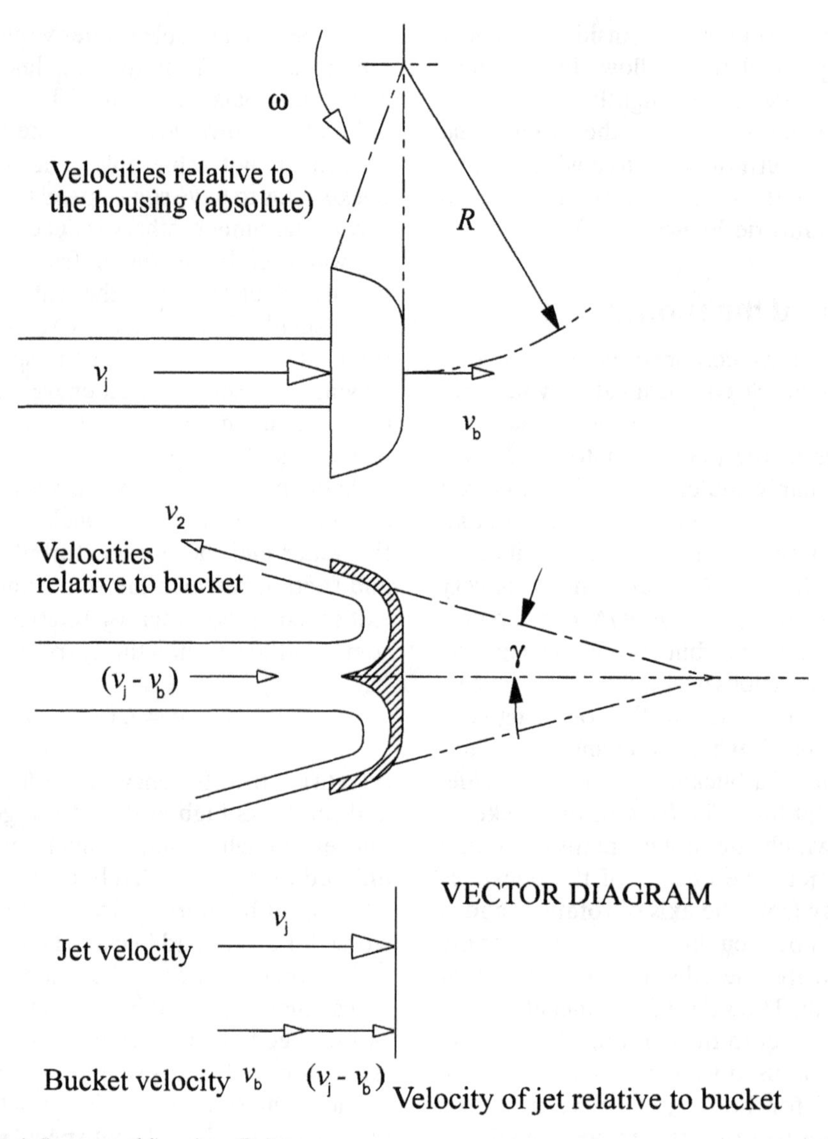

Fig. 2–6: Simplified diagram of flow in a Pelton

the jet. Meanwhile, the cut-off portion of jet catches up with bucket 1. Diagram (c) shows when the rearmost particle of the portion, that was at Y in (b), finally reaches Y' in (c). Bucket 1 now receives no more water until next time round.

Figure 2–7 shows the elapsed time between the various positions if the runner is turning at 1500 rpm. If (a) is at $t = 0$, then (b) is only 1.98 milliseconds later, (c) only another 2.36 milliseconds later. Everything happens very quickly. The runner rotates 7–8° in the time it takes for the water to go around the inside of the bucket.

As the runner speed increases, the cut-off portion takes longer to catch up with bucket 1. There comes a limit at which bucket 1 actually moves off the line of the jet before all the water has hit it. A fraction of the jet is then passing straight through the runner, not touching the buckets. If x is defined as the ratio of the bucket speed at the PCD to the jet speed, then $x = 0$ when the runner is stationary, and $x = 0.5$ when the bucket speed is half the jet speed. For the bucket pattern used in this book (and an 11% PCD jet), a small amount of water at the bottom of the jet starts to pass through at $x = 0.52$. When x exceeds 0.7, the missing water constitutes a major loss of efficiency.

Another effect at high values of x is that the splitter ridge starts to travel faster than the jet. This can be proved mathematically, and a detailed description is given in Section 10.3.2. What it means is that, when x is greater than about 0.6, there are periods when the bucket pushes the jet along. As x increases, these periods become longer, till when x is around 0.8 the buckets are ploughing into the jet when they first come into contact with it. The water is acting as a brake on the runner, rather than helping it turn, and this gives a serious loss of power.

Figure 2–8 shows how the flow within the bucket varies as the bucket moves though the jet. In (a), 15° before the perpendicular, the jet strikes at the end of the splitter, but flows down into the corners near the hub. The extreme edge of the flow may exit from the inner end of the bucket and hit the hub. (b) shows the bucket perpendicular to the jet, with the flow spreading out in the bucket but nevertheless mostly leaving near the centre. Diagram (c)

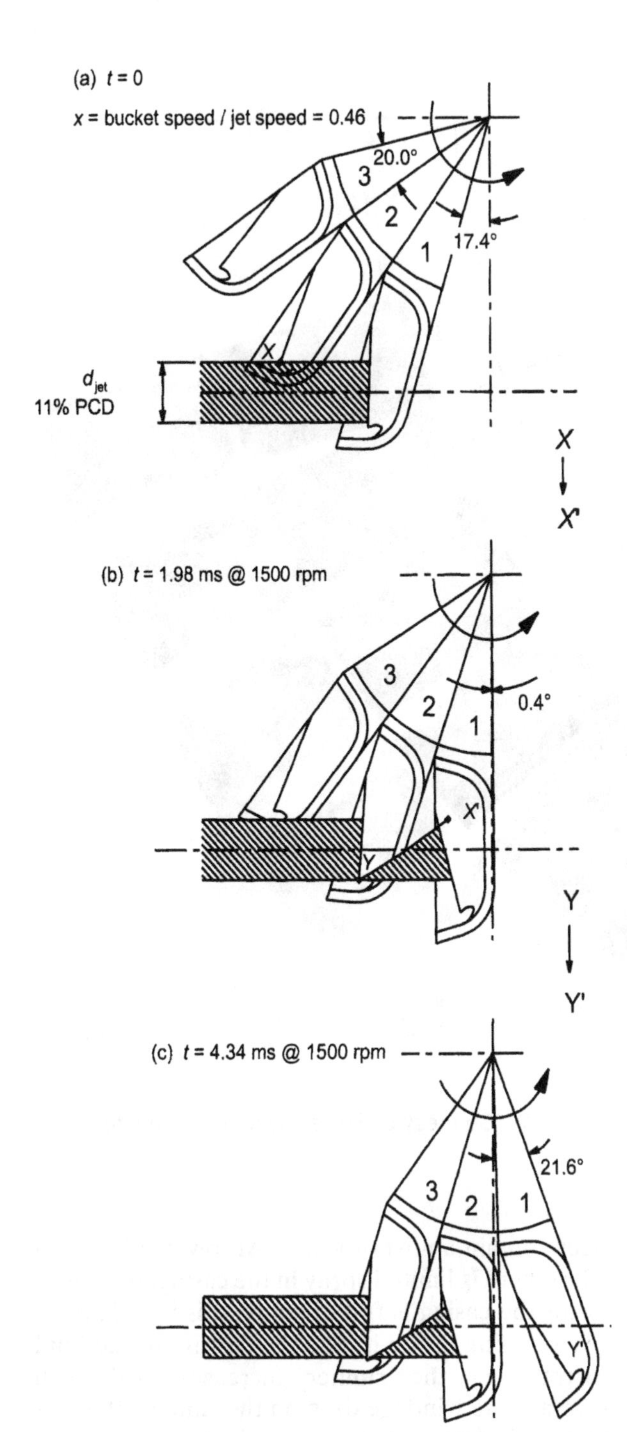

Fig. 2–7: A jet hitting a rotating runner

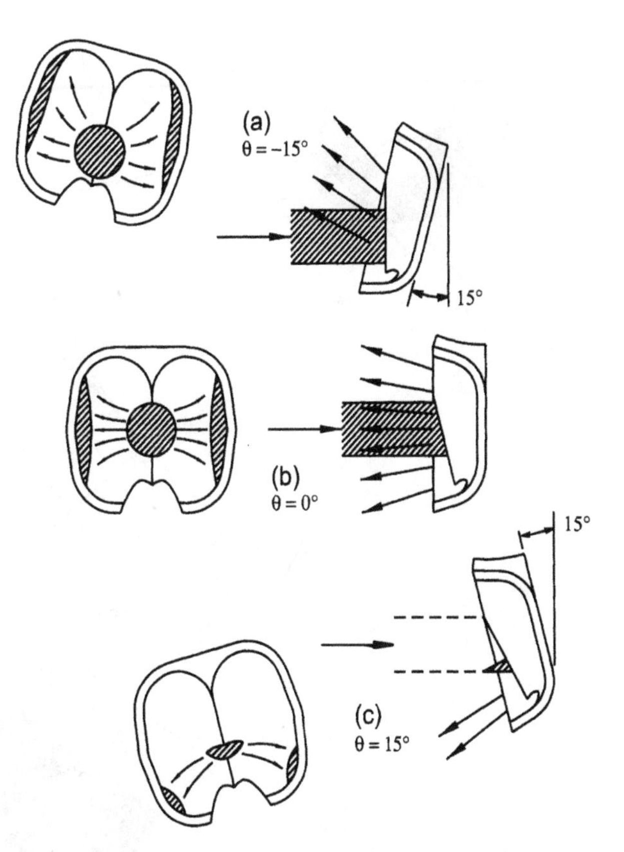

Fig. 2–8: Flow patterns within the bucket at various positions

shows the bucket 15° after the perpendicular. Only a small part of the cut-off portion remains, but this strikes near the end of the splitter, and exits from the outside end of the bucket.

2.3.3 A more complete theory

A Pelton wheel looks deceptively simple, but a complete analysis of its working is extremely difficult. The flow patterns within a bucket are continually changing, so the flow is always transient, never reaching a steady state. This complexity of the flow can be seen in Figure 2–9. The outer ring of figures are high-speed photographs of the jet entering the bucket. The inner ring shows computer generated diagrams of the force on the bucket. The graph shows the variation of the force P_{max} and bending moment in the root, M, as the bucket moves through the jet. The fluid mechanics of two-phase flow – water with a free surface and air above it – is also complex. In the case of a Pelton wheel, it is even worse, because the air is moving at high speed. Even using computers, it has not been possible to produce a complete numerical model of a runner (Kisioka & Osawa, 1972). Some computer modelling has been done of the acceleration of elements of water as they pass over the buckets. Even then, quite a number of assumptions have to be made before the problem can be tackled. The analysis produces traces of the flow paths around the buckets (Brekke, 1984, 1994), and gives many insights into the performance, but it is not a complete mathematical description of the fluid mechanics.

The best Pelton buckets around today have been developed from a mixture of calculation and drawing, backed up by testing. Modern experimental studies use high-speed photography to view the flow in the buckets. However, even though it is not possible to do a rigid mathematical analysis of the flow in a Pelton, the efficiencies being obtained by sophisticated commercial runners are approaching the absolute limit set by geometrical considerations. Further research may lead to efficiency improvements of a fraction of a per cent, but there is not much more to be achieved. It should be noted, in passing, that a fraction of a per cent efficiency,

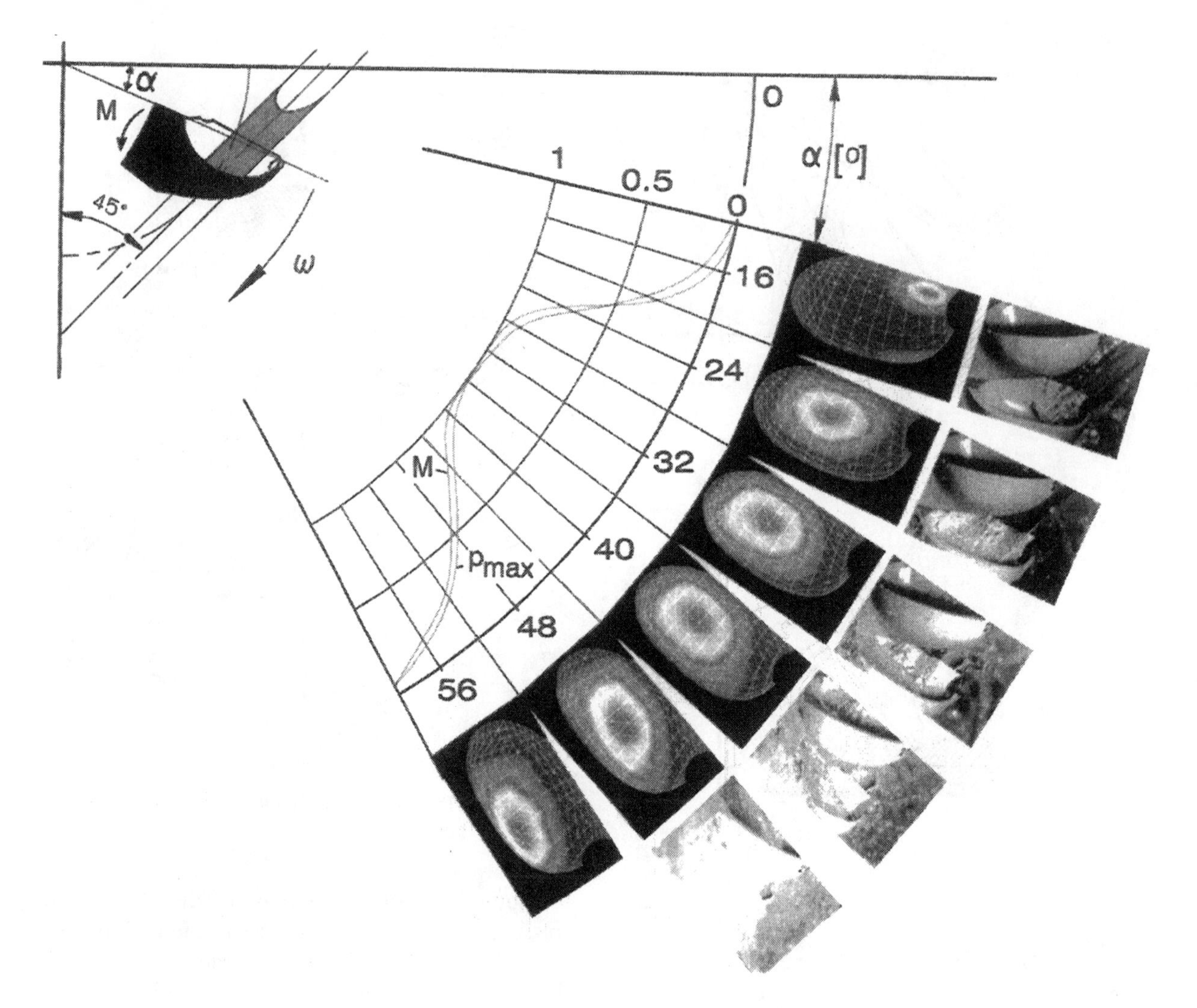

Fig. 2–9: Diagram of various effects on a bucket as the runner rotates (Courtesy of Sulzer Hydro Ltd., Zurich)

while insignificant for a micro-hydro runner, can be important for a very large power-station. While a 1% loss of efficiency in a 10kW micro-hydro plant only equates to 100W, for a 300MW unit it is equal to 3MW – a lot of power.

By making simplifying assumptions, some of the characteristics of a Pelton can be predicted quite well. Even a simple theory shows that the hydraulic efficiency of the runner is reduced by 4–5% by the three-dimensional effects of the runner turning. With a frictional efficiency (η) of 0.98, it gives overall runner efficiency of 91%, which is comparable with the best commercial turbines.

2.3.4 Windage and mechanical friction losses

Some losses arise from the buckets moving through the air and water spray inside the housing; this drag on the runner is termed 'windage'. Windage can be analysed quite simply, as shown in Section 10.3.4. The theory is not very precise, but it does show that the windage power loss is proportional to x^3.

The windage is dependent on the density of the fluid in the casing. The fluid is basically air, but it does contain a mist of water. At low heads (up to 50m) there is limited spray in the casing. At higher heads the casing is full of spray, it is very hard to see anything inside it, and the density of the fluid surrounding the runner increases. This will increase the windage drag on the runner. It is not easy to isolate and measure this drag. The usual way of measuring windage it to drive the turbine externally and measure the power required, but this does not account for the water mist when the turbine is being driven by the jets. In practice, at normal operating speeds, windage is quite small, less than 1%, and so the variation in the fluid density inside the casing is not so important. Brekke (1987) gives a windage loss figure of around 0.5% for 1MW turbines, and states that this can quickly rise to 2% or more away from the design point.

The shaft bearings and, depending on the type, the shaft seals, also contribute to the lost power. The friction torque is approximately constant at all speeds, so the power loss is proportional to speed. Again, it is generally not large, less than 1%.

2.3.5 Overall efficiency

It should be borne in mind that runner losses only represent one part of the losses of the whole system, and this may be a small part. Figure 2–10 shows the components of losses and efficiency within a complete hydropower system.

We are now in a position to discuss the build up of runner losses. Figure 2–11 shows graphically how the final efficiency is built up from the components discussed in the previous sections.

The top line, curve 1, shows the efficiency derived from the simple theory of Section 2.3.1, but ignoring friction and turbulence losses (ζ = 1). The curve is symmetrical, with its maximum at x = 0.5. The loss of efficiency in zone A above the curve represents discharge loss *only*. The slight loss at peak efficiency is in the energy of the water leaving the bucket with a small sideways velocity, and depends only on the bucket side wall angle.

Curve 2 shows the effect of water missing the runner (including the water that is not able to catch up with the runner) at higher speeds, as described in Section 2.3.2. The loss of efficiency in zone B represents energy lost by water that never touches the buckets, so cannot be used. This is not a factor at normal operating speeds.

Curve 3 shows the efficiency calculated using the three-dimensional theory of Section 2.3.2. It is reduced because the momentum force on the bucket does not always act in the most efficient direction, or at the PCD, and because the varying relative velocities of the bucket and jet mean the runner does not take all the power out of the water.

Curve 4 shows curve 3 recalculated introducing skin friction and turbulence into the bucket, with ζ = 0.85. The efficiency loss in zone D represents energy lost into heat in the water.

Finally, curve 5 shows curve 4 with windage and friction losses also subtracted. Efficiency loss in

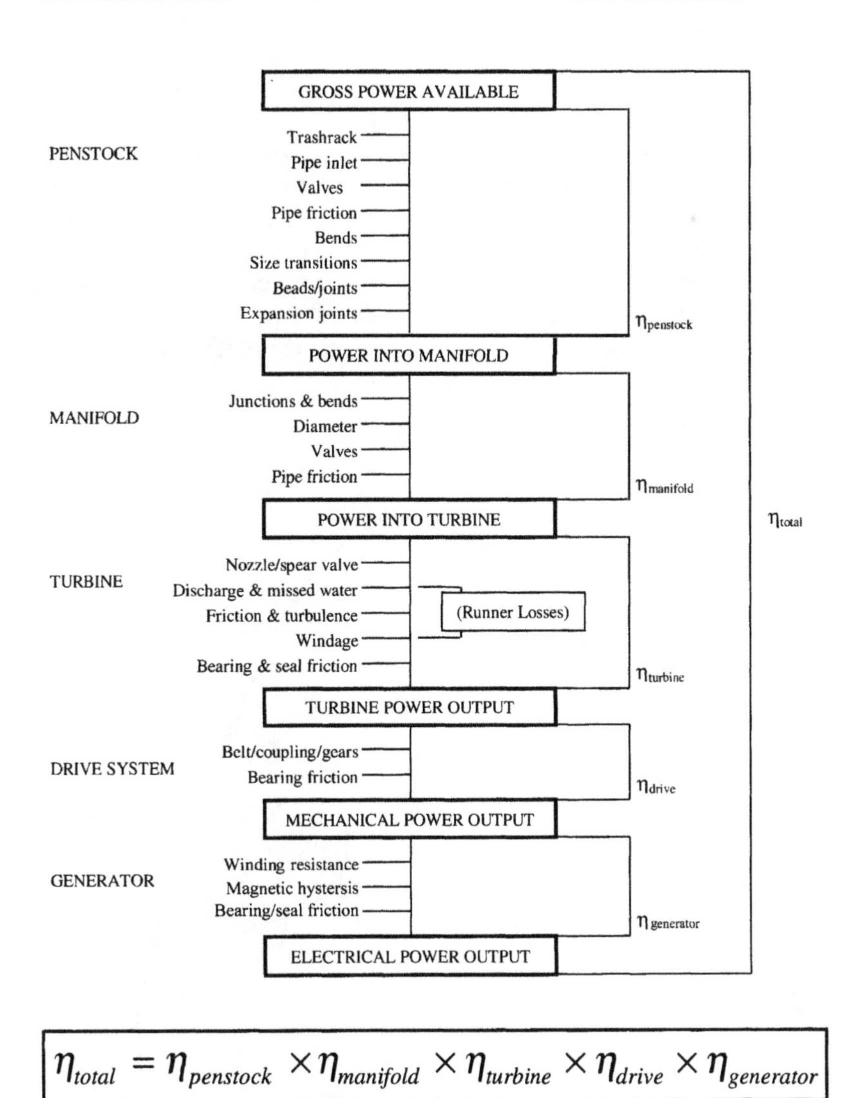

$$\eta_{total} = \eta_{penstock} \times \eta_{manifold} \times \eta_{turbine} \times \eta_{drive} \times \eta_{generator}$$

Fig. 2–10: Power losses and efficiencies within a hydropower system

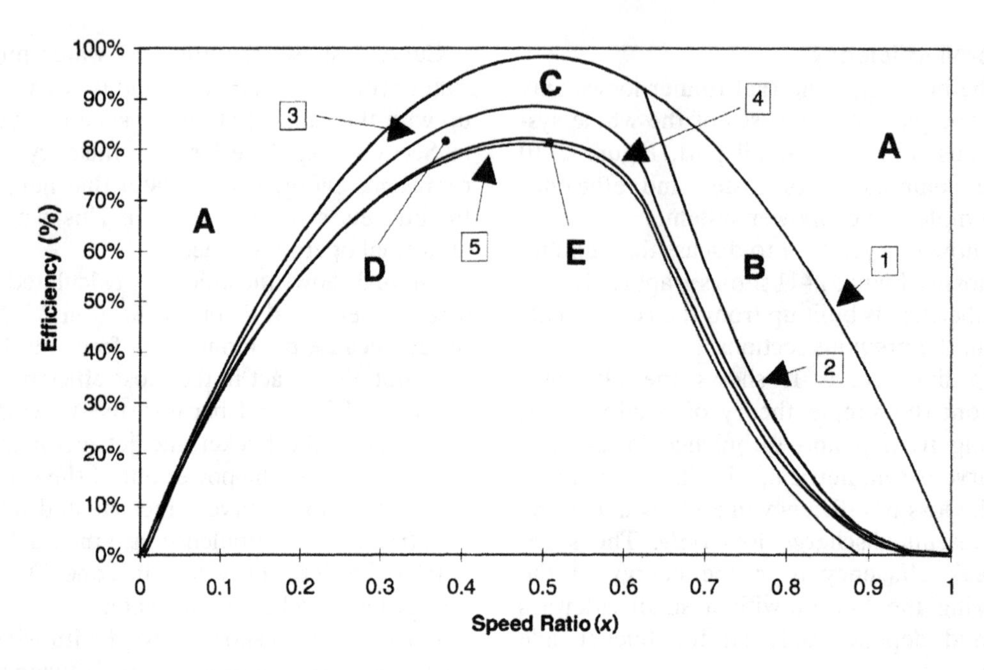

Fig. 2–11: The components of efficiency for a Pelton runner

zone E represents losses to heat in the air in the casing and into the bearings.

Note that the graph is indicative only. It assumes average values for any variables involved, and is the result of simplified calculations. It does help to understand where the power losses occur. The lowest curve, curve 5, is therefore the total efficiency predicted for a runner, and it has most of the features of the efficiency curve for a real runner. An actual curve, such as Figure 2–12, is asymmetrical, falling away faster above the optimum speed than below it.

Note that efficiency varies with the size of the turbine. This happens because the relative effects of gravity, inertia, viscosity, and surface tension vary with scale. This is an important factor in the design of large power-stations. When dealing with power-station turbines that are many metres in diameter, prototype testing cannot be done with full-size runners, and models have to be used. The model may be 10–20% of the full-size turbine, and its efficiency can be 1–2% lower. As yet, there is no accepted way of calculating the scaling effects, though some methods seem to produce quite close

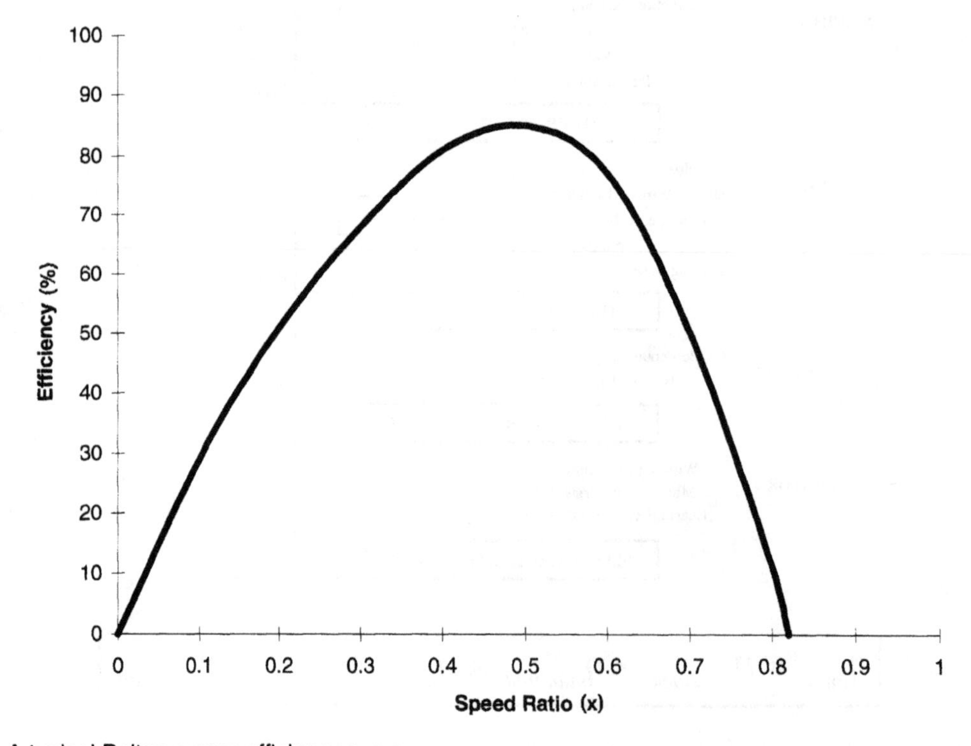

Fig. 2–12: A typical Pelton runner efficiency curve

results (Grein, 1988; Bjerke et al, 1990). Micro-hydro turbines are small enough to be tested full-size, so this is not usually an issue. It is noticeable, though, that very small runners, 100–150mm, have lower efficiencies than runners of 300–400mm or so.

2.3.6 Optimum speed

Earlier it was shown that a simple, two-dimensional treatment of a Pelton gives the optimum bucket speed as half the jet speed, $v_b = v_j/2$, or $x = 0.5$. The optimum speed in Figure 2–11 is just less than $x = 0.5$. This reduction is due to the bearing friction and windage losses, which increase with increasing x and therefore pull down the optimum speed. In fact, the optimum speed is even lower than the simple theory shows, and can be anywhere from $x = 0.42$ to 0.48, 0.46 being a typical value. There are a number of reasons why the optimum speed is less than $x = 0.5$ (Kisioka & Osawa, 1972):

- Windage and friction losses increase with x, so $x_{optimum}$ comes down to keep these a little lower.
- As a bucket cuts into the jet it deflects the remaining part of the jet outwards and retards it. This increases the effective PCD, and reduces the effective jet velocity, so the remaining jet produces its best power at a slightly lower x.
- The flow out of the buckets at $x = 0.5$ has a small absolute velocity, spreads, and hits the following bucket. Away from $x = 0.5$, the outlet velocity is higher, and the efficiency is better.

In summary, the runner produces best efficiency, and highest power output, when the bucket speed at the PCD is less than half the jet speed. This book takes $x = 0.46$ as the optimum speed.

Note carefully that the speed at which a turbine runs is not set by the turbine, but by the load on it. Consider an electrical system. If a turbine/generator is generating 20kW, but is only connected to a load of 10kW, the turbine will speed up. You can only get it to run at its design speed by increasing the load, or by closing the spear valves, until the output matches the load. To predict the actual operating speed, the torque produced by the turbine has to be matched to the torque required by the load, as in Figure 2–13. The point at which the turbine output torque crosses the generator input torque is the actual operating point. In practice, micro-hydro schemes often have Electronic Load Controllers which vary the electric load to keep the speed at the design value.

A turbine does not have to be run at its optimum speed. Moving away ±15% from the optimum speed ($0.39 \leq x \leq 0.52$) only reduces the efficiency by a few per cent, as shown in Figure 2–14 which is the theoretical curve for the efficiencies shown in Figure 2–11. Reducing the speed is actually safer than increasing it, because poor bucket shape, particularly in the notch area, can lead to a rapid fall off of efficiency above the design speed. If the bucket design is proven and tested, varying the operating speed can be useful. It is never done in large power-stations, where every percentage of efficiency gained is important, but in micro-hydro it means you can standardize on a few bucket sizes, and use them over a range of operating conditions.

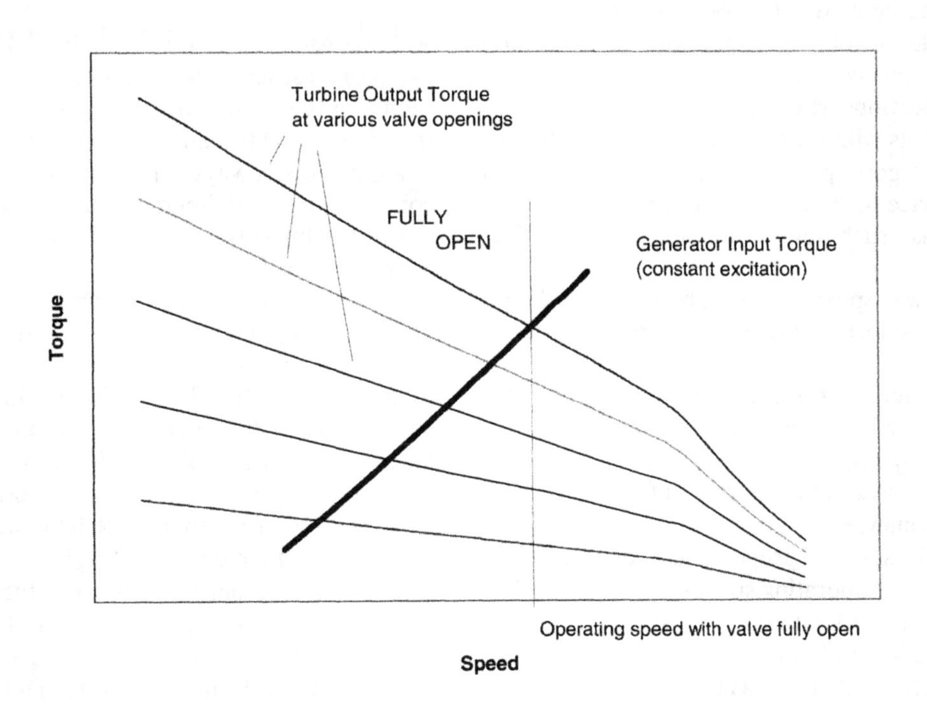

Fig. 2–13: Determining the actual operating speed of a turbine

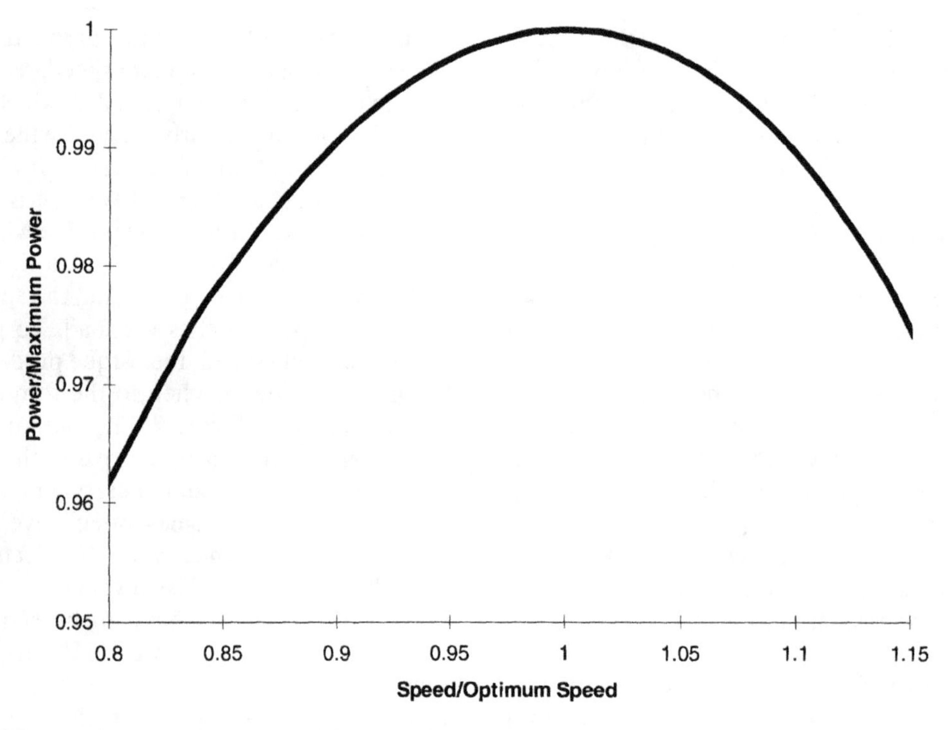

Fig. 2–14: Efficiency/power reduction away from the optimum operating speed

2.3.7 Runaway speed

The runaway speed is the speed the turbine goes to if all the load is removed from it. In a hydroelectric plant, runaway can occur if the generator excitation fails, or the drive belt breaks or falls off. The speed increases rapidly, and it is quite alarming to be in a powerhouse when it happens. The runaway speed is approximately 1.8 times the optimum speed, which means 2700rpm for a turbine normally rotating at 1500rpm – very fast. It is so fast that the windings of a normal generator can be spun off the rotor by the centrifugal force – which makes an unbelievable mess of the generator. (It is common to have a specially strengthened generator for hydro applications where the windings are bonded into their slots with epoxy resin to prevent this occuring.) It is good practice to have an over-speed limiting device to shut down the turbine if it starts to go too fast; such devices are dealt with in a Section 4.4.

The runaway speed is quite a bit less than the jet speed (i.e. $x < 1.0$) for the following reasons:

- Though there is no output power at runaway, some power has to go into the runner to overcome the losses.
- Windage losses are proportional to x^3, and quite large at runaway.
- Friction losses are proportional to x, and are larger than at operating speeds.
- At high values of x, the runner is going nearly as fast as the jet, and much of the water misses the runner. The value of x has to come down before the runner can generate enough power to match the losses.
- At high values of x, the jet actually hits the back of the buckets, braking the runner.

In practice $x_{runaway}$ is about $x = 0.82$, corresponding to a runaway speed 1.8 times the optimum speed. Note that Figure 2–11 slightly overestimates $x_{runaway}$, giving a maximum value of $x = 0.89$ instead of 0.82. This is because only a simplified theory is used, and some of the factors which slow the runner – particularly the last reason given in the list above – are not taken into account.

2.4 Casing design and multiple jets

From the discussion of the flow in buckets, it will be seen that the water emerging from a runner will be travelling slowly, and will be heading nearly straight out sideways, with only a slight backward component. At optimum speed, the water coming out of the buckets has a velocity of about 14–15% of the jet speed.

Figure 2–15 shows the pattern seen in a slow exposure photograph; the inlet is horizontal on the lower right hand side. Most of the water hits the case roughly at 90° to the plane of the area in which the jet hits the runner. It is actually spread out in a diagonal line, and this is typical. The line can often be seen worn into the paint on the inside of a casing. For good efficiency, the casing should catch the water and take it down to the tailrace, preventing it, as far as possible, from falling back onto the runner or the jets.

Multiple jets present further problems. Ideally, all the water from one jet should have left the runner at the point the next jet starts to hit it, but this is not easy to achieve. The bulk of the water leaves the runner over a fairly short angle, but the last bit

Fig. 2–15: Water within a turbine casing

takes quite a while to drip out (Bachmann et al, 1990). Multiple jets also put far more water into the casing, making its design more critical. They make for more spray, more droplets flying around. Mathematical analysis is difficult, but empirical research has lead to a steady improvement in the efficiency of multiple-jet Peltons. The minimum angle used between jets is 60°, so up to six jets can be used, but this is only for vertical axis machines. Horizontal axis machines generally have two jets. The difficulties of making successful multi-jet turbines are such that, unless tests and modifications can be made, the angle between jets should be kept to 80–90° for microhydro.

3
TURBINE SELECTION

For a large power-station the selection of the turbine is a complicated matter. When the output is measured in megawatts, a few per cent extra efficiency amounts to many megawatt-hours of power. It is important to optimize the design to give the best financial return. Occasionally, for a very high-head site, a Pelton turbine is the only possibility, but a given site is often suitable for a number of different types of turbine. (Various alternative types of turbine are shown in Figure 3–2.) Many factors need to be considered, not just the efficiency. Site conditions, such as the amount of excavation needed to place the turbine, are important. Each turbine type has to be evaluated for efficiency at full and part load, the overall installation cost, and operating and maintenance costs.

For micro-hydro, turbine selection is easier. A few per cent difference in overall efficiency is not usually the deciding factor. There may only be one or two types of turbine available anyway. In many countries, micro-hydro is only economical if cheaper, locally-made equipment is used, and this tends to mean cross-flow, propeller or Pelton turbines. Figure 3–1 shows the approximate application ranges of these turbines for micro-hydro. (The diagonal power lines are drawn for an overall efficiency from water to mechanical power of 75%.)

3.1 Pelton turbine selection

Figure 3–3 shows the approximate ranges of head and flow that can be used for a number of sizes of Pelton turbines. Note that the diagonal lines show the shaft output power of the turbines. This chart can be used for initial sizing of a turbine.

A preliminary calculation for Pelton selection can be made quite simply, without needing to know much about the details of any particular manufacturer's turbine. The size of the runner and the number and size of the water jets can be determined from a few simple equations.

3.1.1 Flow equation

The first consideration is the flow. The speed of the water in the jets is dependent only on the head, and the flow is determined by the speed, the area, and the number of jets.

As explained in Section 10.1, the velocity of the jet is given by the equation:

$$v_{jet} = C_v . \sqrt{2g.H_n} \qquad \textit{Eq. 3-1}$$

v_{jet} – jet velocity (m/s)
C_v – coefficient of velocity for the nozzles
g – acceleration due to gravity (m/s²)
H_n – net head at the nozzle (m)

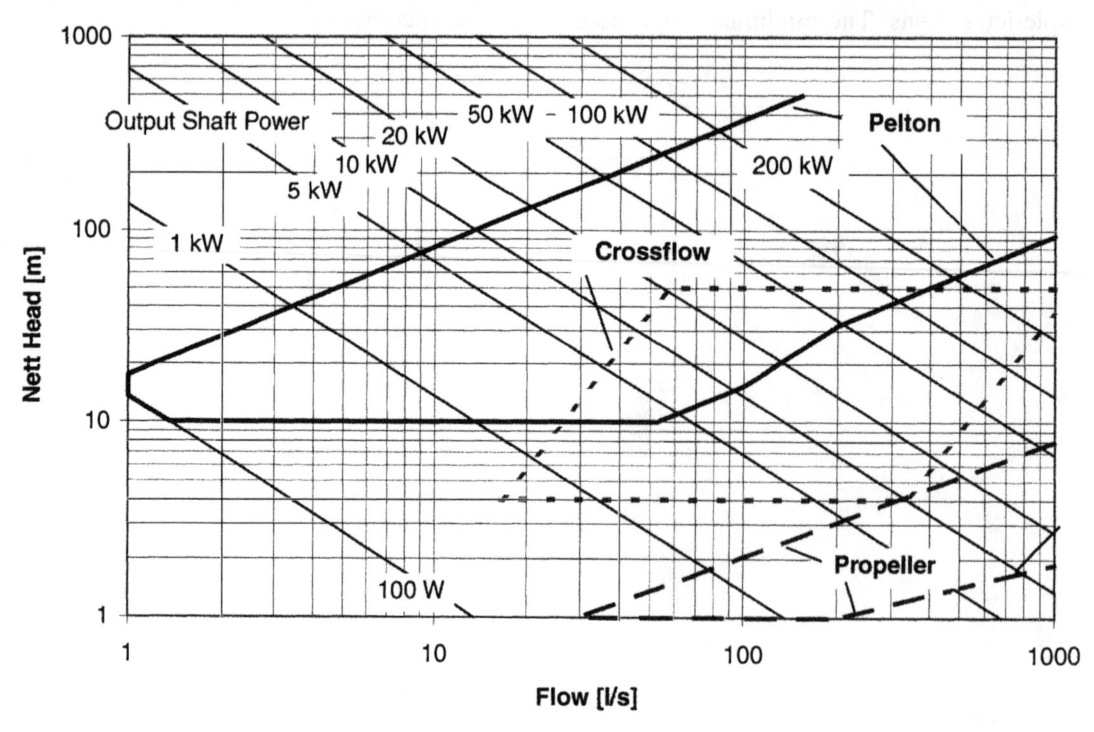

Fig. 3–1: Application ranges of different types of turbine for micro-hydro

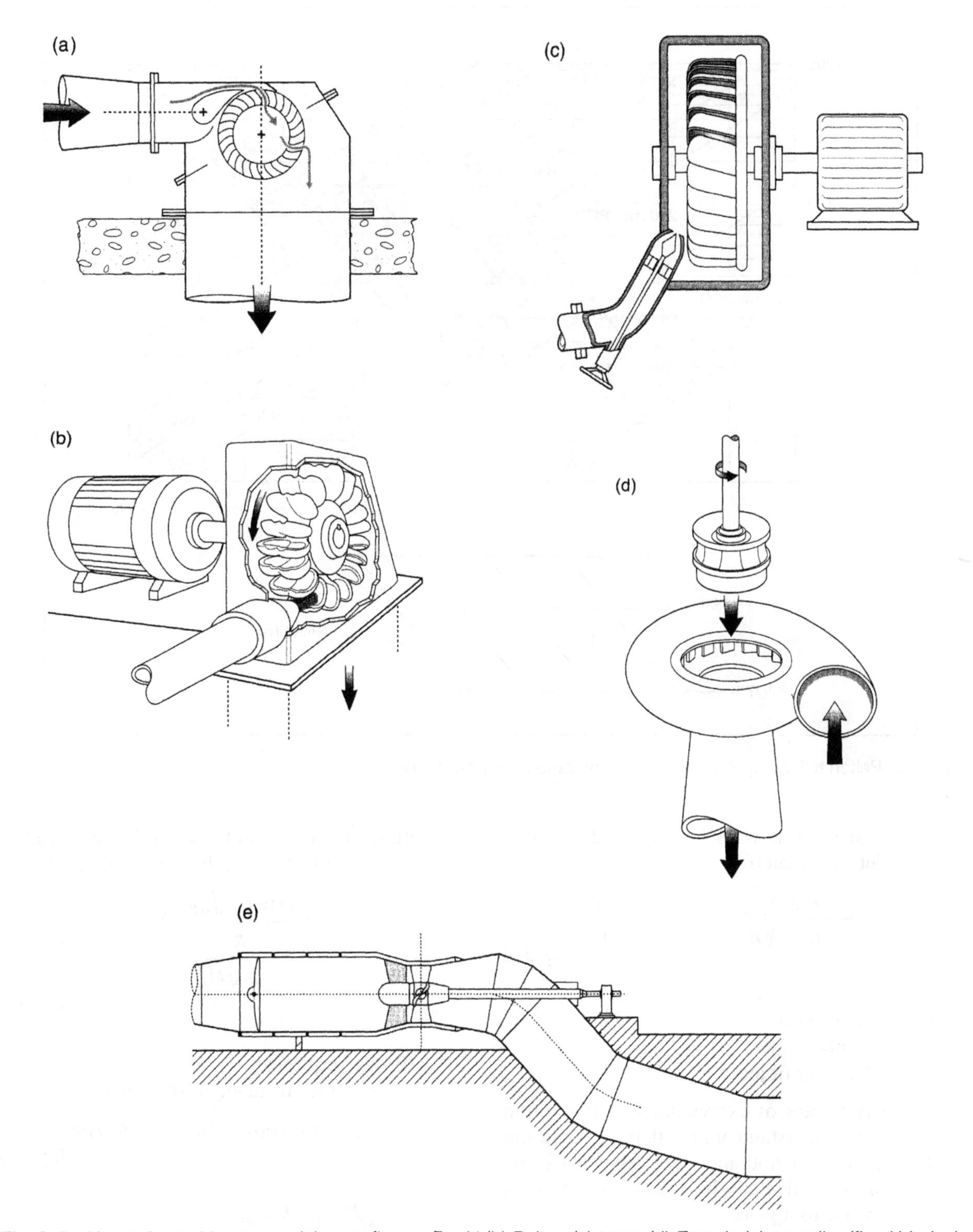

Fig. 3–2: Alternative turbine types: (a) crossflow or Banki (b) Pelton (c) turgo (d) Francis (e) propeller (fixed blades) or Kaplan (adjustable pitch blades)

The flow is then given by this velocity multiplied by the cross-sectional area of the jets:

$$Q = A_{\text{jet}} \times v_{\text{jet}} \times n_{\text{jet}} = \frac{\pi . d_{\text{jet}}^2}{4} . v_{\text{jet}} . n_{\text{jet}} \qquad \textit{Eq. 3–2}$$

Q – flow (m³/s)
A_{jet} – cross-sectional area of each jet (m³)
n_{jet} – number of jets
d_{jet} – diameter of jets (m)

Combining these gives:

$$d_{\text{jet}} = \sqrt{\frac{4}{\pi . C_v . \sqrt{2g}}} . \frac{1}{H_n^{\frac{1}{4}}} . \sqrt{\frac{Q}{n_{\text{jet}}}} \qquad \textit{Eq. 3–3}$$

Using an average value of 0.97 for C_v, this becomes:

$$d_{\text{jet}} = \frac{0{\cdot}54}{H_n^{\frac{1}{4}}} . \sqrt{\frac{Q}{n_{jet}}} \qquad \textit{Eq. 3–4}$$

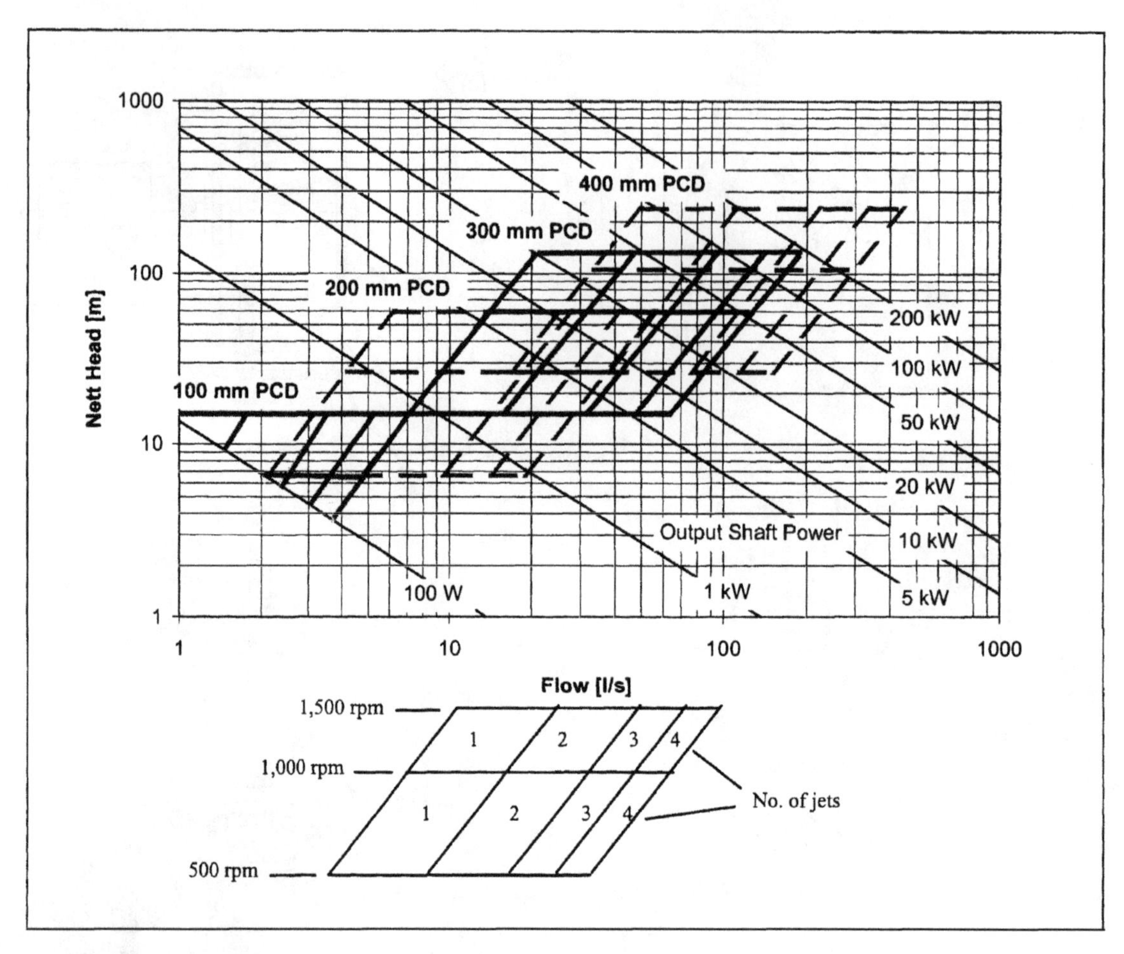

Fig. 3–3: Pelton turbine application ranges for various sizes of PCD

Finally, converting flow to litres/second and diameter of jet to millimetres

$$d_{jet} = \frac{17 \cdot 2}{H_n^{\frac{1}{4}}} \cdot \sqrt{\frac{Q}{n_{jet}}} = \frac{17 \cdot 2}{\sqrt{\sqrt{H_n}}} \cdot \sqrt{\frac{Q}{n_{jet}}} \quad Eq.\ 3–5$$

Q – flow (litres/s)
H_n – net head (m)
n_{jet} – number of jets
d_{jet} – diameter of jet (mm)

Note the two ways of expressing $H_n^{\frac{1}{4}}$. It is often more easy to understand $\sqrt{\sqrt{H_n}}$ than $H_n^{\frac{1}{4}}$, and the double square-root notation shows that it can be worked out on a calculator by entering H and then pressing square-root twice.

3.1.2 Speed equation

The basic speed equation for a Pelton is given by:

$$2\pi \frac{N}{60} \cdot \frac{D}{2} = \frac{\pi N.D}{60} = x.v_{jet} \quad Eq.\ 3–6$$

N – speed of the runner (rpm)
D – pitch circle diameter (m)
v_{jet} – jet speed (m/s)
x – ratio of runner speed at the PCD to the jet speed

Substituting for v_{jet} from Equation 3–1, and using a typical value of 0.46 for x, this equation becomes:

$$N = \frac{60x.C_v.\sqrt{2g}}{\pi} \cdot \frac{\sqrt{H_n}}{D}$$

$$= 37 \cdot 7 . \frac{\sqrt{H_n}}{D} \quad Eq.\ 3–7$$

3.1.3 Power equation

The power output from any turbine is given by

$$Power = Pressure \times Flow \times Efficiency \quad Eq.\ 3–8$$

Or:

$$P = p.Q.\eta_{total}$$
$$= (H_{gross} \times \rho_{water} \times g).Q.\eta_{total} \quad Eq.\ 3–9$$

P – power (W)
p – pressure (N/m^2)
Q – flow (m^3/s)
η_{total} – total system efficiency (unit-less, fraction)
H_{gross} – gross hydraulic head (m)
ρ_{water} – density of water (m^3/s)
g – acceleration due to gravity (m/s^2)

This equation is discussed in more detail in Section 9.4.1.

3.1.4 General procedure

To choose a turbine, the net head and flow for the site need to be known. If the net head is not known, a first approximation can be made from the gross head with penstock losses taken as 5%. Net head is then 0.95 times the gross head (where gross head is the height difference between the water level in the forebay tank and the turbine centreline).

The basic procedure is to calculate the size of jet that is needed to deliver the flow for single or multiple nozzles. From the jet size you work back to the runner diameter, and then use this to calculate the speed of the turbine. This completes the initial calculations. You then have to make some judgements about the number of jets you are going to use, and the most appropriate speed.

Step 1 – Number of jets and jet diameter

Horizontal-axis Peltons very rarely have more than two jets. It is possible to build vertical-axis turbines with up to six jets, but this requires very careful design and manufacture. For micro-hydro, four jets is a realistic maximum number of jets.

To start selection, put the maximum design flow into Equation 3–5 to calculate the jet diameter for 1, 2, 3 and 4 jets.

Step 2 – Runner PCD

A given PCD of runner has a limited size of jet that it can accept. For most bucket designs, this is 10–11% of the PCD, and depends on the bucket shape. For the design of bucket used in this book, the design jet diameter is 11% of PCD.

$$D \leq \frac{d_{jet}}{0{\cdot}11} \qquad \textit{Eq. 3–10}$$

The jet diameters calculated for 1, 2, 3 or 4-jet turbines above can be used in Equation 3–10 to calculate the minimum possible PCDs for these configurations.

If a slight reduction of efficiency is acceptable, the buckets can be used with jets up to 12% of PCD (see the discussion on jet divergence in Section 2.2.3). As a rough guideline, for heads up to about 50m, the loss of efficiency with a 12% PCD is hardly noticeable. Above 100m, the efficiency for a 12% jet can be several per cent less than for an 11% jet. Nevertheless, if plenty of water is available, using a 12% jet can be a good option even at higher heads. A 12% jet has nearly a fifth more flow than an 11% jet, so the power it gives is much higher, even with lower efficiency. If you are prepared for a much larger loss of efficiency, of perhaps 10% or even more, then jets as large as 15–20% PCD can be used, and this may be appropriate in special circumstances. It is advisable to test turbine performance with such large jets before committing to using them.

Step 3 – Available runner PCD

In practice, not every size of runner will be available. Manufacturers will standardize on a series of PCDs, say 200, 225, 250, 300, 350, 400 mm. The calculated figures should be rounded up to the nearest size available.

Step 4 – Runner speed

Equation 3–6 can now be used to find the speed at which each of the chosen PCDs will run.

Step 5 – Gear ratios

Finally, the required gear ratio to convert this turbine speed to the driven equipment speed can be calculated.

3.1.5 Worked example

A site has a net head of 44m and a flow of 40litres/s. The turbine is to drive a 1500rpm generator. Runners are available in 25mm increments from 100mm to 400mm.

No of jets	1	2	3	4
Step 1				
Jet diameter (mm)	42.2	29.9	24.4	21.1
Step 2				
Runner PCD (m)	0.384	0.272	0.222	0.192
Step 3				
Available PCD (m)	0.400	0.275	0.225	0.200
Step 4				
Turbine speed (rpm)	625	909	1111	1250
Step 5				
Gear ratio	2.40	1.65	1.35	1.20

All these are possible. Staying with the easiest options of 1 or 2 jets, the best solution is probably the 2-jet turbine, because it is substantially smaller than the 1-jet version and therefore likely to be cheaper. A 2-jet turbine also allows more flexibility that a single jet, because 2-jets can be used for full flow, but only one need be opened at small flows.

3.1.6 Making a final choice

Having done the calculations, the final choice of turbine may be obvious, but often some experience is required to say which is best.

Direct coupling

Direct coupling of the turbine to the driven machinery has many benefits, and should be aimed for when possible. This means running the driven machinery at the same speed as the turbine. So if a 1500rpm synchronous generator is being used, the turbine must also run at 1500rpm. Eliminating belt drives or gearboxes saves the mechanical power losses associated with them, usually makes the installation cheaper, and reduces the maintenance. Note that Pelton runners need to be carefully balanced to operate at these high speeds; the procedure is given in Section 5.1.4. While standard generators are 4-pole (running at 1500rpm for 50 Hz, 1800rpm for 60 Hz), 6-pole generators are available that run at slower speeds (1000rpm, 1200rpm), and these can be chosen for direct coupling to a slower turbine. However, 6-pole generators are bulkier than 4-pole ones, and cost more. It is also possible to get 2-pole generators that run at 3000rpm or 3600rpm, but these speeds are rather high for a Pelton, particularly at runaway.

One way of achieving direct coupling for electrical installations is to mount the Pelton runner on the generator shaft, eliminating the turbine bearings and a shaft coupling too. If this is done, good attention has to be paid to the shaft seal, because any leakage will obviously go straight into the generator; shaft seals are discussed in detail in Section 4.9. It is well worth checking first whether direct-coupled speed is achievable with the sizes of runner available.

Operating away from optimum speed

Note that Pelton turbines do not have to be run at exactly the optimum speed shown by the calculations. The power/speed curve for a Pelton is fairly flat around the optimum speed, and the turbine can be operated a little way away from the optimum with only a small loss of efficiency. Typically a micro-hydro turbine could be run +10% to –15% away from the optimum speed, with only a 1–2% reduction in power. So, to direct couple to a 1500rpm generator, Peltons with optimum speeds from 1275–1650rpm could be used – see Figure 2–14.

An exception to this rule is turbines with many jets. There is a phenomenon called *falaise* observed with 6-jet, vertical-axis machines, where the turbines are highly sensitive to overspeed and become unstable. If a 6-jet machine is being used, it should not be run above its optimum speed.

Multijet turbines

Horizontal-axis turbines usually have one or two jets. Large commercial power-stations commonly use six-jet vertical-axis machines, but this requires careful design and accurate manufacture. Poor design leads to interaction between the jets and greatly reduced efficiency. Water from one jet needs to leave the bucket before the next jet hits it, and the casing needs to take the exhaust water efficiently into the tailrace. If the turbine is made incorrectly, the water leaving the buckets can hit the next jet, and the exhaust flow and spray can choke the casing. Even commercial turbines suffer a slight loss of efficiency from using six jets. For the above reasons, micro-hydro turbines generally have one or two jets, and hardly ever have more than four.

Small, vertical-axis turbines can be readily made with three or four jets. This arrangement is very common below 5kW, and with a little effort can be used for turbines up to about 20kW. Larger than this, it becomes quite a problem to fit the turbine manifold into the powerhouse and retain reasonable service access. It is not advisable to attempt to use five or six jets unless you have a proven design of turbine or are prepared to do substantial development to make it work.

A horizontal-axis layout is easier to use for larger turbines. The most common layout is to have one or two jets, and standard units made by the larger commercial companies never have more than two jets. This keeps the manifold short and compact, and ensures that the exhaust water is discharged straight into the tailrace. It is possible to make horizontal-shaft turbines with 4-jets that work well, but there are difficulties. For example, Perera (1995) describes a Pelton turbine at Tawakelle, Sri Lanka. This is a 4-jet turbine with two jets horizontal, on opposite sides of the casing, and the other two jets around the top of the casing, so that the angle between the jets is 60°. This machine produces 120kW, the casing stands over 2.4m high, and the manifold reaches to the ceiling of a large tea-factory building. There is no apparent interference between the jets. However, this demonstrates that the manifolds for 3-jet or 4-jet machines tend to be huge and unwieldy, and have to be supported high over the turbine. The exhaust water also needs to be considered. The water from jets that hit the runner on the side where the buckets are going upwards is flung into the upper part of the casing, and care has to be taken that it does not fall back on to the runner or other jets. As a general rule, only use one or two jets for horizontal axis machines.

In summary, up to four jets can readily be used on vertical-axis turbines for schemes of a few kilowatts, and at higher power it is possible with experience. Horizontal-axis machines should normally have one or two jets, and can be used to generate up to several hundred kilowatts or more.

Speed ratio limitations

For direct coupling, the turbine speed has to be the same as the driven machinery. If a belt drive or a gearbox is used, the turbine speed can be greater or less than the driven machinery, but the speed ratio must be within the limits of the drive system chosen.

Typical input speeds for common types of driven machinery are given in Table 3–2 overleaf. These general values are for guidance only, and the recommended speeds should be obtained from the equipment manufacturers before going too far with the scheme design.

The most common types of drive for micro-hydro systems are flat belts or V/wedge belts. Table 3–1 overleaf gives some recommendations for the type of drive that can be used for different sizes of schemes, and the approximate speed ratios that can be safely achieved. These are guidelines only, and manufacturers information should be consulted before using belts.

Table 3–1: Recommended belt types for different scheme sizes

Power	*Up to 8kW*	*8 to 20kW*	*20–50kW*	*Over 50kW*
Recommended type	Any	Any flat belt Wedge belt	Synthetic flat belt Wedge belt	Synthetic flat belt only
Maximum ratio	1:5	1:4	1:3.5	1:3

Table 3–2: Typical input speeds for common types of driven machinery

Driven machinery	*Input power (kW)*	*Typical input speed (rpm)*
Flour grinding mill, twin stone type	4–5	550–750
Rice hulling mill	5–8	900–1500
Oil expeller	4–6	175–250
Saw mill	2–12	300–550
Synchronous generator, 50Hz, 4-pole		1500
Synchronous generator, 60 Hz, 4-pole		1800
Induction generator, 50 Hz, 4–pole		1550
Induction generator, 60 Hz, 4-pole		1860

3.1.7 Further examples

The following examples further illustrate how to choose a turbine. Example 1 is straightforward, and a turbine can be chosen readily. Some of the other examples are more difficult, in that there are no easy choices. They illustrate the limits of the Pelton application range, and the problems of trying to use a Pelton in sites that are not really appropriate.

For all the examples, assume the following:

- The Pelton runner PCD sizes available are: 100, 125, 150, 175, 200, 250, 300, 350, 400, 500, 600 mm.
- All these runners accept a maximum jet size of 11% PCD.
- The minimum recommended jet size is 8% PCD, but it can be taken a bit lower.
- The turbines can be run +10% to –15% away from the optimum speed with little reduction in efficiency.
- Belt drives are available for ratios of 1:4 for less than 20kW and 1:3 for larger powers.
- The overall mechanical efficiency of a micro-hydro plant from water power to shaft power is 70%.
- The overall efficiency from water power to electrical power is 60%.

Example 1

A site has a net head of 38m and minimum flow of 5l/s.

The turbine is to drive an induction generator at 1550rpm.

1a. Select the most suitable runner size, jet diameter and number of jets.
1b. Estimate the electrical output power.
1c. This power is not thought to be enough for the small village, even though they only want a few lights. Further searching reveals another stream that can be channelled from a little way away and added to the existing stream. The net head remains the same but the flow is increased to 15l/s. Select a Pelton for this new data.
1d. What is the electrical output for the increased flow?

Solution to Example 1a

For $H_n = 38$ m, $Q = 5$l/s.

Using Equation 3–5, for 1 jet:

$$d_{jet} = \frac{17 \cdot 2}{\sqrt{\sqrt{38}}} \cdot \sqrt{\frac{5}{1}}$$
$$= 15 \cdot 4\text{mm}$$

The results for 2, 3 and 4 jets can be obtained by using the formula again with $n_{jet} = 2$, 3 and 4. A short-cut is to multiply the value of d_{jet} for 1 jet by $1/\sqrt{2}$, $1/\sqrt{3}$ and 1/2 respectively. The results are tabulated below.

No of jets	**1**	**2**	**3**	**4**
Jet diameter (mm)	15.5	11.0	8.9	7.7
Runner PCD (m)	0.141	0.100	0.081	0.070
Available PCD (m)	0.150	0.100	–	
Turbine speed (rpm)	1549	2324		
Gear ratio	1.00	0.67	–	–

Since 0.1m is the minimum size PCD available, there is no point in filling the last two columns.

A single jet, 150mm PCD Pelton is clearly the best choice. The turbine can be direct coupled to the generator (it would probably be mounted directly on the generator shaft) as it runs at just the right speed.

The two jet, 100mm PCD Pelton runs too fast.

Solution to Example 1b

The only problem here is to guess the various efficiencies. The overall efficiency of 60% is given above, but in this example the *net* head is given, not the gross head, so the penstock/manifold efficiency has been allowed for. The easiest way to tackle this is to guess that the penstock/manifold efficiency is, say, 95%, and obtain the overall efficiency from the net head as 0.6/0.95 = 0.63. Then the electrical output power is (from Equation 3–9):

$$P = 38 \times 1000 \times 9.8 \times 0.005 \times 0.63$$
$$= 1173W$$

Or 1.2kW.

Alternatively, the efficiency for each component could be guessed. Realistic values would be:

Penstock:	95%
Manifold:	98%
Nozzle:	94% (= $C_v^2 = 0.97^2$)
Runner:	80%
Drive:	100% (Direct coupled)
Generator:	80%
TOTAL:	56% (All the above multiplied together.)

This would give an output power of 1043W.

Solution to Example 1c
For H_n = 38m and Q = 15l/s:

No of jets	1	2	3	4
Jet diameter (mm)	26.8	19.0	15.5	13.3
Runner PCD (m)	0.244	0.172	0.141	0.122
Available PCD (m)	0.250	0.175	0.150	0.125
Turbine speed (rpm)	930	1328	1549	1859
Gear ratio	1.67	1.17	1.00	0.83

The 3-jet, 150mm PCD looks the favourite. For a small installation like this it is quite easy to incorporate three jets, and again the runner could be mounted on the generator shaft. (Note that the head has not changed, so a 150mm runner is still required for a speed of 1550rpm.)

Solution to Example 1d
This is trivial. Assuming the efficiencies have not changed (which would need to be checked before a final design is proposed), the flow has increased threefold, so the power should increase likewise, to 3 × 1210 = 3630W or 3 × 1043 = 3129W, depending on the efficiencies assumed.

Example 2
Select a Pelton turbine for the following site:
Net head 120m.
Design flow 144l/s.
To drive a synchronous generator at 1500rpm.

Solution to Example 2
For H_n = 120m and Q = 144l/s:

No of jets	1	2	3	4
Jet diameter (mm)	62.4	44.1	36.0	31.2
Runner PCD (m)	0.567	0.401	0.327	0.283
Available PCD (m)	0.600	0.400	0.350	0.300
Turbine speed (rpm)	688	1032	1180	1377
Gear ratio	2.18	1.45	1.27	1.09

All the gear ratios are possible with belt drives, even for the power here (around 110kW).

The 4-jet, 300mm PCD turbine could be direct coupled, running at 1500/1377 = 1.09 or 9% over its optimum speed. However, for a turbine of this power making a successful 4-jet turbine requires considerable expertise, so this is probably not an option. The same applies for a 3-jet turbine.

The single-jet solution is possible, but a 600mm PCD runner makes a very big turbine. The best solution is the 400mm, 2-jet turbine.

Example 3
Select a Pelton turbine for the following site:
Net head 75m.
Design flow 20l/s.
To drive a variety of milling equipment with a range of input speeds from 200rpm for an oil-expeller to 600rpm for a rice-huller.

Solution to Example 3
For H_n = 75m and Q = 20l/s:
The output power is around 12kW mechanical.

No of jets	1	2	3	4
Jet diameter (mm)	26.1	18.5	15.1	13.1
Runner PCD (m)	0.238	0.168	0.137	0.119
Available PCD (m)	0.250	0.175	0.150	0.125
Turbine speed (rpm)	1306	1866	2177	2612
Gear ratio to 200 rpm	0.15	0.11	0.09	0.08

All of the speeds are too fast for the oil-expeller. A 4:1 reducing drive could only bring 1306rpm down to 327rpm, which is well over the 200rpm required. The others are much too fast. The turbines with 3 or 4 jets are unlikely to be able to withstand the operating speeds, and at overspeed would fly apart.

This shows the limit when the flow is too low for the head. One way round it is to use a single-jet turbine but with an even larger runner to keep the speed down. A 400mm PCD runner would have a speed of 816rpm, requiring a gear ratio of 0.25, or 4:1 reducing. This is possible, but not very elegant. The jet size would be 0.026/0.400 = 0.065 or 6.5% of PCD, which is really too small.

An alternative would be to use a *layshaft*: a separate shaft, on its own bearings, with two pulleys of different diameters on it. The speed reduction can then be achieved in two stages, the first in a belt drive from the turbine to the layshaft, and the second from the layshaft to the oil-expeller. The large reduction ratio would then be easily achieved, but be careful of the extra power losses in the drive system. Another alternative is to drive an electrical generator and run the mill machinery from an electrical motor. This is much more expensive, and has power losses in the generator and the motor. It might allow you to put the mill in the village instead of down by the river.

Example 4

Select a Pelton turbine for the following site:

Net head 75m.
Design flow 415litres/second.
To drive a synchronous generator at 1500rpm.

Solution to Example 4

For H_n = 75m and Q = 415l/s:

The output power is around 200kW.

No of jets	**1**	**2**	**3**	**4**
Jet diameter (mm)	119	84.2	68.7	59.5
Runner PCD (m)	1.082	0.765	0.625	0.541
Available PCD (m)				0.600
Turbine speed (rpm)				544
Gear ratio to 200rpm				2.76

Far too much flow! A 600mm, 4-jet is the only possibility, and yet this for a very high power installation. Be very wary of trying a 4-jet Pelton here! The head is probably too high for crossflow or propeller turbines, too. What can be done?

Using two, 2-jet, 600mm Peltons, would work, but again this is not very elegant. Use a Francis turbine? Or look for another site!

4
DESIGN

This chapter describes how to design the various components of a Pelton turbine. It assumes that the runner PCD, the number of jets, and the size of the jets have been chosen, as described in Section 3.1. It includes recommendations on the optimum shapes and sizes, and calculation procedures.

4.1 The buckets

A Pelton bucket can be designed by drawing and calculation, but the process is tedious. As discussed in Section 2.3.2, the flow through the bucket is complex, with each part of the jet following a different trajectory around the bucket. The design has to consider these different flow paths, and ensure that all the water enters and leaves the buckets at the correct points and in the correct directions. It is complicated, and approximations have to be made. This means that a design done on a drawing board still needs to be tested experimentally, to fine-tune the shape. Even using computers, to date it has not been possible to analyse a Pelton runner adequately, and experimental proving is still required.

In practice, hardly anyone designing a bucket starts from scratch now. Experimental work done over decades has led to a sort of 'natural selection' among designs, and though there are many different manufacturers, the buckets they produce are remarkably similar in shape. Most new designs are based on an existing pattern, modified a little to suit the manufacturer's production techniques, or in line with their most recent test results.

Because bucket design is so dependent on testing, no attempt has been made here to describe calculation and drawing methods for designing a bucket, though some of the general principles are given; more detailed theory is given in Nechleba (1957). This manual considers one particular design of bucket which has reasonable efficiency, but was chosen primarily because it is a simple to manufacture. If the reader possesses a proven pattern, and the correct orientation and diameter for fitting the buckets to a runner are known, that bucket design can obviously be used as an alternative.

4.1.1 Bucket shape

The first part of a bucket that the jet should hit is the splitter ridge (see Figure 2–7). This should be sharp and smooth, so that it cleanly splits the jet into two halves. The force on the bucket comes from it catching the water and taking out as much of its momentum as it can. Getting a good torque from this force needs the force to act at as large a radius as possible. To achieve this, the position of the end point of the splitter ridge is important. The optimum position will have most of the water hitting the bucket when it is nearly at right angles to the jet. The angle of the splitter ridge is set so that it is approximately at right angles to the jet when the full cross-section of the jet is hitting it.

The rest of the internal shape is designed to allow the water to flow freely round, and out at the edges. The water should emerge at as favourable an angle as possible, to give the maximum moment from the momentum change. It is obvious that to do this, the surfaces must be smoothly curved, with a reasonably large radius. In the pattern used in this book, the curves are spherical, but they are usually elliptical.

Ideally, the bucket would take the water around a complete 'U', removing all the energy from the jet, and leaving the water stationary. This is not achievable in practice, as the next bucket would hit the water left hanging in mid-air. The edges of the bucket are consequently angled outwards slightly, just enough to make sure that the water clears the next bucket.

The end of the bucket, between the end of the splitter and the outside edge, it cut away to form a notch, and the shape of this is important. The notch is there to allow the bucket to straddle the jet before the water starts flowing into it. This means that the bucket is at a much better angle to the jet when the water hits it, giving a good flow pattern, and a greater force on it. If there were no notch (as was the case in Pelton's original buckets – see Figure 1–11) the jet would initially hit the outside lip of the bucket and flow straight in towards the hub, where it has no adequate escape route. This significantly reduces the power of the runner. The jet should always strike the splitter ridge first, never the edge of the bucket. The surface between the splitter ridge and the outside end of the bucket has to be shaped so that it directs all the water from the ridge into the bucket cups. If the notch is too deep, it can lead to part of the jet passing straight through, instead of flowing around the insides of the bucket. At very high heads, making the notch too shallow can lead to cavitation problems in the region behind it and lead to stress fatigue, but this does not occur in micro-hydro.

The back of the bucket should be cut away behind the splitter ridge to stop it slapping the top of the jet. This can be seen in Figure 2–7a, where bucket 2 is just entering the jet; unless the bucket is cut away, parts of the splitter to the right of X could hit the top of the jet. The path of the jet relative to the bucket can be easily drawn, and a groove that follows this path can, in theory, be cut into the back to avoid the water. In practice such a groove is too deep, and leaves the bucket very thin and weak, so a compromise has to be reached between slapping and strength. If there is no groove, slapping can cause a marked loss of power. Under very high heads, cavitation can cause pitting in the groove area on the back of the bucket. This is due to high-velocity flow running along the end of the splitter ridge and then being pulled away from it, leaving an area of low pressure. Water vaporizes on the surface, creating small bubbles. As pressure returns to the region, the bubbles collapse rapidly in miniature 'implosions' which can damage the surface, causing pitting. However, since cavitation is generally associated with high heads of 500m or more, it is not a problem for micro-hydro. This is fortunate, as design solutions for this problem are difficult.

Apart from the groove in it, the shape of the back of the bucket is not particularly important. There must be adequate wall thickness near the end and sides, and a good strong attachment to the hub, with no discontinuities to act as stress concentrations.

A detailed theoretical analysis shows that the ratio of the width of a bucket to the runner PCD should vary with head. (More correctly, width/PCD is inversely proportional to the specific speed per jet.) Under lower heads, the water velocities are smaller, and the buckets need to be wider to handle all the water. As the head increases, the buckets should be made narrower, and this makes the runner slightly more efficient. For this reason, the width of buckets for large power-stations are tailor made for the running conditions. However, for heads up to a few hundred metres, the difference in width is quite small. For micro-hydro, it makes sense to standardize on an average width to PCD ratio. The loss of efficiency is negligible.

4.1.2 A bucket design

Figure 4–1 shows a drawing of a Pelton bucket. Its primary advantage is that it is easy to manufacture. It is constructed from simple geometric shapes, and a casting pattern can readily be made from the drawing, as described in Section 5.1.1. The bucket can be made at any size. All dimensions are given as a percentage of the runner PCD. So, for example, if a Ø400mm PCD runner is required, the length of the bucket would be 34% × 400 = 136mm. This means that a bucket could be tailor-made to suit a given site. In practice, it is easier to

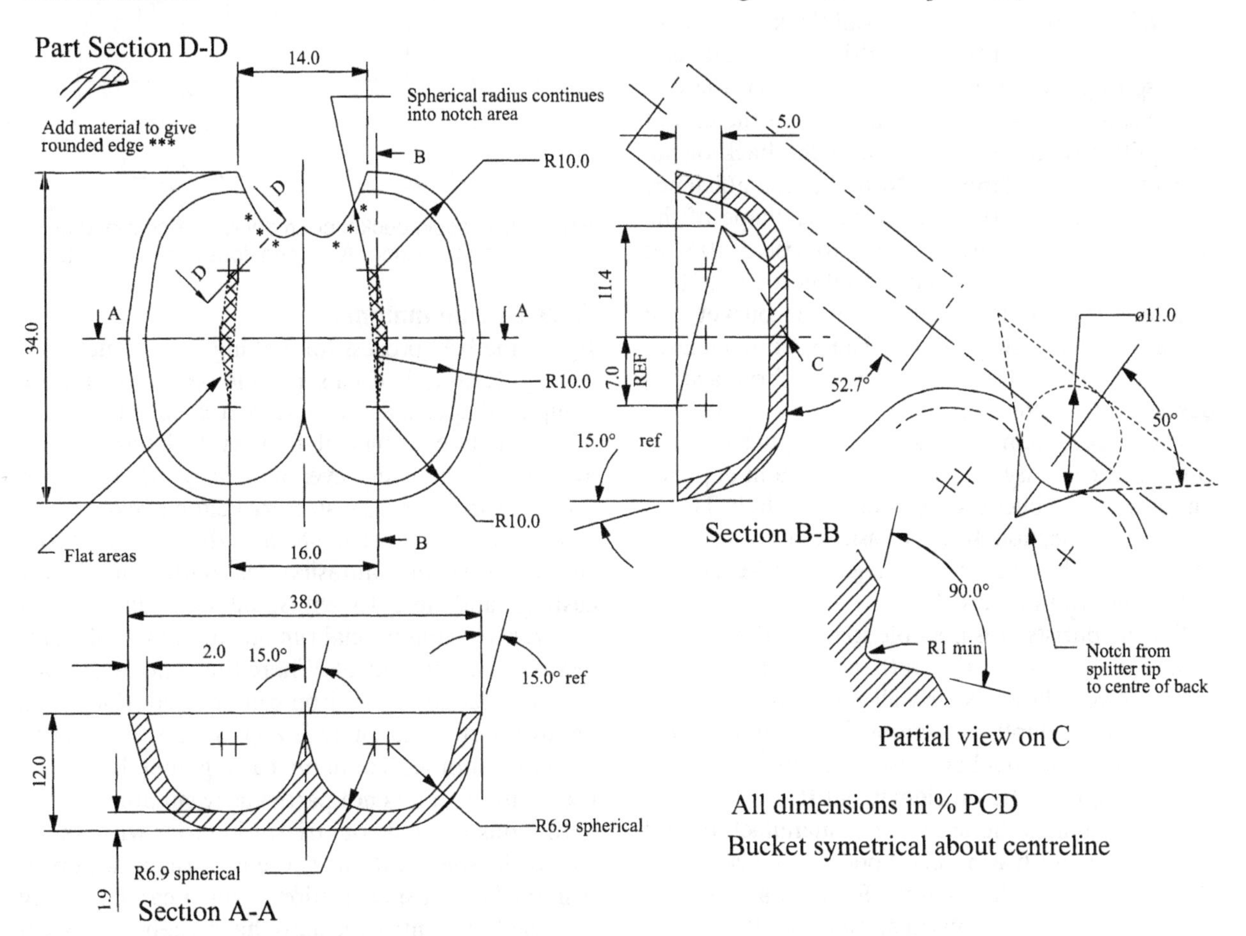

Fig. 4–1: A scaleable Pelton bucket. All dimensions are in % PCD

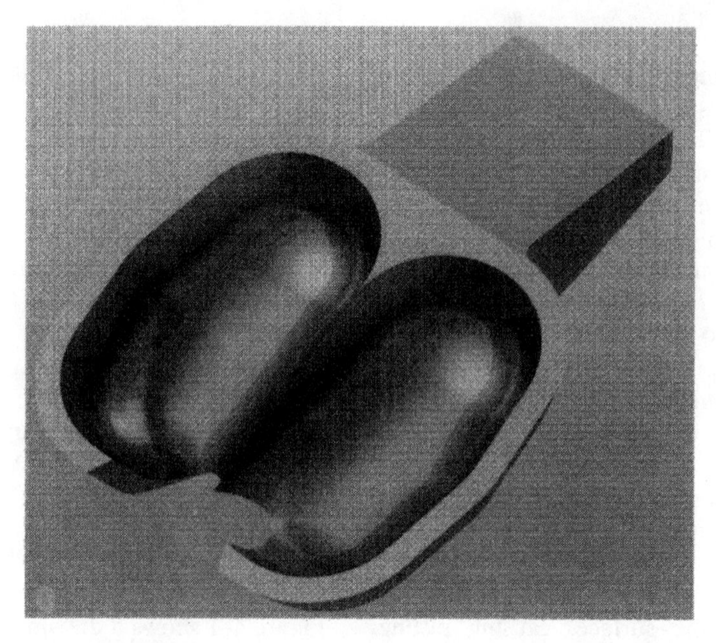

Fig. 4–2: A CAD drawing of the bucket shown in Figure 4–1

standardize on a range of sizes in, say, 25mm steps, and keep these patterns in stock.

The bucket has a flat back and a flat top, which makes it easy to hold in a fixture for machining. In the middle of each cup there is a flat area, shown cross-hatched in the drawing. From this flat, the sides rise up around a radius and then continue on an outward slope of 15°. The wall thickness is constant. The notch is formed by driving an imaginary radiused wedge through the pattern from the end of the splitter ridge. A 90° groove from the end of the splitter ridge to the middle of the back of the bucket prevents 'slapping'. Note that the efficiency of the bucket is very sensitive to the shape of the notch and groove, and it is easy to lose 5–10% of the bucket power with poor detail here. Note that Figure 4–1 shows the final size of the bucket, not the size of the casting pattern. The pattern needs to be made slightly larger to allow for shrinkage, as described in Section 5.1.1.

The bucket stem is not shown in Figure 4–1, because this can vary; the shape of the stem depends on the way the buckets are fixed to the hub. This is discussed in Section 4.2.2. A basic stem for machining, used for bolting or clamping the buckets to the hub, is shown in Figure 4–3.

For the purists, it is possible to make this pattern using an elliptical section rather than the radiused section given above. Nearly all commercial buckets are based on elliptical sections. Using an elliptical section on the bucket design given here may slightly improve the efficiency, but this will not be noticeable unless the accuracy of manufacture and the surface finish are very good. It is also more difficult to make the pattern. Readers are recommended to use the radiused section, but the elliptical section is also shown in Figure 4–4.

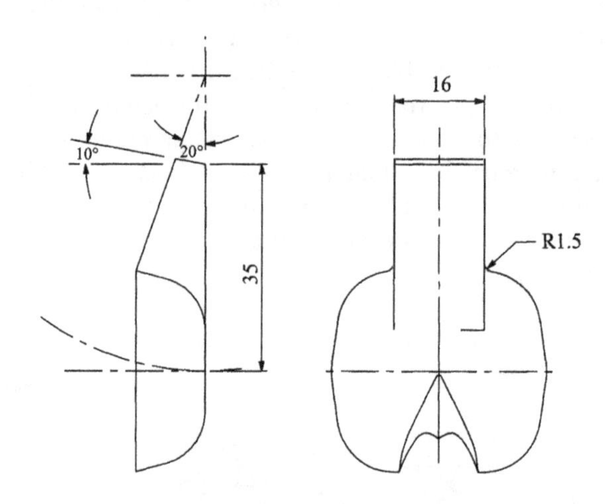

Fig. 4–3: A basic bucket stem design for clamped or bolted fixing. All dimensions are in % PCD

4.1.3 Bucket materials

By far the best process for making Pelton buckets is casting. Casting is cheap, reliable, and can produce complex shapes quickly. Indeed, it is difficult to produce a Pelton turbine if foundry facilities are not available. Fortunately, even a basic, poorly-equipped foundry can make a satisfactory Pelton bucket.

The chief requirements are that the material should be strong, abrasion resistant, suitable for casting, and able to withstand extended use in water. Most commercial runners are made of stainless steel, which is ideal. Where this is not available, another steel or a cast iron can be used. For small, light-duty runners, brass is a good substitute. Aluminium, another common casting metal, can be used, although it is not particularly suitable.

In industrialized countries, foundries will ask for a material specification. There is a horrendous profusion of metal specifications, which can be highly confusing. Countries usually have their own steel composition standards and naming systems, and

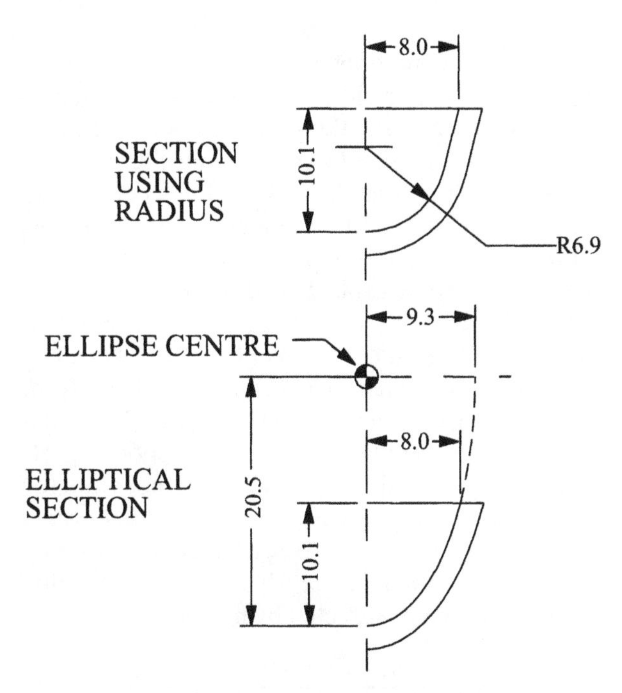

Fig. 4–4: Standard radiused bucket section and the equivalent elliptical section. All dimensions are in % PCD

there is little common ground between them. A good foundry should be able to cross-refer between the systems, and find an equivalent to any given specification.

Small, local foundries in less-developed countries often do not work to specifications at all. They melt down any scrap they can find to form 'brass', 'cast iron', or whatever, of highly variable composition. There is not too much point quoting a standard specification to such a foundry. The best that can be done is to ensure that they use reasonable quality scrap. Try to use scrap from a high-quality product. For example, melting down cylinder blocks from volume production trucks can produce an acceptable cast iron. If you are not sure what material is being used, it is advisable to test the strength of a sample piece, as described in Section 5.1.2. The bucket design given here is quite strong, and is tolerant of poor material.

Aluminium

Aluminium has been used by a number of small manufacturers to make small, low-head runners. Its great advantage is that it is very easy to cast, and relatively cheap. It is by far the easiest material with which to start making runners. The problems with it are that it corrodes in water, it creates a *galvanic couple* with steel parts (including bolts) which accelerates its corrosion (seizing the bolts), and it is quite soft. It is just about acceptable for very small, low-head runners, but even then, its wear rate will be higher than the more normal brass or bronze. It can be argued that its cheapness and simplicity of manufacture more than make up for its shorter life, because the replacement cost is that much less than it would be with other materials. A grade of aluminium that has been used for runners is LM6 to BS 1490.

Copper-based alloys: brass and bronze

Copper-based alloys are probably the best materials with which to start. They are usually called brasses if the primary alloying element is zinc, or bronzes if another element is used. Other metals may be added in small quantities, such as lead to improve machinability. The melting point of brass is low (around 904°C for 60/40 brass), so it can be cast using a simple furnace or fire, making it particularly appropriate for small village workshops. It has good corrosion resistance in water, and is readily machined. Its strength is reasonable, but nothing like as good as iron or steel. It has two main disadvantages. The first is that, unlike ferrous metals, it has no fatigue limit. This means the stresses within it have to be kept very low to make a bucket. Secondly, it is relatively soft. If there is abrasive material in the water it will wear away quickly. Because of these restrictions, brass is really only suitable for suitable for small runners (say up to 200mm PCD).

A basic brass for micro-hydro buckets is 60/40: 60% copper, 40% zinc. This is cheap, readily available, and easy to work. There are some health hazards involved in working with such high-zinc alloys, as the zinc tends to evaporate from the crucible into the foundry atmosphere. The corrosion resistance of this brass in normal water is reasonable, but salinity will eat away the zinc, making the metal porous.

Brasses are covered by the international standard ISO 1191 Part 1, and the usual national standards (e.g. British BS 1400:1985 and the American Unified Numbering System – UNS in ASTM-B30). The addition of a small quantity of tin improves both the corrosion resistance and the strength, and so a better material than plain 60/40 brass is ISO G-CuZn36Sn (BS 1400:1985 Grade SCB4), which is 60–63% copper (Cu), 0.5% max. lead (Pb), 1.0–1.5% tin (Sn), 0.01% max. aluminium (Al), and the remainder zinc (Zn). This is a naval brass suitable for sand-casting with an ultimate tensile strength of 250–310N/mm^2 and an 0.2% proof stress of 70–110N/mm^2.

Bronzes are also used for small Pelton buckets. Silicon bronze is reported to give excellent life and few problems (Shaeffer, 1991). A suitable grade is UNS C87200, which is 89% min Cu, 1% max. Sn, 0.5% max. Pb, 5% max. Zn, 2.5% max. iron (Fe), 1.5% max. Al, 1.5% max. manganese (Mn), 4% silicon (Si) (between 1–5%). It has a yield strength of 172N/mm^2 and a tensile strength of 380N/mm^2.

Grey cast iron

The next step up is cast iron. This is stronger, harder and better wearing than bronze, but can still be readily machined. There are several types of cast iron (grey, SG – also called nodular, malleable etc.), but only grey is considered here. Grey cast iron is the simplest and most common form. Its melting point is around 1450°C, which is attainable by most foundries. It normally has 1.7–4.5% carbon, 1–3% silicon, and low amounts of sulphur, manganese and phosphorous. Excess carbon takes the form of graphite flakes in the metal matrix, and this gives the iron the colour from which its name derives.

It is not advisable to weld grey cast iron. While it is possible, it is difficult to do it well, and requires careful preheating and control. Poor welding leads very easily to weak welds and cracking. For this reason, it is recommended that cast iron buckets are machined to fix them to the runner.

Standards for grey cast iron (for example, the British Standard BS 1452:1977 or the American National Standards Institute ANSI/ASTM A48–76) classify cast irons by their properties rather than their composition. Thus, the BS grade numbers define the tensile strength, in N/mm^2, of a 30mm diameter cast bar tested in the specified fashion. It is up to the foundry to meet the specification. A reasonable strength to use for micro-hydro would be 220N/mm^2 or 260N/mm^2, corresponding to grey cast iron to BS 1452 Grade 220 or 260. Cast iron has low ductility and yields just before it ruptures, so a proof stress is usually quoted instead of a yield stress. An advantage of grey cast iron is that it is relatively insensitive to notches, the endurance limit for a notched sample being only slightly less than for a polished specimen.

Steels

Steel is much harder to cast than iron, and small foundries may not be able to offer it. Its melting point is higher, and it may need to be taken up to 1600°C or more for pouring, which cannot be achieved by basic foundries. It is harder to make clean, sharp-edged castings in steel, and the capabilities of local foundries must be investigated before choosing to use it.

There are a huge number of steels, and the properties of steel can be tailor-made for Pelton runners. Plain carbon steels have only small quantities of alloying elements in them. Alloy steels have elements such as manganese, chromium or nickel in them, and may be further divided into high-alloy or low-alloy steels depending on the amounts of other elements they contain. Steel castings are usually normalized by holding them at a raised temperature for an hour or so, and air cooling. This helps to eliminate hard spots and relieves residual stresses.

Some grades of cast steel are weldable and can be used if the buckets are to be welded to the hub. However, even if the buckets are bolted to the hub, a weldable grade of steel can be useful to weld the buckets for repair. Large commercial runners are commonly reconditioned by building up worn or pitted areas with weld and grinding them back to shape. Generally speaking, the harder and more corrosion-resistant and wear-resistant the steel, the more difficult it is to weld. Weldable steels can be welded with standard equipment and electrodes. Difficult steels require preheating, low-hydrogen electrodes, and controlled conditions.

One method for estimating the weldability of non-austenitic steels is given in Kempe (1989). The percentage equivalent of carbon is calculated from the following formula, where C, Mn, Mo, V, Ni and Cu are the percentages of the corresponding elements in the metal. The formula does not work for austenitic steels, but it is possible to check which type of steel you have with a magnet: austenitic steels are non-magnetic, whereas non-austenitic steels are strongly attracted by magnets.

$$C_{eq} = C + \frac{Mn}{6} + \frac{Cr + Mo + V}{5} + \frac{Ni + Cu}{15}$$

Eq. 4–1

For	
$C_{eq} \leq 0.41\%$	weldable
$0.41\% < C_{eq} \leq 0.45\%$	weldable without preheating but with low-hydrogen electrodes
$0.45\% < C_{eq} \leq 0.60\%$	weldable with preheating and low hydrogen electrodes
$C_{eq} \geq 0.60\%$	welding inadvisable

A suitable basic plain carbon steel (i.e. a steel with only small amounts of alloying elements) for casting would be BS 3100:1991 Grade A1. This has maximum percentages of 0.25% carbon (C), 0.60% silicon (Si), 0.90% manganese (Mn), 0.06% phosphorous (P), and 0.06% sulphur (S). The ultimate tensile strength (σ_{ult}) for a standard test piece is 430N/mm^2 minimum, and the lower yield strength (σ_{yield}) is 230N/mm^2. A stronger weldable steel, but one which may need preheating before welding, is BS 3100:1991 Grade A3 (0.45% C, 0.60% Si max., 1% max. Mn, 0.06% max. P, 0.06% S max.), which has σ_{ult} = 540N/mm^2 min. and σ_{yield} = 295N/mm^2.

Stainless steel is ideal for buckets and runners, if local foundries can produce it. It can have corrosion resistance, wear resistance, machinability, weldability, and strength that are more than adequate for micro-hydro. A suitable grade is 304C12 to BS 3100:1976, which is a low-carbon austenitic steel. This has a composition of 0.03% max. C, 17–21% Chromium (Cr), 8% min Ni, 1.5% max. Si, 2.0% max. Mn, 0.04% max. P, and 0.04% max. S. Its 0.2% proof stress is 215N/mm^2, and its ultimate tensile strength is 430N/mm^2. This is readily welded without preheating or post-welding heat treatment, and it can be machined, though it is prone to work harden. Stronger, higher carbon grades of stainless may need preheating for welding and/or post heat treatment, and can give corrosion cracking from the root of the weld. A commercial grade of stainless steel that has been used successfully for Peltons is 3CR12, which is a cheaper version of 304C12 (with less nickel and chromium). Its corrosion resistance is slightly less than the 304 steel.

Large commercial runners are nearly always made of a chrome-nickel steel similar to ASTM:A743-A2b or DIN 17445: No. 1.4313, which is a 13% Cr, 4% Ni steel. Foundries use an advanced de-oxidizing process to keep the levels of impurities, such as hydrogen, nitrogen, oxygen and sulphur, to a minimum. Specialist foundries may have their own proprietary versions of this steel for runners.

Plastic

An alternative production process to casting is plastic injection moulding. This can be used to make either individual buckets, or complete runners and the components can be very accurate, with excellent surface finish. The unit cost of making each part is low, but injection moulding machinery is expensive and the tooling costs for making metal moulds are high. Consequently, this process is good for high volume production – for making thousands of runners. Micro-hydro manufacturers usually make individual turbines, or maybe even a few dozen at a time, but the volume is far too small for injection moulding to be economic. Suitable materials are acetal and polyurethane.

It is possible to make plastic components in low volumes. This can be done, for example, using urethane resins in silicone rubber moulds (Cunningham, 2000). High quality individual buckets can be made, and it is possible to produce complete runners, though the complex shape of a runner sometimes leads moulders to simplify the shape, giving reduced efficiency.

The main problem with plastic is its softness. It is very quickly eroded, and even in quite clean water a runner will have to be replaced regularly. Plastic is worth considering for small turbines and low heads, particularly if runners or buckets are readily available and cheap.

Sheet steel

Casting can be cheap and repeatable, and can produce excellent buckets. There is something in the engineering mind, though, that looks at a Pelton bucket and wants to fabricate it. Surely by using a press to make the cups, bending a few other parts, and then welding them all together, an excellent runner can be made? This seems an especially attractive option when there are no local foundries, or when the quality of local castings is poor.

Many attempts have been made to fabricate buckets from mild steel. It is quite possible to make a reasonable-looking bucket, but it tends to take a lot of time, and may work out more expensive than using castings. The real problem is that fabricated runners are weak and very prone to fatigue cracking. If you are looking for a small, slow-speed, low-head turbine, with a small jet for the bucket size, a fabricated runner might work. Any attempt to get maximum power from a fabricated runner will probably result in flying buckets.

4.1.4 Stress calculations

If a micro-hydro bucket is going to break, it usually does so at the stem. It is possible to estimate the stresses in the stem using the procedure given in Section 10.4. These calculations are straightforward, but rather time consuming. The general results of these calculations, for the standard designs given in this book, are presented below. If the reader is using a modified stem or bucket design, a different attachment detail, or different material, it is advisable to re-do the stress calculations for your particular case.

There are basically two load cases to check. The first is the centrifugal load at runaway, and the second is the bending moment due to the jet force. As discussed before (in Section 2.3.7), if the load is removed from a Pelton turbine, it quickly accelerates to its 'runaway speed', which is about 80% more than its optimum operating speed. The centrifugal forces on the bucket are proportional to the square of the rotational speed, and so increase dramatically at runaway. This puts a strong tensile load into the bucket stems, which can cause them to snap. Runaway is basically a fault condition, which only occurs when something goes wrong. Nevertheless, it usually does occur at some stage in the life of a micro-hydro turbine, and the runner must be able to withstand the runaway centrifugal forces. Even though automatic jet deflectors, described in Section 4.4.2, may be fitted to stop the runner at, say, 25% overspeed, it is possible for such safety devices to fail to work. Peltons without overspeed protection will invariably go to runaway occasionally.

While runaway is an occasional condition, bending due to the jet force is part of the normal operating cycle. The jet hits the bucket at approximately its mid-point, and this causes a bending moment in the stem. There is no bending moment when the buckets are out of the jet, and the moment builds up and falls back to zero each time a bucket passes through a jet, as illustrated in Figure 4–5. This means that the bending stresses in the stem must be treated as a cyclical load, and the number of cycles can be very high. Fatigue happens very quickly in Pelton turbines. A 2-jet Pelton turbine running at 1500 rpm reaches 10^6 fatigue cycles in under 6 hours. If there is going to be a failure, it can be expected to happen soon after commissioning.

Once the stresses have been calculated, they need to be compared with the allowable stresses for

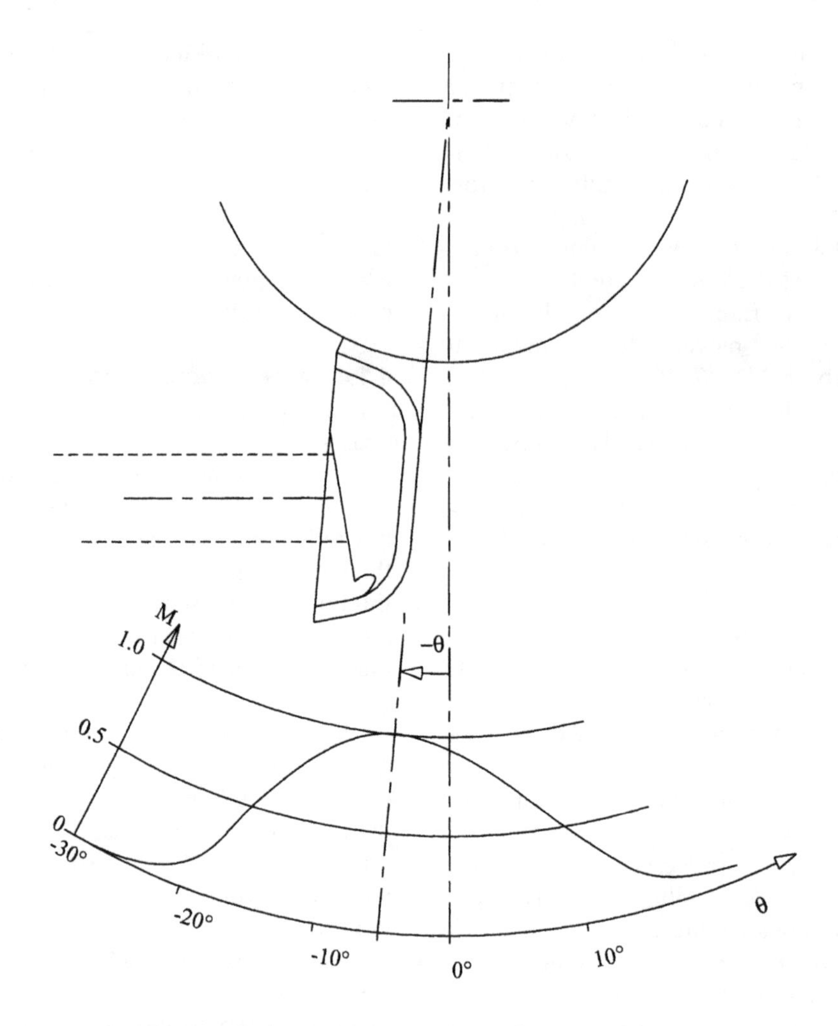

Fig. 4–5: Bending moment in the stem of a bucket due to jet force. Based on Figure 2–9

the bucket material. The allowable stresses for a variety of common Pelton bucket materials are shown in Table 10–1 in Section 10.4.3.

For a given design of bucket, it is found (using the calculations shown in Section 10.4) that both the centrifugal and bending stresses are dependent only on head, irrespective of the size of the runner. This means that a 150mm PCD runner and a 600mm PCD operated under a head of 50m will both experience the same stresses. This unexpected result means that a simple table can be drawn up for limiting heads for a given bucket design. Table 4–1 shows the maximum heads that can be used for the standard bucket design and attachment details used in this book. It is based on the calculations and material properties given in the Section 10.

Table 4–1 shows the maximum allowable heads for both runaway and jet force, calculated for a jet diameter of 12% of PCD, and typical values of other variables. The limiting head for each case is shown in bold. In all cases here, the limiting head is set by the fatigue load. The table shows why high-head runners are always cast in one piece. Clamping and welding are both reasonable for most micro-hydro heads, but care needs to be exercised for higher heads. Note that Table 4–1 only applies to the Pelton design used in this book for the single-piece, clamped, and welded attachment details shown in Figure 10–9 and Figure 10–10 in Section 10.4, and for the material specifications shown in Table 10–1 in the same section. While they may be indicative for other designs, buckets

Table 4–1: Limiting nett heads for various attachment methods for tensile stress at runaway and fatigue stress due to the jet force

Attachment method	*Brass – 60/40*		*Cast Iron – Grade 220*		*Steel – Grade A1*	
	Runaway (tensile)	*Jet force (fatigue)*	*Runaway (tensile)*	*Jet force (fatigue)*	*Runaway (tensile)*	*Jet force (fatigue)*
Single-piece casting	726m	**225m**	1517m	**282m**	1488m	**395m**
Clamped	222m	**112m**	463m	**139m**	455m	**195m**
Welded	–	–	–	–	717m	**191m**

which are different in any way should have calculations done for them.

Large commercial runners are almost always made of 13% Cr, 4% Ni steel. Fatigue life curves for this material were established for this material in a Norwegian research programme in the 1960s, leading to design guidelines used by a number of large turbine manufacturers. The production of this steel is carefully controlled, and finished runners are machined and ground to eliminate surface defects above a certain size. Dye penetration, magnetic-particle, ultrasonic, and radiographic crack detection methods are then used to check that there are no surface defects larger than 2mm × 1mm, and no internal defects larger than 2mm × 2mm. The stress is analysed using finite element analysis, and is kept down to 35–45N/mm^2 (and in some cases as low as 30N/mm^2). The fatigue life is then predicted to be in excess of 10^{10}–10^{11} cycles. Even then, runners are inspected annually, and any defects found are ground out, welded and finished. This level of quality control is not possible for micro-hydro, but neither is it needed. The heads in micro-hydro are usually quite low, the life expectancy is not so high, and it is not too expensive replace or repair a runner after some years if necessary.

The calculations given in this book only consider the bucket stem, but buckets for commercial, high head Peltons are also stressed around the notch area. This is a structurally weak area of the bucket because the notch itself cuts away the support from the outside corners of the cups. If the stresses are too high, fatigue cracks appear, starting from the notch and working their way outwards until the corners snap off. Commercial Pelton manufacturers check these stresses using computer programs which first predict the flows in the buckets and then feed the forces determined from those calculations into a finite element stress analysis. Such techniques are beyond the scope of this book. There are occasions when micro-hydro buckets show cracks developing in the notch; this only happens at higher heads, and is usually because the notch edge has been left too sharp, acting as a severe stress concentration.

4.2 The runner

4.2.1 The number of buckets

As a general rule, a small runner is cheaper to make than a large one. It takes less material to make it, and the housing and associated components can be smaller. The cheapest Pelton turbine for a given site is usually the smallest one that can take the flow. For a given size of jet, a Pelton bucket has to be of a certain minimum size. If it is too small, it cannot handle all the water. The starting point for selecting a turbine is to use as many jets as possible, to keep the jet size small, and then to choose the smallest size of bucket that will accept that size of jet, as described in Section 3.1.

Having chosen a bucket size, it might seem best to make the runner as small as physically possible and, in order to keep the speed high, to use as few buckets as possible and fit them as close as possible to the shaft. The problem is that if there are too few buckets, some of the water in the jet will not be caught. It travels along between the buckets, but never hits any of them. Careful design of the splitter ridge and notch can improve the amount of water caught, but there is a minimum number of buckets required to catch all the water. The exact number depends on the bucket design, but is generally between 18 and 22. This minimum number can be found by calculation or drawing. The calculation procedure follows the general method given in Section 10.3.2.

It is possible to use a much larger number of buckets. In extreme cases, this gives the appearance of small buckets fixed on to the edge of a large disc. The resultant runner turns relatively slowly, but may be appropriate for a special application. For the bucket design used in this manual, the minimum number of buckets is 18, with each bucket 20° apart. All the subsequent discussion will assume an 18-bucket runner using this pattern, giving the cheapest and most efficient runner, although it should be remembered that it is possible to make a runner using this pattern with more than 18 buckets.

4.2.2 Attaching the bucket to the hub

There are a number of way of mounting Pelton buckets on the runner hub. The whole runner can be cast as one piece. Alternatively, the buckets can be made as separate items and bolted, clamped or welded to the hub.

Single piece castings

With the improvement of casting techniques in the 1960s, it became possible to cast runners in one piece, even for turbines many metres in diameter (Figure 4–6). Runners for large power-stations are nowadays usually made in this way. (The largest runners are made in sections, with several buckets on each. The sections are then welded together to make the runner.) The advantages are that the runner is stronger, and the buckets can be mounted close together on the hub to give the optimum PCD for a given bucket size. Single-piece casting does, however, require specialist skills. (There are only a handful of foundries in the world that can make really large Pelton runners.) A good foundry should be able to make smaller, micro-hydro-sized runners, but by no means all foundries are capable of doing so.

If single-piece casting is available, it is a good technique for micro-hydro. A 'monobloc' runner is usually cheaper than a runner with clamped or bolted buckets, because no machining is required to fix the buckets to the hub. A single-piece runner

Fig. 4–6: Three sizes of Pelton runners cast as single pieces (Evans Engineering Ltd.)

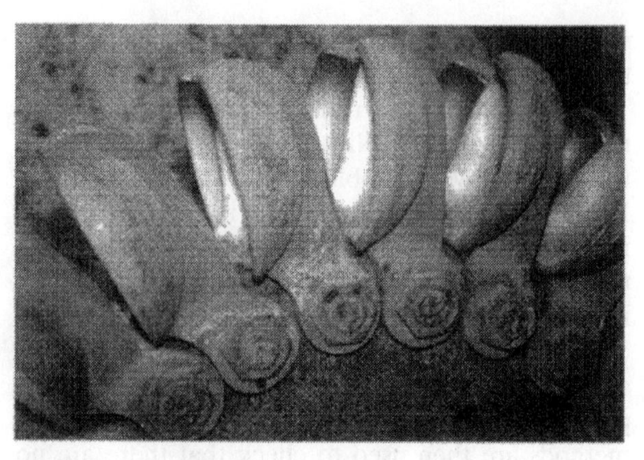

Fig. 4–7: An old Pelton runner showing buckets bolted to the hub

is often cheaper than a welded runner, too. The casting process needs to be controlled well, as one weakness in the material – a blow-hole, crack, or slag inclusion – can cause a bucket to shear off. (Even the major Pelton manufacturers, using runners from experienced, top-quality foundries, still have occasional breakages.) It is a good idea to make the runner from weldable steel, so that a breakage can be repaired if necessary, without having to scrap the whole runner.

Bolted buckets

Prior to the 1960s, it was conventional to bolt individual buckets to the hub (Figure 4–7). The buckets were first machined, and then carefully fitted on to the runner. To improve the strength, and to maintain a constant pitch, each bucket would also be attached to the buckets on either side of it (see Figure 4–8(e)). Machining and assembly was time-consuming and costly. The bolting arrangement took up a lot of space, and meant that the buckets were spaced further apart than was ideal. While, in theory, it was possible to stock spare buckets and replace one if it got damaged, in practice each bucket was individually machined and hand-fitted, so replacing one was a skilled job. All these factors led to the switch to monobloc castings.

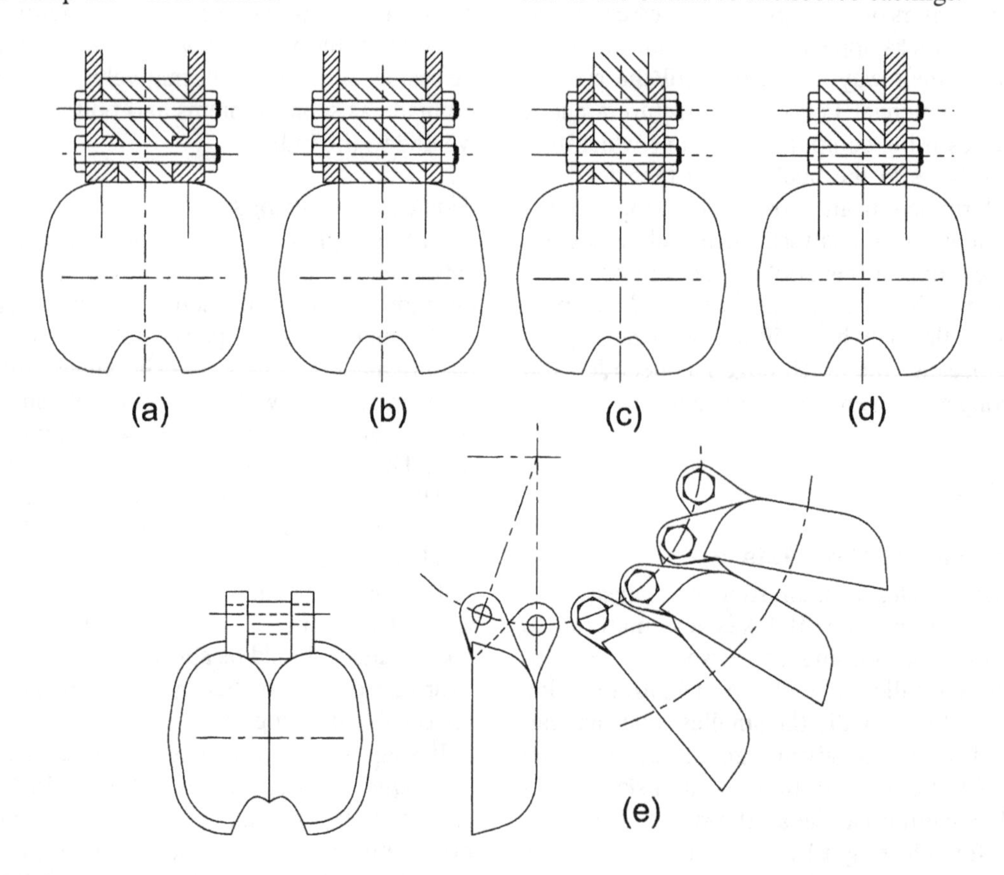

Fig. 4–8: Various means of bolting Pelton buckets to the hub

There are numerous ways that a bucket can be bolted to a hub, some of which are shown in Figure 4–8. Arrangements (a) to (d) all locate the bucket with two bolts through the stem. Arrangement (e) is similar to Figure 4–7, where each bolt locates two buckets. The loads on the bolts can be quite high in any of these arrangements. Historically, bolted attachment has always been done with precision or 'dowel' bolts that are a tight fit in their holes. The load is then carried in shear by the bolt material. There must be no clearance in the holes, and the bucket is held rigidly by the bolts. High strength material is usually used for the bolts. How the bolt loads can be calculated is discussed below. These bolted arrangements can be reliable up to quite high heads, but the machining has to be accurate. If numerically-controlled machining centres are available, this sort of arrangement can be produced relatively easily and it is possible to make the buckets interchangeable for servicing. Without such machinery, skilled fitting is required to make a precision-fit bolted arrangement work.

If the bolts are not a precision fit, then the forces have to be carried by the friction generated by the bolts' pre-tension. Unfortunately, there is only room for quite small bolts in the stem of a bucket, and it is difficult to get enough clamping force. Friction clamping alone can only be used for very low heads. Calculation of the bolt loads and clamping forces is discussed below.

Arrangement (a) in Figure 4–8 can be used as a compromise. This design uses the basic stem from Figure 4–3. The bolts can be fitted in clearance holes, but additional support is provided from the hub sides. These are made a good press fit into the recesses in the sides of the stem, so that the bucket is primarily located by this fit. This is much easier to machine than a precision-fitted arrangement, and has been successfully used for micro-hydro up to 100m head or so. If the hub is not a press fit into the bucket, then the load has to be carried by friction, and the allowable head is much lower. Note carefully too that the bolted bucket stress calculations in Section 4.1.4 assumed this press fit. If the buckets are loose the allowable head is not so high.

The basic theory for calculating bolt loads is given in Section 10.5. This gives the general formulae so that any bolt arrangement can be tackled, but also derives the specific formulae for the two-bolt arrangement in Figure 4–9. The force on each bolt is made up of two components, one which is the direct shear force from the jet force, S, and the other, T, which comes from the moment generated by the jet force. The shear component in the bolts is

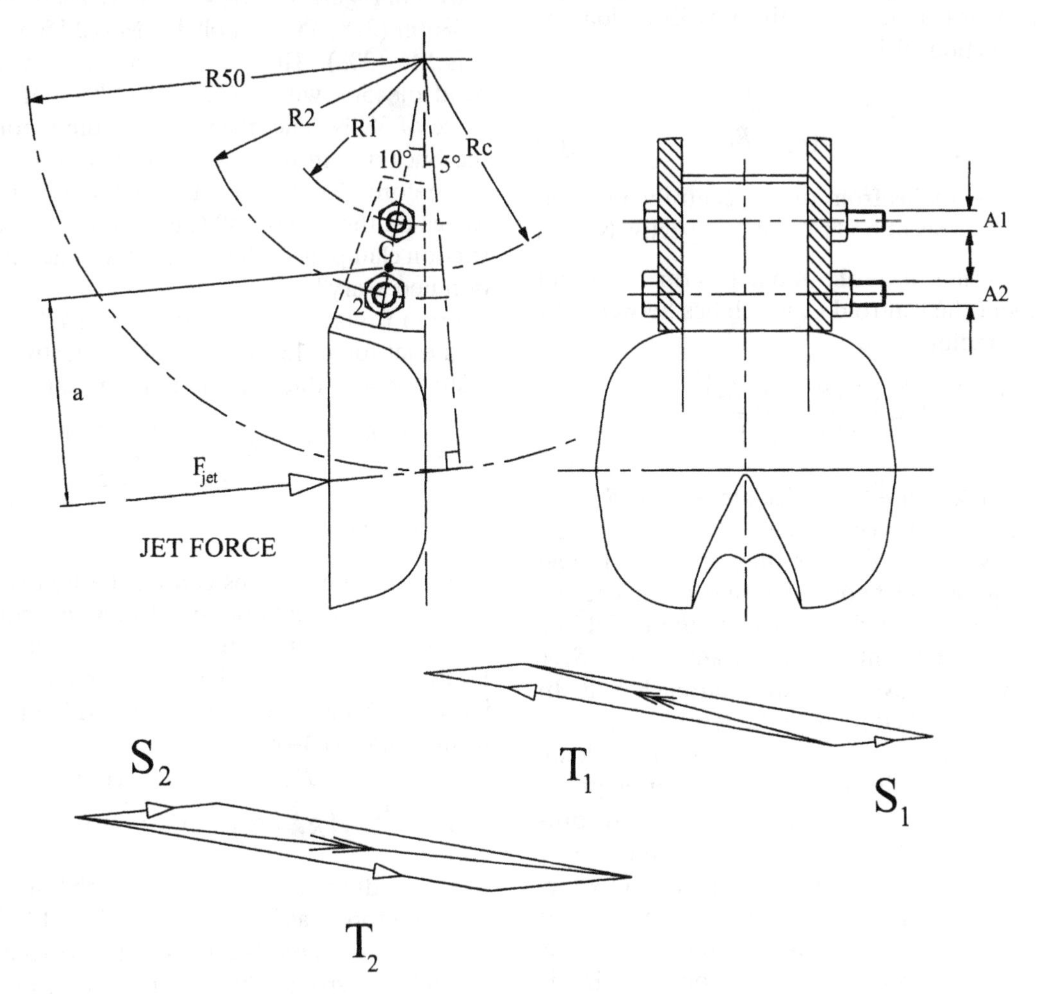

Fig. 4–9: A two-bolt clamping arrangement

Table 4–2: Bolt and bolted joint properties

		Grade 4.6 black mild steel				*Grade 8.8 high tensile steel*			
Bolt dia.	*Pitch*	*Pre-tension*	*Torque*	*Friction load*	*Shear load*	*Pre-tension*	*Torque*	*Friction load*	*Shear Load*
(mm)	(mm)	(kN)	(Nm)	(kN)	(kN)	(kN)	(Nm)	(kN)	(kN)
M5	0.80	2.7	2.6	0.5	3.5	6.9	6.8	1.2	7.0
M6	1.00	3.8	4.5	0.6	4.9	9.8	11.6	1.7	9.8
M8	1.25	6.9	11.0	1.2	9.0	17.8	28.2	3.0	18.0
M10	1.50	10.9	21.7	1.9	14.3	28.1	55.9	4.8	28.7
M12	1.75	15.9	37.9	2.7	20.9	40.9	97.5	7.0	41.9
M16	2.00	29.6	94.4	5.0	39.5	76.0	242.9	12.9	79.2
M20	2.50	46.2	184.4	7.9	61.8	118.8	474.4	20.2	123.7
M24	3.00	66.5	318.7	11.3	89.0	171.1	819.7	29.1	178.2
M30	3.50	105.8	634.7	18.0	142.4	272.1	1632.5	46.3	285.2

given by the following formula, which is the same as Equation 10–78 in Section 10.5.2:

$$S_1 = F_{jet} \cdot \frac{A_1}{(A_1 + A_2)}$$

$$S_2 = F_{jet} \cdot \frac{A_2}{(A_1 + A_2)} \qquad \textit{Eq. 4–2}$$

A_1, A_2 – cross-sectional area of bolts 1 and 2 respectively

The moment component is given by Equation 10–81 from Section 10.5.2:

$$T_1 = T_2 = \frac{F_{jet} \cdot a}{(R_2 - R_1)} \qquad \textit{Eq. 4–3}$$

R_1, R_2 – radius from runner centreline to centres of bolts 1 and 2 respectively

The dimension '*a*' is the moment arm of the jet force about the centroid C, which lies between the bolts at a radius:

$$R_c = \frac{(A_1 . R_1 + A_2 . R_2)}{(A_1 + A_2)} \qquad \textit{Eq. 4–4}$$

Equation 4–2 is used to calculate S_1 and S_2, and these are drawn as vectors parallel to the jet force. Equation 4–4 is used to plot the position of C, and then Equation 4–3 finds the values of T_1 and T_2, which are plotted at right angles to the radial line between the bolt centres. Vector addition of $S_1 + T_1$ and $S_2 + T_2$ gives the force on each bolt. In practice, T_1 and T_2 dominate the result.

The bolt must be able to take the resultant load. For precision fit bolts, the effective 'bearing pressure' on the bolt hole (the force divided by the projected area of the hole) must be less than the yield stress of the material, with an adequate safety factor. (Obviously, the bolts must not fail in single or double shear – whichever is relevant – but this is unlikely to be a problem.) If friction clamping is being used, the friction load must be greater than the resultant force, plus a safety factor. Approximate friction loads, with other properties, are given for various bolts in Table 4–2. The shear load is the failure load for single shear across the core diameter of the thread, assuming the shear stress is 0.7 UTS. If friction clamping is being used, the bolts *must* be torqued up as in the table.

Numerical example

A 200mm PCD Pelton wheel is bolted to the hub as shown in Figure 4–9. R_1 = 40mm (20% of PCD), R_2 = 58mm (29% PCD), bolt 1 is M5 (2.5%), and bolt 2 is M6 (3%). The turbine operates under a net head of 45m, with a Ø24mm (12%) jet, giving a force of 389N. The maximum bending moment on the bucket is in the position shown in Figure 4–9. (5° before Bottom Dead Centre). What are the forces on the two bolts? Can this be carried by the friction clamping of the bolts, or is some other location necessary?

From Equation 4–2, *S*1 and *S*2 are 41% and 59% of the jet force, 159N and 230N respectively. From Equation 4–4, the centroid is at a radius:

$$R_c = \frac{(A_1 . R_1 + A_2 . R_2)}{(A_1 + A_2)} = \frac{(\frac{\pi}{4} \times 5^2 \times 40 + \frac{\pi}{4} \times 6^2 \times 58)}{(\frac{\pi}{4} \times 5^2 + \frac{\pi}{4} \times 6^2)}$$

$$= 50 \cdot 6\, mm$$

(Note that the π terms cancel, the bolt diameters can be in any units provided they are consistent, and R_c comes out in the same units as R_1 and R_2.) By drawing (or calculation) the moment arm of the jet force about the centroid is a = 51.2mm. Finally, from Equation 4–3:

$$T_1 = T_2 = \frac{F_{jet} \cdot a}{(R_2 - R_1)} = \frac{389 \times 51 \cdot 2}{(58 - 40)} = 1{,}106\ N$$

Vector addition of the forces on each bolt gives the resultant forces as F_1 = 953N and F_2 = 1330N.

Now, from Table 4–2, the friction loads are 500N and 600N respectively for Grade 4.6 M5 and M6 bolts. This is less than the bolt load, so the clamping is insufficient by itself. For Grade 8.8 bolts the figures increase to 1200N and 1700N, which are OK, but with a safety factor of only about 1.3. Friction clamping alone would be acceptable for Grade 8.8 bolts, but it would be advisable to go up to Grade 10.9, and even then the bolts should be torqued up carefully.

These calculations can be generalized a little. As with other forces on Pelton wheels, the maximum acceptable head for clamping is independent of the actual size of the runner, provided everything is scaled equally. By making a few assumptions about bolts (that the stress area is 73% of the nominal bolt area), the maximum heads for friction clamping to work – for Peltons which are exact scale copies of the above example – are given in Table 4–3. A safety factor of 1.5 is included in these figures.

Table 4–3: Maximum allowable head for friction clamping for the arrangement shown in Figure 4–9

Bolt grade	*Maximum head (m)*
4.6	13
8.8	33
10.9	48
12.9	56

These heads are not very high, and show that it is usually better to use some other location method – precision-fit bolts or press-fit hub sides.

When using bolts for fitting buckets, it is usually best to locate the components on the shank of the bolt, not on the thread. International standards for metric bolts relate the length of thread to the nominal bolt diameter as:

$$L_{thread} = 2 \times d + 6 \quad \text{for } L \leq 125\text{mm}$$
$$L_{thread} = 2 \times d + 12 \text{ for } 125\text{mm} < L \leq 200\text{mm}$$
$$L_{thread} = 2 \times d + 25 \text{ for } L > 200\text{mm}$$

Welding

Welding is potentially a good means of attaching buckets to a hub for low to medium heads, but it needs to be done carefully. To achieve a reasonable fatigue life, the stress in the weld needs to be kept low, so it is not appropriate for high-head sites. Large power-station turbines have never used welded runners, and nowadays, single-piece castings are the obvious choice because they are even cheaper than welded assemblies. Welding is still used for repairing buckets, even for the largest runners.

For micro-hydro runners, where the heads tend to be quite small, welding is a possibility. Where single-piece castings are not available, it offers a significantly cheaper construction method than machining and bolting, allowing a runner to be completed in a matter of hours rather than days. It requires an understanding of the welding process, the correct equipment, careful preparation, and a skilled welder. It is advisable to use dye-penetration testing on larger or more highly-loaded runners. The bucket material must be a weldable steel; cast iron will not work. The detailed procedure is given in Section 5.2.2.

Figure 4–10 shows different ways of attaching the bucket. Arrangement (a) is the most common. Here a double V-notch is created by chamfers on the bucket and the hub. There is a flat area in the middle to help locate the bucket at the correct diameter, but this is only a few millimetres wide so that it fuses when the base of the notch is welded. The disadvantage with this arrangement is that it takes a lot of weld metal, and the large amount of heat put in can lead to cracking and distortion.

The weld area in arrangement (a) can be minimized by making the stem narrower, but a reasonable width is require to provide adequate sideways support. The angle of the V-notches can be reduced, but it has to be wide enough for a welding electrode to get into it. It might seem possible to increase the flat land in the middle, between the notches, but this is fatal. This area must fuse during welding. If it does not, a sharp-ended crack is left cutting across the stress lines in the weld, and this can quickly lead to fatigue failure.

The arrangement in Figure 4–10(b) shows an alternative that requires less weld. A central tongue on the bucket goes into a groove in the hub. Although this leaves an un-welded, U-shaped gap in the centre, the edges of this gap are in line with the prevailing stress, and the arrangement does not lead to cracking.

Figure 4–10(c) is a variant of the tongue and groove design, with the tongue on the hub, and the groove in the bucket. While this gives a lighter bucket, it may give problems during casting as foundries usually pour the metal into the mould through the stem area of the bucket. If the volume of the stem is reduced too much by a groove, the flow of metal is impeded, and a poor-quality casting may result. Consult the foundry before using this design.

While the attachment of the bucket to the hub is important, the buckets also need to be welded to each other. The basic principle here is to provide a

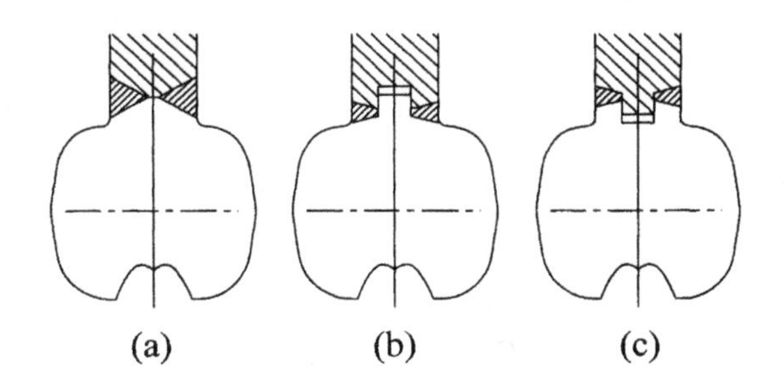

Fig. 4–10: Methods of welding buckets to the hub

weld preparation to weld into, and to keep poor weld details away from stress concentrations. The weakest point, and the place where cracking usually begins, is where the base of the splitter ridge on one bucket joins the back of the next bucket. A suggested stem design is shown in Figure 4–11.

Whatever the welding arrangement, great care needs to be taken over the welding process. In particular, the first, root weld needs to be laid down carefully. See the Manufacturing Section 5.2.2.

4.3 Spear valves and nozzles

4.3.1 Nozzles or spear valves?

Choosing how many jets to use for a turbine has been dealt with in Section 3.1.6. The recommendation was that one or two nozzles are used for horizontal-axis turbines, and up to four for vertical axis turbines. Having selected the number of jets and the jet size, the choice needs to be made between plain nozzles and spear valves.

Spear valves have a number of advantages:

- The power output can be adjusted.
- It is easy to turn the flow on or off.
- The maximum power can be set after testing.

The advantages of nozzles are:

- They are simple to make.
- They are cheaper than spear valves.
- They can be a couple of per cent more efficient than spear valves.

The choice of whether to use a spear valve or a nozzle is usually determined by the overall valve arrangement for the plant.

The simplest possible system would be one where there are only nozzles, no valves on the manifold or penstock, and the flow is controlled at the inlet to the penstock. This is adequate for low-head sites, but can be rather exhausting if one has to run several hundred metres uphill for an emergency shutdown.

It is possible to have nozzles only, no valves, and use deflectors to start and stop the turbine. The starting procedure is then to put the deflectors in front of the nozzles, run up the hill and let the water into the penstock, run down again and lift the deflectors to start the turbine. This system is safer than the basic one above, but still lacks sophistication.

It is more usual to have valves to control nozzles. A gate valve, butterfly valve, or ball valve could be fixed at the end of the penstock, or on each branch of the manifold for each jet. This arrangement is reasonable for small pipes and small heads, but the force required to open a valve under a high pressure can be considerable. The author knows a site that has a nozzle with a 200mm gate valve under a 125m head. It takes four people to open it, and it can only be a matter of time before the valve spindle breaks. Butterfly valves and ball valves take less force to open them, but one should check that they can be opened under pressure. As a rule-of-thumb, simple nozzles should not be used above about 60m head; spear valves are a better option.

Note that gate, butterfly, or ball valves should be either open or closed. They cannot easily be used to

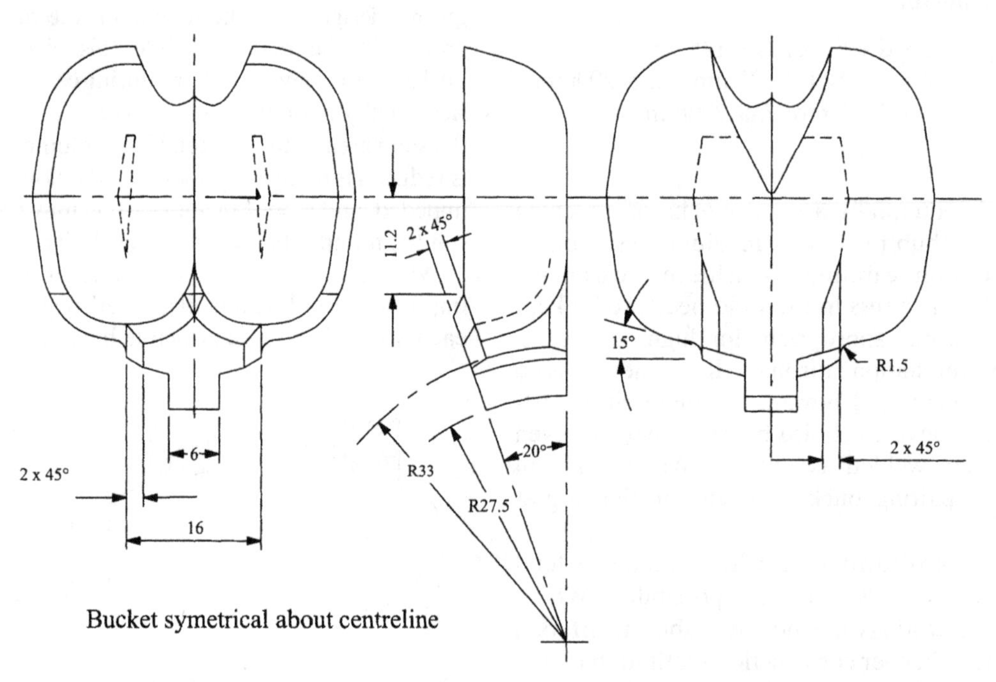

Fig. 4–11: Bucket stem details for welded attachment. The basic bucket is as in Figure 4–1. All dimensions are in % PCD

vary the output power. If they are only partially opened, they create tremendous turbulence in the flow, and can give big losses.

Coarse flow control can be achieved on multi-jet turbines by not having all the nozzles working. On single jet Peltons, some measure of flow control can be achieved by having a number of nozzles of different sizes. For example, on a single-jet Pelton, a large diameter nozzle might be used in the wet season, and a small diameter nozzle in the dry season. Some small Peltons have been designed with easily-changed orifice plates in the nozzles to control the jet diameter.

Spear valves can easily be opened under high pressure, and can be cracked open just a little to start the turbine under very little power. The flow can be adjusted from virtually nothing to full flow steadily and easily. It is common to use another valve behind the spear valve, so that the spear valve can be emptied to unblock it or service it, or for extra safety if one needs to work inside the turbine. In this case, the shut-off valve may still have to be opened against pressure, but it is not as bad as with a nozzle. The spear valve should be kept closed, and as soon as the shut-off valve is cracked open, water rushes in and fills the spear valve, equalizing the pressure. The shut-off valve can then be opened easily. On some very big, high-head sites, where the force required to open the shut-off valve is still too high, a small by-pass valve is used to fill up the spear valve before the main shut-off valve is opened. Spear valves are more expensive than nozzles, but they are the only option for high-head turbines.

Good control over the power can be achieved on multi-jet Peltons by having one spear valve and the rest nozzles. The basic power range is set by opening a number of nozzles, and then fine control is achieved with the spear valve. If any sort of flow-control governing is to be used, then spear valves must be used, and they have to be designed to be actuated by the governor. This is not covered in this manual, but the general principles are discussed in Section 4.14.4.

In summary, on small schemes where flow control is not needed, use nozzles to keep the cost down. When nozzles are not appropriate, use spear valves.

4.3.2 Nozzles

Various types of nozzle are shown in Figure 2–2. Orifices give good performance when they are new and sharp, but deteriorate rapidly as they wear. They can be used for small turbines, but need to be replaced regularly. Tapered nozzles give better performance, and better life. The most efficient nozzle shape tapers slowly at about 7° (14° inclusive angle). This shallow taper makes the nozzle very long, and awkward to fit into the turbine housing. It is more normal to use a steeper taper, of say 20–30°. Try to use as shallow a taper as fits in the space available. Alternatively, a rounded nozzle, as shown in Figure 2–2, can be used.

The calculations in Section 3.1.1 on turbine selection give the jet size to be used. *Note carefully that this is the size of the water jet, not the nozzle size.* As discussed in Section 2.2.1, the jet emerging from a nozzle first contracts down to a narrow point called the vena contracta, and the jet size is calculated at this point. The nozzle has to be larger than the design jet size.

$$d_{nozzle} = \frac{d_{jet}}{C_c} \qquad \textit{Eq. 4–5}$$

C_c – nozzle contraction coefficient

C_c can be found for various nozzles in Section 2.2.1. C_c for different nozzle designs can be extrapolated from the values shown.

The lip of the nozzle can be made sharp, or have a slight flat on it. A sharp lip gives the best jet, and is the easiest to make, but it will be worn away more quickly by the water. The nozzle will last longer if a small flat is machined around the orifice – see Figure 4–12. The end of the nozzle should be carefully machined, with a good surface finish, so as not to disturb the flow.

Nozzles can be made in a variety of ways. The whole nozzle can be machined out of a single block, but this is very wasteful of material. Pieces can be welded together to make a blank for machining. One way of doing this is to roll a conical section from plate – an example is given in Figure 4–12. If a number of large nozzles are being made, consider using a casting. Nozzles are generally made of steel or cast iron to give good erosion resistance. With high heads the jet is very fast, and soft metals can erode quite quickly.

The nozzle can be directly welded on the end of the manifold pipe – a cheap and simple arrangement – but it is then very hard to unblock the nozzle, or to replace it if it wears. For small nozzles, a screw thread can be used to fit it on to the manifold, but for larger sizes it is best use a flanged joint. If flanges are used, always make a spigot to align the nozzle with the pipe behind it. If you just rely on bolts to line up the pipes, there will be some misalignment, leading to turbulence.

After the vena contracta, the jet starts to expand again. By the time the jet hits the runner, it will have expanded and fragmented somewhat. For this reason, the nozzle should be kept as close as possible to the runner. It may be necessary to leave space for a deflector mechanism to be fitted to the nozzle, as described in Section 4.4.2, but otherwise a clearance of a few millimetres is all that is necessary between the nozzle and the runner.

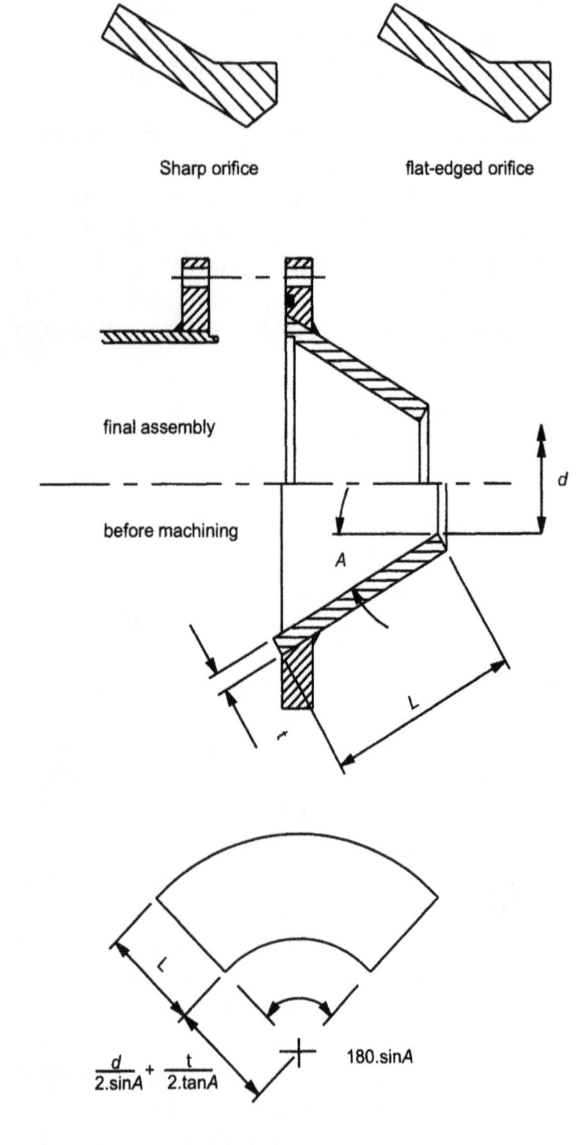

Fig. 4–12: Nozzle details

4.3.3 Spear valves

As mentioned before, spear valves are an excellent invention for controlling the flow into a Pelton runner. They give continuous adjustment of the jet from almost nothing up to maximum flow, and yet give only small losses. Note that spear valves are also called *injectors*, *injector valves*, or *needle valves* by some manufacturers.

Basic dimensions for a spear valve are given in Figure 4–13. It is interesting that many publications give 'optimum dimensions for a spear valve', but they are all completely different from each other. It appears that a wide variety of shapes work reasonably well. Figure 4–13 should be viewed as giving the reader a starting point from which to design, rather than being the only dimensions that will work. The diagram shows a 55° spear in an 80° nozzle. Major manufacturers use 60°/90°, 50°/90°, and many other variations, but they all seem to work. The nozzle angle has to be somewhat larger than the spear angle to make sure the jet converges and does not break away from the spear, which can lead to pitting and erosion.

The orifice diameter, d_{nozzle}, needs to be larger than design jet diameter, because the jet continues to contract after leaving the valve. In Figure 4–13 it is made 25% larger, large enough to ensure that the correct jet size can be achieved. The spear valve has a design travel of $0.42 \times d_{nozzle}$, which should give the jet diameter shown. The maximum allowable spear movement is $0.6 \times d_{nozzle}$, which will give a 14% larger jet. The jet diameter can be calculated from Equation 2-5.

The conical section of the spear should sit in the orifice, so the radius should start further back. As with nozzles, a sharp edge can be used for the orifice, but it is better to have a small flat. This needs to be at the same angle as the spear (55°) to seal

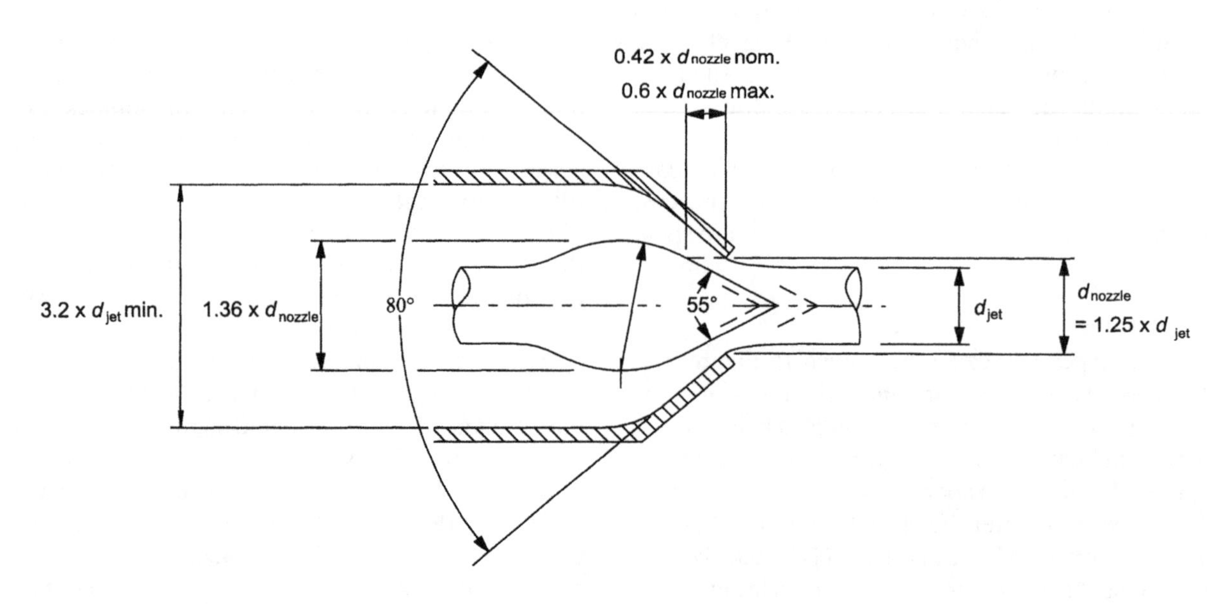

Fig. 4–13: Recommended spear valve dimensions

against it. A well-machined spear valve will not leak at all even under very high head.

Both the nozzle and the spear can wear, particularly under high heads or if there is abrasive silt in the water. If erosion is expected, the spear should be made so that it is replaceable. The orifice at the nozzle can also be put in as a replaceable plate, as in the detail in Figure 4–14. If a small amount of clearance is allowed in the location of the orifice plate, say 1.0mm, the orifice can be positioned so that it is exactly concentric with the spear, and seals perfectly. It should then be bolted tightly in place, and ideally located with a couple of spigot pins.

The velocity of the water in the body of the spear valve needs to be kept down to avoid large losses over needle valve guide fins, bends etc. As a rule of thumb, the area should be about ten times the jet area (corresponding to a diameter $3.2 \times d_{jet}$). It is a good idea, if possible, to keep the water velocity roughly constant through the penstock, the manifold branches, and the spear valve, in which case the spear body should be the same size as the manifold branch.

There are numerous ways of putting together a spear valve, and two are shown in Figure 4–14. The first design shows a simple spear valve with a supporting bearing near the nozzle. The spear has the hand-wheel fixed directly to its outer end, and it screws into a threaded portion of the bearing. As the handle is turned, the spear is screwed in or out of the nozzle. The spear is screwed on to the end of the shaft, so that it can be removed to allow the shaft to be withdrawn from the back of the valve. To prevent the spear from becoming unscrewed, it is pinned to the shaft. In normal operation, the pin stays inside the bearing, so that it cannot fall out. It is possible to make the spear head and shaft in one piece, but then the spear can only be withdrawn forwards, and the runner has to be removed to do this.

The seal is a packing seal. The best form of this is a square section rope, often of an asbestos-based material. At least three full turns of rope should be used. Alternatively, it is available as a thread, and many turns are used to fill the seal cavity. Packing is cheap, widely available, easy to use, and tolerant of poor surface finish or corrosion. The housing does not have to be machined to tight tolerances, just to the nominal size of the seal. Because the area behind the packing is sealed from the water, it can be greased, and a grease nipple should be arranged to allow the bearing to be re-greased. The pressure in the seal can be adjusted by tightening the bolts that hold the rear bearing. The bolts should be tightened just enough to stop water leaking out. If the bolts are too tight, the spear valve will become stiff and hard to turn. Seal manufacturers make a variety of proprietary packing seals. One type of lip seal is shown in the detail in Figure 4–14. The seal shown has two end pieces, and a variable number of sealing sections. Like the packing rope, it needs to be compressed to push the lips of the seal against the shaft and the housing to seal.

The front bearing needs to run in water. The best combination of materials for this bearing is a phosphor bronze bearing running on a stainless steel shaft. This works well with water lubrication, and does not corrode. Any general grade of phosphor bronze and stainless steel will work (for example BS 1400:1985 Grade PB1, equivalent to ISO 1191 Part 1 Grade G-CuSn10P phosphor bronze, and BS970:Pt.1:1983 Grade 304S15 stainless steel). It is possible to make the whole shaft out of stainless steel, but this is expensive. A cheaper solution is to use mild steel and put a stainless steel collar on it under the bearing. If stainless steel is not available, use a bronze sleeve on the shaft as well as a bronze bearing. Various attempts have been made to use plastic bearings, and these can work well. The problem with them is that many plastics, such as nylon, swell after extensive submersion in water. You can find that the spear valve locks solid after the turbine has been running a few months. Only use plastic if you are sure of the quality and know its water absorption characteristics.

The second design in Figure 4–14 does away with the front supporting bearing. This simplifies the construction, and removes the troublesome front bearing. The rear bearing has to be longer than before to give better support to the shaft. This design is easily dismantled, as the spear diameter is less than the bearing diameter, and the whole spear shaft can be withdrawn out of the back.

This second design shows the use of O-ring seals instead of the packing type. O-rings are again cheap, widely available, and easy to fit, but they do require a corrosion-resistant housing and shaft, accurate machining of the housing, and good surface finish on the sealing surfaces. For information on how to use O-rings see Appendix 12. O-rings are not adjustable like packing seals, and this simplifies the design of the spear valve a little. It is possible to use sophisticated lip seals instead of O-rings, but it is often not necessary. O-rings are used by major Pelton manufacturers on very large spear valves.

In mechanically operated spear valves, such as the one shown in Figure 4–15, the spear moves straight in and out of the nozzle, without rotation, and some designers try to achieve this on hand-operated valves too. The arrangement shown in Figure 4–14 takes a different approach, and rotates the spear in order to move it. While this leads to marginally more rubbing contact between the spear and the nozzle lip, experience has shown that the extra wear caused is negligible, while the spear design is considerably more simple.

In both designs, the pipe has to emerge from the housing on a bend to give room for the spear valve hand-wheel. The temptation is to make this bend too tight in order to keep down the overall length

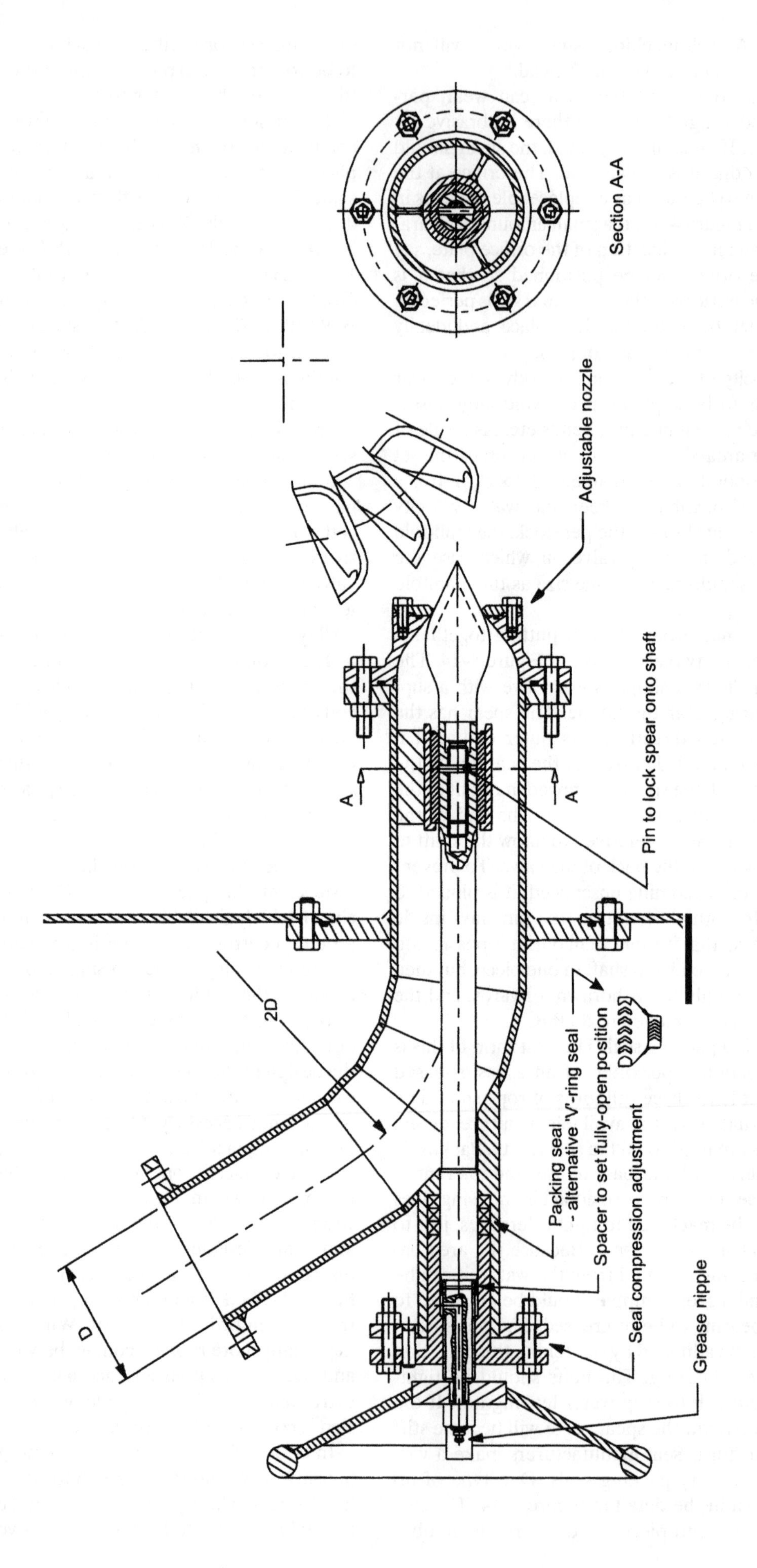
Section A-A
Adjustable nozzle
Pin to lock spear onto shaft
A
A
2D
D
Packing seal
- alternative 'V'-ring seal
Spacer to set fully-open position
Seal compression adjustment
Grease nipple

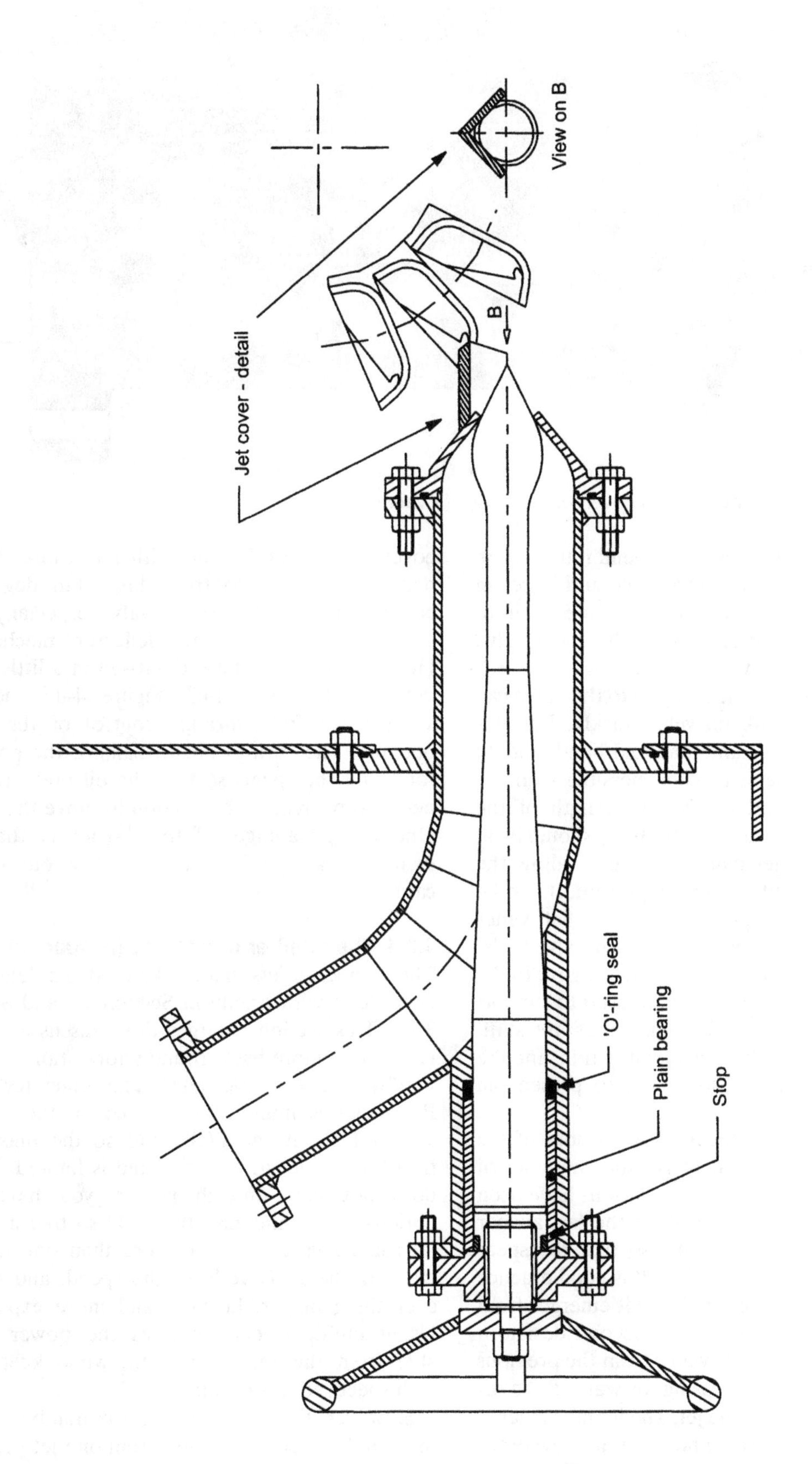

Fig. 4-14: Spear valve layouts

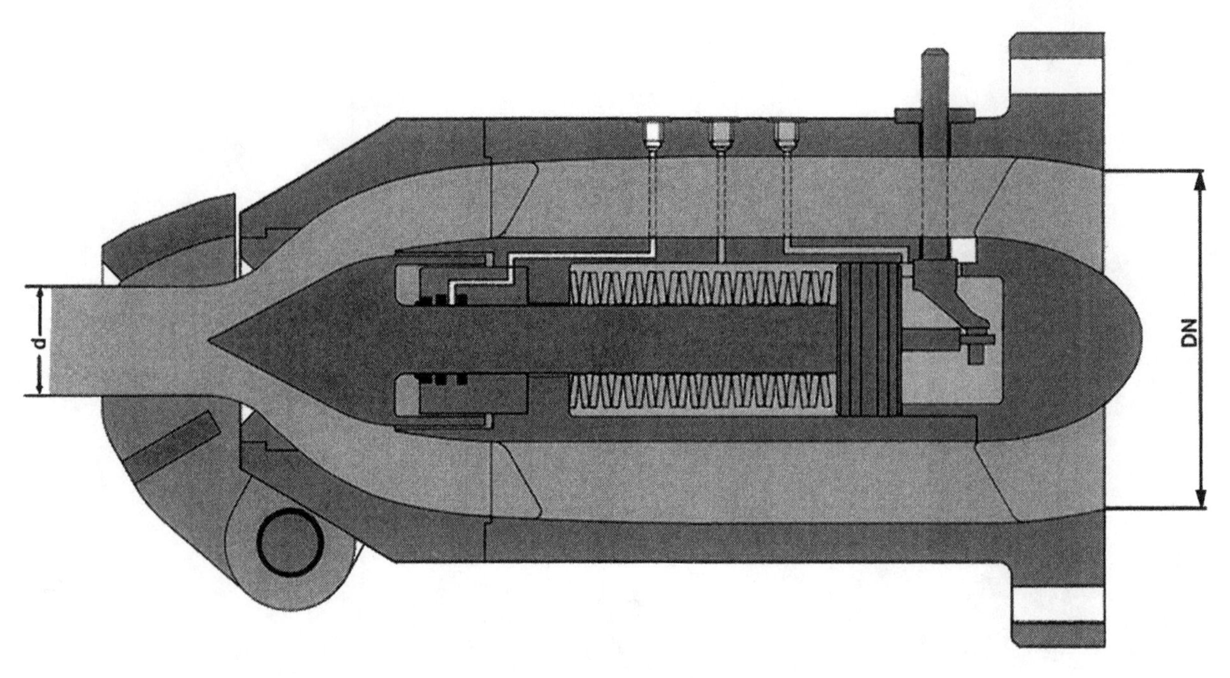

Fig. 4–15: Spear valve for use with a governor (courtesy of Kvaerner Energy)

of the spear valve. However, the bend must not be too sharp or it will cause turbulence and losses in the valve. As a general rule, the centreline radius of the bend should be at least twice the spear valve body diameter, as shown in the drawings.

Swirl in the pipework leading up to the valve can cause the jet to break up very quickly. For this reason the spear valve and manifold bends should be kept in one plane as much as possible – do not have close compound bends. The length of the straight section of the spear valve body should be at least 10 times the jet diameter long to allow the flow to sort itself out. If swirl is present, it can be straightened out by putting in long, flat vanes around the straight section of the spear valve body. It is good practice to put these in as standard. In the first arrangement the vanes can be used to support the front bearing. Three or four vanes are sufficient. They should be kept reasonably thin, and the ends should be tapered so as not to present an obstruction to the flow.

When designing the spear, make a note of the withdrawal room required to get the shaft out of the back of the valve. Many installations have been built with the powerhouse wall or the last anchor block too close to the turbine, so that the spear cannot be removed for servicing! It is good practice to put a small cover over the jet as it emerges from the nozzle. This is shown in the second design in Figure 4–14. This prevents water from the previous jet, water droplets in the casing, or water flung off the runner, from hitting the jet. The jet may emerge absolutely smooth from the nozzle, but a few droplets of water can cause it to break up. The cover prevents this.

It was stated in the foreword to this manual that is is for micro-hydro Pelton turbines, and would not cover the use of Peltons with governors. This is because governor control adds a fair degree of complication, to the spear valve especially, and sometimes also to the deflector mechanism. However, governors are discussed in a little more detail in Section 4.14.3. Figure 4–15 shows a spear valve for hydraulic control of the valve opening. The springs are to balance the pressure forces on the spear, so that the oil pressure only needs to overcome the friction to move the spear. One big advantage of this layout is that the manifold can go in a straight line out of the casing.

4.3.4 The number of jets and jet spacing

The number of jets that can be used in a Pelton has been dealt with briefly in Section 2.4 and Section 3.1.6. This section enlarges those discussions, and give a little more background information.

Why use multiple jets? The diameter of a Pelton jet is limited by the size of the bucket, say to 12% of the PCD, and so the maximum flow for one jet at a given head is limited. If this does not give enough power, you have two options; either increase the PCD so that a larger jet can be used, or have more than one jet. Increasing the PCD reduces the speed, and makes everything bigger, heavier, and more expensive. Using multiple jets increases the power available from the same size unit, while keeping a high speed, so it is cheaper.

However, there is a limit to the number of jets that can be used. The water from one jet needs to be clear of the runner before the water from the next jet enters. This is because the jets have to be placed at intervals around the outside of the runner, and the jets must hit the runner at different angles. If two jets were allowed to play on one bucket at the same time, the flow paths around the bucket would be different, the two flows would deflect each other, and there would be turbulence and a loss of power.

A runner rotates somewhat more than 70° from the first droplets hitting a bucket to the last droplets leaving it (see, for example, Figure 2–9). It is therefore not surprising that some loss of efficiency is found if jets are put closer than 75°. Jets can be as close as 60°, but it must be accepted that there will be some interaction and a certain reduction in efficiency. At 60° spacing, it is obvious that the maximum possible number of jets is six.

Six-jet turbines suffer from a phenomenon known as *falaise*, which results in instability if the unit is operated above its optimum speed. This is due to the water being carried around in the buckets for a greater angle at the higher speeds than at lower speeds, resulting in more interference between the jets. Consequently, the efficiency curve for a 6-jet Pelton falls away very quickly at higher speeds, and the runaway speed is lower.

Most manufacturers limit the angle between jets on horizontal-axis machines to 70–75°, and rarely have more than two jets. It is possible to make a horizontal turbine work with up to four jets and with only 60° between the jets, but under these conditions there can be problems. It is not advisable to try it unless you have facilities to test and develop the design. Vertical-axis machines with six jets are common but can give difficulties to inexperienced manufacturers. It is safer to limit the number of jets to four or fewer until experience has been gained.

4.3.5 The effect of the nozzles or spear valves on the penstock

The nozzles or spear valves fitted in a turbine affect the thickness of the penstock needed for the site. Water hammer occurs when the flow in the penstock is stopped suddenly. The water has momentum, and if the speed of the water is reduced quickly this momentum generates pressure. The size of the surge head is determined by the amount of flow being stopped, the time it takes to stop it, and the penstock size and material.

The full calculation procedure for penstocks is given in both Harvey et al (1993) and Inversin (1986), and the formulae are given in the spreadsheet in Section 11–4. The surge head depends on whether the valve closure or blockage occurs in a time greater or less than the critical time – the time it takes for a pressure wave to travel the length of the penstock and back. If a spear head breaks off, or if a soft object blocks a nozzle, the closure is instantaneous (and therefore less than the critical time), and the surge pressure comes from the simple formulae:

$$h_{surge} = \frac{V_{wave} . \Delta V}{g}$$

$$V_{wave} = \sqrt{\frac{1}{\rho . \left(\frac{1}{K} + \frac{d}{E.t} \right)}} \qquad \textit{Eq. 4–6}$$

h_{surge} – surge head (m)
V_{wave} – pressure wave velocity = speed of sound within the penstock (m/s)
ΔV – change in velocity of flow in penstock (m/s)
g – acceleration due to gravity (9.8m/s^2)
ρ – density of water (1000kg/m^3)
K – bulk modulus of water (2.1×10^9N/m^2)
d – pipe bore (m)
E – Young's modulus for penstock material (N/mm^2)
t – thickness of penstock (mm)

Suppose a Pelton turbine has three nozzles, and one of them is blocked suddenly. If the velocity in the penstock is 3m/s, after the blockage the flow will be 2m/s, and ΔV is 1m/s. Plain nozzles can block instantly, especially if they are small. Frogs are particularly good for blocking nozzles, though it does them no good; if they can get round the trashrack, they are just the right shape and consistency to completely seal a nozzle shut. Other debris – balls of leaves and sticks, stones – can also block nozzles. Surge calculations should be based on one nozzle blocking instantaneously.

For multijet Peltons, a first approximation for ΔV is:

$$\Delta V = \frac{V}{n_{jet}} \qquad \textit{Eq. 4–7}$$

V – initial velocity in the pipe (m/s)
n_{jet} – number of jets on the penstock

However, the detailed calculations will show that the worst case is having only one jet of a multi-jet Pelton open, and letting that jet block. The reason is that when only one jet is operating, the flow in the penstock is less than its design value, and the penstock losses are less than the design losses. This means there is a higher flow through one nozzle on its own than if it is operating with any of the others open. As an example, consider a two-jet Pelton. Suppose the design water velocity in the penstock when both jets are open is 3.0m/s. When only one jet is open the penstock losses will be somewhat less, and the velocity be slightly more than 3.0/2 = 1.5m/s, perhaps 1.6m/s. If this one nozzle blocked, the whole of the 1.6m/s would be stopped. ΔV in this instance is 1.6m/s, not the 1.5m/s that would be expected from Equation 4–14. So the surge pressure will actually be 1.6/1.5, or roughly 7% more, than would be expected from simple calculations. In most cases the difference is not significant, and Equation 4–14 can be used. When working with high heads, or using small safety factors on the penstock, full calculations should be done to determine the actual ΔV. This can only be worked out by calculating the penstock losses and flow for only one jet being open. Note that water hammer is much less severe in plastic penstocks than in steel because the speed of the pressure wave, V_{wave}, is much lower.

Spear valves are not easily blocked by debris. The opening is annular, and even frogs cannot extrude to fill round the whole area. It is possible to completely block the nozzle if the spear comes loose, or if its head drops off. This is the worst case, and it is recommended that surge is calculated for this case. A multiple spear valves turbine is treated in the same way as one with multiple nozzles above.

If there is no chance that the head of the spear valve could ever fall into the nozzle, maximum surge calculations may be reduced. To calculate the surge pressure in this case, the quickest possible valve closing time has to be estimated. Simplified equations are given in Section 11 and in Inversin (1986); more detailed Allievi charts are in Arter et al (1990).

4.4 Overspeed protection: deflectors and other mechanisms

4.4.1 The need for overspeed protection

If the load is removed from a Pelton turbine, it speeds up very quickly. For example, if the turbine is driving a generator and the circuit-breakers for the consumer load trip, the turbine is left generating full power when the generator cannot dissipate it. The turbine will accelerate to its runaway speed in a matter of seconds. As discussed in Section 2.3.7, runaway speed is about 1.8 times the optimum speed, so for a turbine running normally at 1500rpm, runaway can be 2700rpm – very fast.

Some machinery may be able to cope with such overspeed, but some cannot. The windings on a synchronous generator rotor can be flung out of their slots if the speed is too high, and this sort of failure causes a tremendous amount of damage. Even if the windings are suitable for high speeds, the generator bearings may not cope with running at high speed for extended periods. If the driven machinery cannot cope with runaway speed, the turbine must be fitted with an automatic mechanism to stop it in the event of overspeed.

Some generators can withstand high speeds. Squirrel-cage induction motors are generally able to withstand very high speeds, and can be used as induction generators without overspeed protection. Check with the manufacturer's specifications. Some manufacturers of synchronous generators will also supply generators able to withstand overspeed. If the windings can withstand high speeds, but the bearings are limited, the manufacture may put a time limit on the overspeed: say ten minutes at 3000rpm. The overspeed requirement needs to be specified when the generator order is placed. You should assume that a generator cannot withstand overspeed unless the specifications say it can. Large power-stations are never allowed to go to runaway speed. The maximum overspeed is commonly set at 33%, with overspeed trips set to operate 25% higher than normal speed.

4.4.2 Deflectors

The simplest, most common way of stopping a Pelton in an emergency is to use deflectors on the jets. These are shaped plates that are moved into the jets to deflect the water away from the runner. Even a small deflector can make all the water miss the buckets. Without any power going into it, the turbine will slow down until it stops. Deflectors can work in two ways, as shown in Figure 4–16. The first type, Figure 4–16a, normally sits outside the jet, and has a sharp leading edge. When it is rotated into the jet it slices off the bottom part of the jet, leaving the upper part of the jet still hitting the runner. Only when the deflector has been moved fully into the jet does the runner stop.

The second type, Figure 4–16b works in a rather different way. It normally sits between the runner and the jet, and is rotated into the top of the jet. It deflects the top layers of the jet, but these in turn deflect the lower portion of the jet. By the time the deflector is about half-way across the jet, the whole jet will be deflected away from the runner. At first sight it seems quite surprising that this deflector works, because it does not go the whole way across the jet. The principle can be verified by using a spoon to deflect the water from a tap. To deflect all the water in the jet away from the runner reliably, the deflector has to come down about three-quarters of the way into the jet. The exact depth needed depends on the geometry. The top edge should remain a little way outside the jet, or else water will be deflected backwards into the runner, causing it to go into reverse. Almost all commercial turbines use this type of deflector. This is because it is more compact, and requires less force to operate it.

The deflectors act on the water after it has left the nozzle, and so make no difference to the pressure in the manifold or penstock. The deflector can be moved very quickly, deflecting the jet in a fraction of a second, and there will be no surge pressure in the penstock. For an emergency shut-down, it is best to use the deflectors first, and then slowly close the spear valves or the manifold valves.

Pelton governors often control both the spear valves and the deflectors. When the load on the turbine changes, an initial, fast adjustment is made with the deflectors. The spear valves are then adjusted slowly, and the deflectors are backed off out of the jet again (see Figure 4–16(c)). Thus the penstock does not have to cope with the repeated surge pressure that would result from shutting the spear valve quickly. Both types of deflector can be used for this system.

In principle, the force and torque on the deflector can be calculated quite easily. The jet hits the deflector with a velocity v_{jet}, and leaves with a velocity $k \times v_{jet}$, as in Figure 4–17. A vector diagram can be drawn to find the change in velocity, Δv. The force F on the deflector is the rate of change of momentum in the water, and will be in the opposite direction to Δv. The torque on the deflector pivot is this force times the moment arm about the pivot, r.

$$F = -\rho.Q.\Delta v$$
$$T = F.r \qquad \textit{Eq. 4–8}$$

F	– jet force on the deflector (N)
ρ	– density of water (kg/m³)
Q	– flow in jet (m³/s)
Δv	– vector change in velocity (m/s)
T	– torque on the deflector shaft (N.m)
r	– moment arm of F about the deflector pivot (m)

The difficulty with this approach is working out where the jet leaves the deflector. When it hits the plate, it is not neatly bent down as in Figure 4–17,

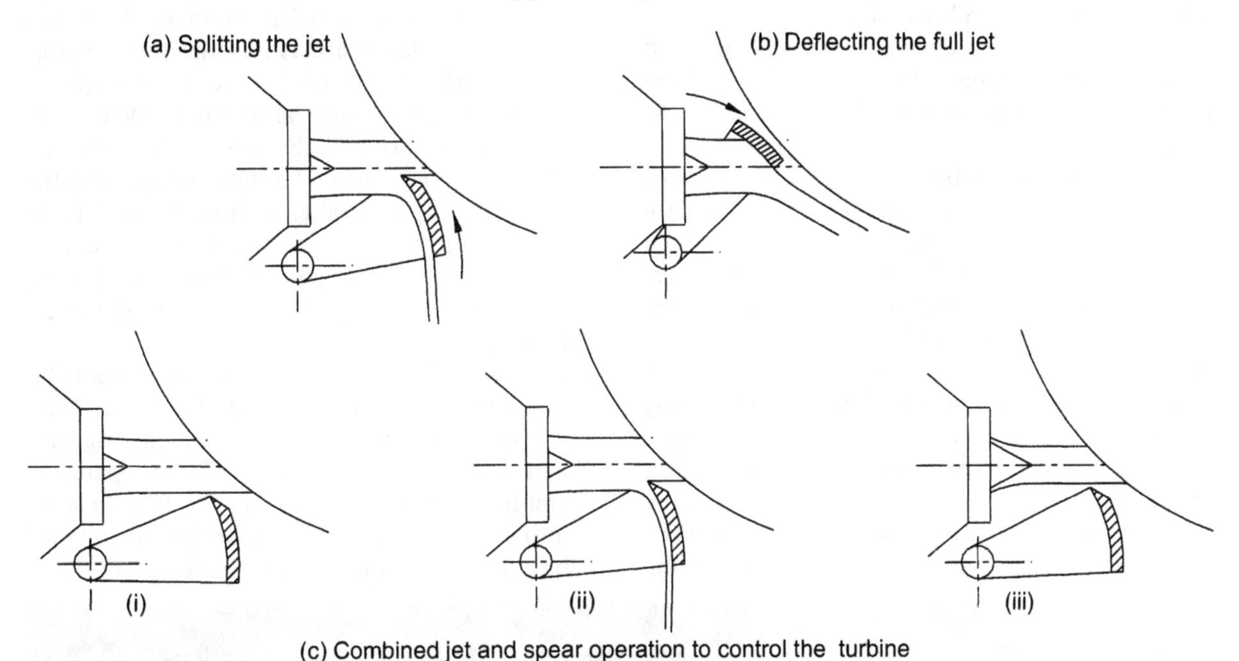

Fig. 4–16: Principles of deflector operation: (a) Splitting the jet. (b) Deflecting the full jet. (c) Combined jet and spear operation by a governor to prevent surge pressure in the penstock

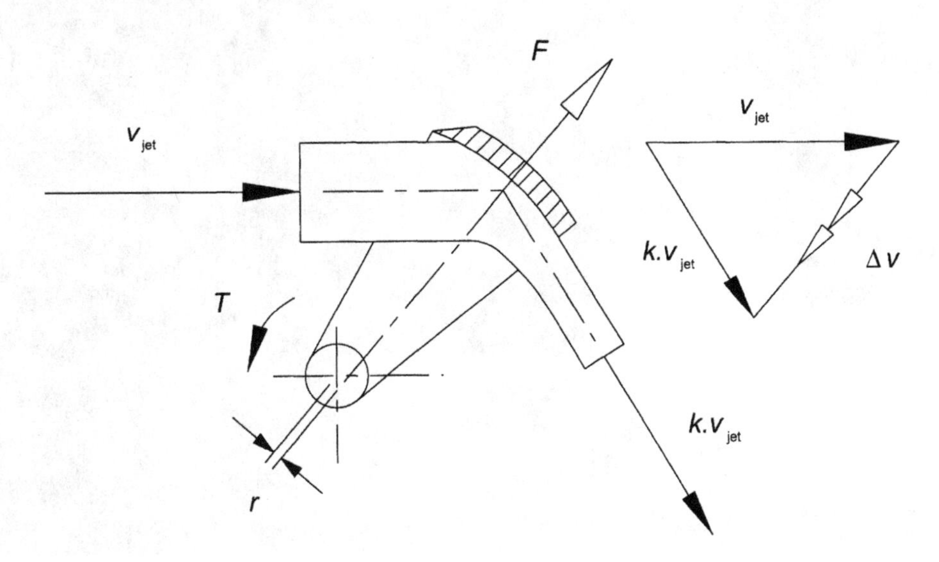

Fig. 4–17: Calculating the force and torque on a deflector

but it spreads out. Some water is deflected out of the top of the deflector, some spreads sideways and hits the sides of the deflector, and only a proportion is bent downwards as shown. There are two unknowns: firstly, how much of the water leaves the jet with momentum, and secondly, in which direction does the water leave.

The author has found that the force can be estimated as follows. For jet-splitting deflectors (Figure 4–16a), the water jet comes out of the deflector tangential to the trailing edge, as shown in the figure. If a value of 0.65 is used for k, the forces come out as measured For jet-bending deflectors it is much harder to determine the line of the water leaving. The simplest approach is to draw the jet so that it looks right. The value of k for the jet-bending arrangement is larger, because the water hits the plate at a shallower angle; a figure of k = 0.75 can be used. The deflector bearings and shaft must be able to cope with the force F, and the operating mechanism must be able to overcome the torque T plus friction.

The deflectors can be operated in many ways. One of the simplest is to join all the deflectors together with a lever system, and put a dead weight on the end of the lever. When the lever is released, the weight falls, the deflectors are moved into the jets, and the turbine stops. One arrangement is shown in Figure 4–18. Here the lever is held up by a d.c. solenoid. A relay in the solenoid power supply cuts off the current if there is a fault in the system, particularly if the turbine overspeeds. The deflectors could also be moved by electric motors, by hydraulic rams from a pressurized oil system, or by cylinders using the pressure of the water in the penstock. Whatever method is used, it should be fail-safe. For example, it you use an electric actuator to move the deflectors, there needs to be a battery backup system to power it, for the occasion when the generator fails and the turbine starts to overspeed.

4.4.3 Other shut-down mechanisms

Deflectors are by far the most common shut-down mechanism. They are simple, reliable, and easy to make, and it is recommended that the reader uses them unless there is a compelling reason for doing otherwise. Some other mechanisms that have been tried are:

- Brakes: putting a disk or hub-type brake somewhere in the drive system. This can work for small turbines, but it is hard to make the system reliable. Its advantage is that the turbine stops quickly, whereas with a deflector the turbine can keep turning for several minutes. Ideally, the brake should normally be in the locked position, requiring power to make them release the turbine (like the air-brakes on a lorry) so that any failure stops the turbine. A brake can be used in conjunction with deflectors if it is important that the turbine is stopped quickly when there is a failure.
- Penstock inlet valves: an automatic valve at the top of the penstock, shutting off the water. There is a long delay between the valve being triggered and the turbine stopping while the penstock empties. The penstock *must* have a tube to let air into it when the valve closes, or else the pipe will be collapsed by the vacuum created in it.

Fig. 4–18: A deflector mechanism held up by a solenoid and operated by a dead weight (KMI, Nepal)

- Automatic valves in the manifold: electrically or hydraulically-operated valves that close themselves when a fault occurs. Beware of surge! The penstock and manifold must be able to cope with the surge caused by shutting the valves.
- Governor-controlled spear valves: similar in principle to valves in the manifold. The spear valves must not be closed too fast.
- Vanes to cause swirl in the jets: an old design of turbine twisted the flow-straightening vanes inside the spear valve to make the flow swirl as it entered the nozzle. This gave a wide, conical spray from the nozzle, which was not strong enough to turn the runner. This is unnecessarily complicated compared with a deflector, and no-one uses it any more.

4.4.4 Overspeed sensing

The speed of the turbine can be sensed in a number of ways. The most direct, and reliable, is to have a mechanical switch, operated by centrifugal force, that triggers when the speed exceeds a preset limit. Generator manufacturers can often supply these built-in to the generator. Very large generating sets will have two of these switches, in case one fails. Another alternative is to have an electronic speed sensor that measures the shaft speed, and to fit a trip circuit to this. A third method is to measure the frequency of the generator output, because this is directly proportional to the speed. In designing safety circuits, care must be taken that the overspeed trip is operated for all conditions that lead to overspeed, including a broken drive belt or complete electrical failure.

4.5 General shaft and bearing layout

A Pelton turbine can be made as a separate unit, with the runner on its own shaft and bearings, completely self-contained. It is also possible to integrate the turbine with the driven machinery, for example by putting the runner on the shaft of the generator.

4.5.1 Direct shaft mounting

If the speed of the turbine can be made the same as the speed of the driven machinery, then one should consider an integrated design. When choosing a turbine, it is good practice to try various diameters of runner, and changing the number of jets, to see if it is possible to make the speeds the same – see the discussion in Section 3.1.5. The advantage of an integrated design is that no drive system or couplings are required, and it eliminates at least two bearings. This makes the unit more compact, and it is usually cheaper and more reliable.

The most common use of this system is for generators, where the runner is mounted on the generator shaft as shown in Figure 4–19(a). For micro-hydro, it is quite possible to get Pelton turbines to run at the 1500rpm required for a 50Hz, 4-pole, synchronous generator (or 1800rpm for 60Hz). It is also possible to purchase 50Hz, 6-pole generators that run at 1000rpm (or 1200rpm for 60Hz). Low speed generators are, however, more bulky and more expensive, so it is not always economical to use them.

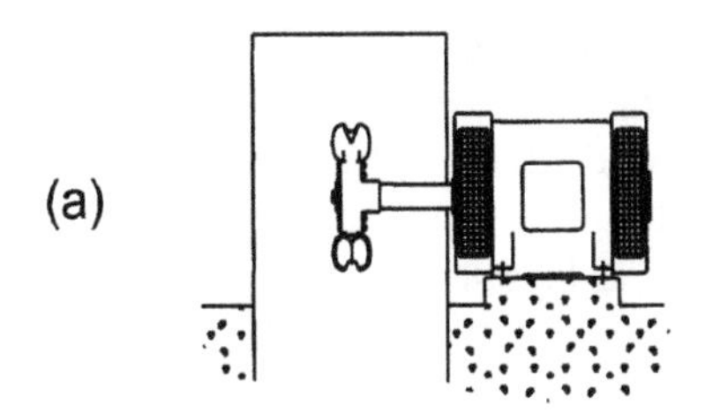

Runner mounted on a generator shaft

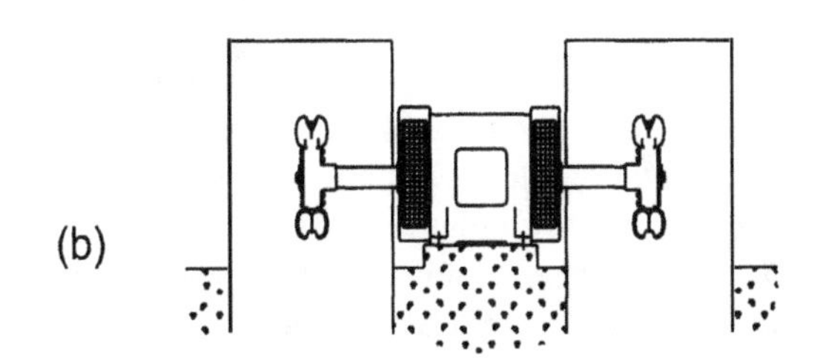

Two runners mounted on either end of a generator shart

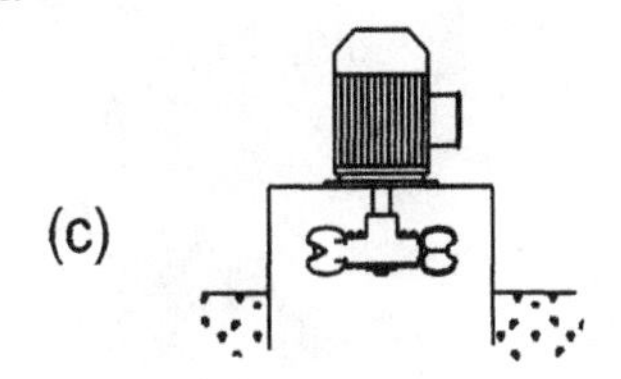

A Peltric set

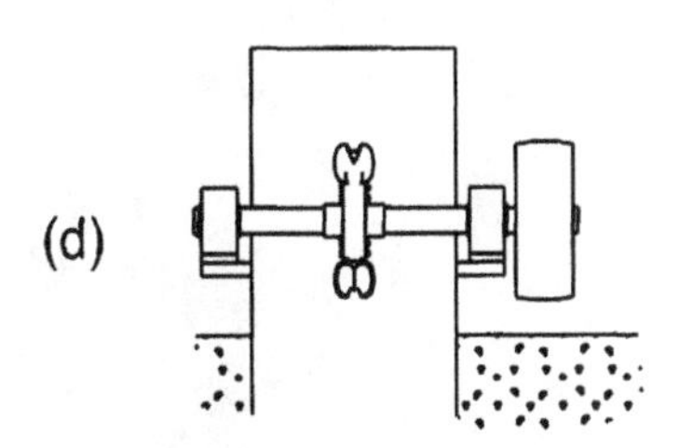

Stand-alone unit with the runner on its own shaft and bearings

Fig. 4–19: Various turbine layouts

Before mounting a runner directly on to a generator shaft, check with the generator manufacturer that the unit can withstand the load. The manufacturer will need to know the load on the shaft (from the jet forces and the weight of the runner) and the distance from the front face of the generator to the load, and the life required. If the manufacturer cannot be contacted, one can look to see what bearings are in the generator and calculate using the equations in Section 4.7.3.

Another option is to get a generator with a shaft that comes out of both ends (Figure 4–19(c)).

These are not usually stock items, but are available from specialist manufacturers. With two 2-jet turbines on each end, this produces a lot of power from a compact unit. It is not so easy to service the generator with this layout.

4.5.2 Peltric sets

Peltric sets are simply a special case of direct shaft mounting (see Figure 4–19(c)), but they have become so popular that they merit a mention in there own right. The name *Peltric* was coined in Nepal for small, self-contained, Pelton generating units. They use induction motors as generators, and generally come complete with an *Induction Generator Controller*, or IGC (Figure 4–20).

Because they are self-contained units, Peltrics can be sold as 'do-it-yourself' units for the customers to install for themselves. Kits for units of, say, up to 5kW, can be sold complete with IGC, ballast tank, pipework and valves. Detailed instructions need to be given to help the user survey the site head and flow, choose the correct pipe for the penstock (often HDPE plastic pipe) and install the Peltric.

Fig. 4–20: A Peltric set (made by KMI, Nepal)

Induction motors for Peltric sets are much cheaper than synchronous generators. They generally have good, large bearings, that are designed for heavy belt drive loads, and they can easily cope with the runner forces. They can also usually withstand overspeed without any special modification, so the turbine does not need to have deflectors or an automatic shutdown arrangement. Some induction motors are splash-proof, which is particularly appropriate for hydro use.

An induction motor needs capacitors fitted across its terminals for it to work as a generator. These capacitors provide the magnetizing current. Some simple calculations have to be performed to select the correct size of capacitors. It is difficult to get single-phase induction motors to excite, and 3-phase motors are almost always used. For small schemes, a 3-phase delta-connected motor is wired to produce a single-phase output. For details of induction motor selection, choice of capacitor, and the use of IGCs, see Smith (1994).

Drinking water supply systems often have break-pressure tanks in line from the water source to keep the pressure within the limits of the pipe. Peltrics can readily be fitted just before a break-pressure tank, using energy otherwise wasted, and making a very cheap installation because the water system is already there.

4.5.3 Separately mounted runners

There are many cases when it is not possible to mount the runner on the generator shaft directly:

- When the generator bearings cannot take the runner forces.
- When the generator shaft cannot withstand the bending moment from the runner.
- When the turbine speed is different from the generator.
- When the powerhouse layout or other design considerations dictate having separate turbine and generator units.
- When there is no generator! The turbine may be for a mechanical drive only.

In these cases the turbine needs to be self-contained, with the runner shaft mounted on its own bearing (Figure 4–19(d)).

Various drive options for separately mounted runners are shown in Figure 4–21. The simplest arrangement is to have a bearing on each side of the runner, as in arrangements (a), (b), (c), (e) and

(a) Direct coupling with flexible shaft coupling

(b) V-belt drive (© ITDG)

(c) Flat-belt drive

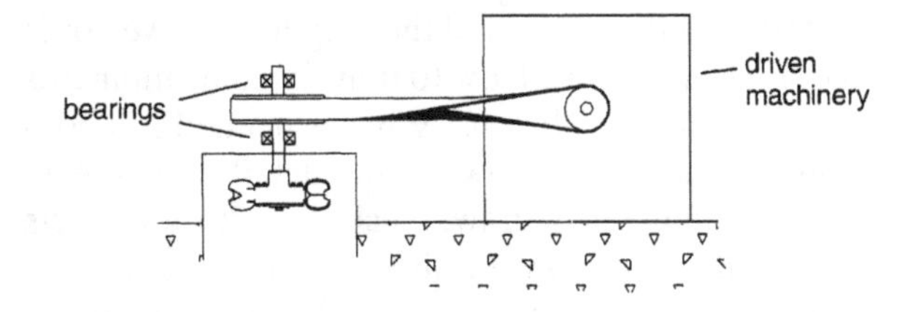

(d) Flat-belt drive with shafts at 90°

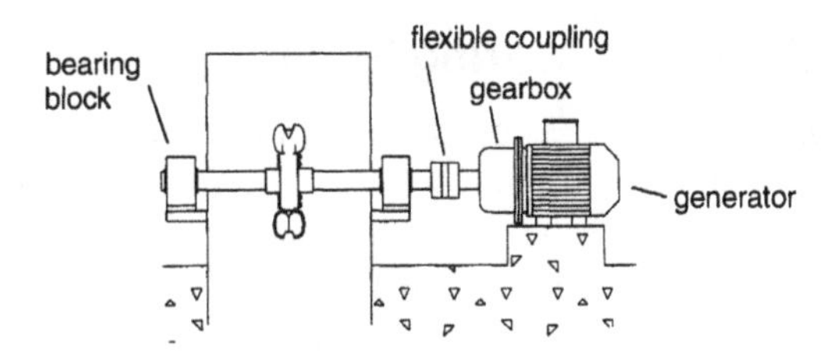

(e) Gearbox coupling

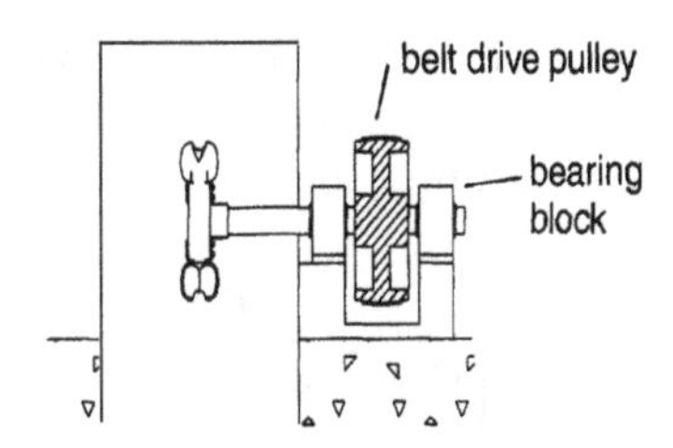

(f) Overhung arrangement with a drive pulley between the bearings

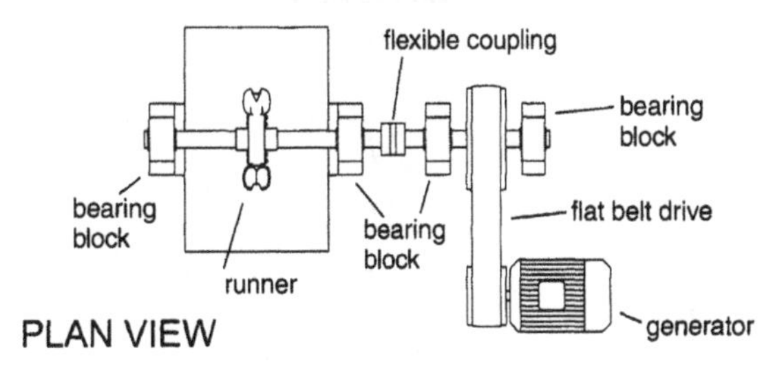

(g) Flexible coupling and layshaft

Fig. 4–21: Drive options

(g). These bearings will also have to cope with the forces from the external drive system. Flexible couplings and self-contained gearboxes put very small or no forces on the shaft, but belt drives put quite large forces on it.

Another option is to have two bearings on one side of the runner, so the runner is on an overhung or cantilevered shaft, as in arrangements 4(d) and (f), and also, of course, in Peltrics. This is similar to putting the runner on a generator shaft. This arrangement is the only one practical for vertical-shaft turbines. It is also useful for horizontal-axis machines, as only one shaft seal is needed, and the casing can be designed to give clear and easy access to the runner from the unsupported side.

Whatever the arrangement, the bearing forces and driveshaft bending must be calculated, as in Section 4.6. The bearings and shafts must then be sized as in Section 4.7.3 and Section 4.8.2 respectively. Sometimes the bearing forces are so large that the bearings become massive and expensive, or difficult to fit. In such cases the runner and drive system forces can be separated by using a layshaft, as in arrangement (g). The twisted belt, arrangement (d) is only suitable for low-power applications; and many belt manufacturers do not recommend this arrangement. The gearbox arrangement (e) looks neat but is quite rare as such gearboxes are relatively expensive.

4.6 Shaft and bearing forces

Calculating the forces on the shaft in a Pelton turbine is relatively straightforward. The main forces come from the water jets, the runner and shaft weight, the bearings, and the drive system. The water jet force can be calculated in two ways. One method is to work from the basic principles of the jet acting on the bucket, and this is given in Equation 10–25. The other way is to work backwards from the mechanical output power, using basic relationships between torque and power:

$$F_{jet} = \frac{T}{n_{jet}.\left(D/2\right)} = \frac{2T}{n_{jet}.D}$$

$$= \frac{2\left(P_{mech}/\eta_{force}.\omega\right)}{n_{jet}.D}$$

$$= \frac{2P_{mech}}{\eta_{force}.n_{jet}.D.\omega}$$

$$= \frac{2P_{mech}}{\eta_{force}.n_{jet}.D.\left(2\pi.N/60\right)}$$

$$= \frac{60P_{mech}}{\eta_{force}.\pi.n_{jet}.D}$$

Eq. 4–9

F_{jet} – jet force (N)
T – jet torque on runner (N/m)
n_{jet} – number of jets
D – PCD (m)
P_{mech} – runner mechanical output power (W)
η_{force} – proportion of the force that produces output torque
ω – runner angular speed (rad/s)
N – runner speed (rpm)

Some of the torque from the jet force is used to overcome windage and bearing losses, so the jet force is actually higher than the amount calculated from the output power. The efficiency η_{force} takes this into account, and a figure of 0.95 can be used.

The forces from belt drives etc. depend on the belt tension. Refer to the manufacturers catalogue to find the forces that come from V-belt or flat belts. It is interesting to note that a flat belt exerts the same force on a pulley when it is transmitting no load as when it is under full load; the force comes purely from the belt tension. Knowing the jet forces, the weights, and any belt forces, the bearing forces and the bending moment on the shaft can be calculated. Treat the forces acting vertically on the shaft separately from the horizontal forces, and then add the forces and moments vectorially at the end. The example below shows the procedure.

Example

Consider the turbine and drive system shown diagrammatically in Figure 4–22. It is a horizontal-shaft, 2-jet Pelton with one jet vertical and one jet horizontal. Each jet acts on the runner with a force of 2000N. The shaft is supported on two bearings. The weight of the runner is 800N, the weight of the pulley is 1000N, and the weight of the shaft can be ignored. The turbine drives a generator through a flat belt drive system, which acts on the turbine pulley with a force of 6000N, 15° below horizontal. Dimensions are given in mm. This example is based on a real microhydro plant, generating 100kW of electrical power from a head of 125m.

Solution

The first step is to draw a side elevation force diagram. The shaft and runner are drawn as a free body, as if they were hanging in space. The forces are drawn as they act on the shaft and runner. Since the shaft is rotating at constant speed, and stays where it is fixed in the turbine, everything must be in equilibrium. The forces and bending moments on the free body must balance, cancelling each other out. The forces at the pulley are its weight and the vertical component of the belt tension. The forces at the runner are the runner weight and the vertical jet force. The unknowns are the two vertical forces on the bearings, V_1 and V_2. These are found by taking moments around the bearings. For the moments around bearing 2:

$$(1000 + 6000 \times \sin 15°) \times (0.4 + 0.4 + 0.175) + (800 + 2000) \times (0.4) = V_1 \times (0.4 + 0.4) \therefore V_1 = 4511\text{N}$$

V_2 is found in a similar way by taking moments about bearing 1. Having found V_1 and V_2, the bending moment diagram can be drawn for the end view, as shown. It is important to be clear about the directions of V_1 and V_2. These are the forces of the bearing on the shaft, and are upwards in this case. The force of the shaft on the bearings is downwards, equal and opposite to the bearing force on the shaft.

Next, the plan view is drawn. Using the same method as above, the forces are calculated, and the bending moment diagram drawn.

Finally, the various force and bending moment vectors on each bearing are added together as in the bottom two diagrams. Drawing force vectors is quite straightforward. The direction of the vector is the direction of the force, and the length of the vector is the size of the force. How to draw bending moments vectors is not so obvious. What is needed is a convention, and this is shown in the detail in Figure 4–22. The moment is drawn such that the force acts around it in a clockwise direction, in the same direction one would turn a screwdriver to tighten a screw. For a given point on the shaft, only consider

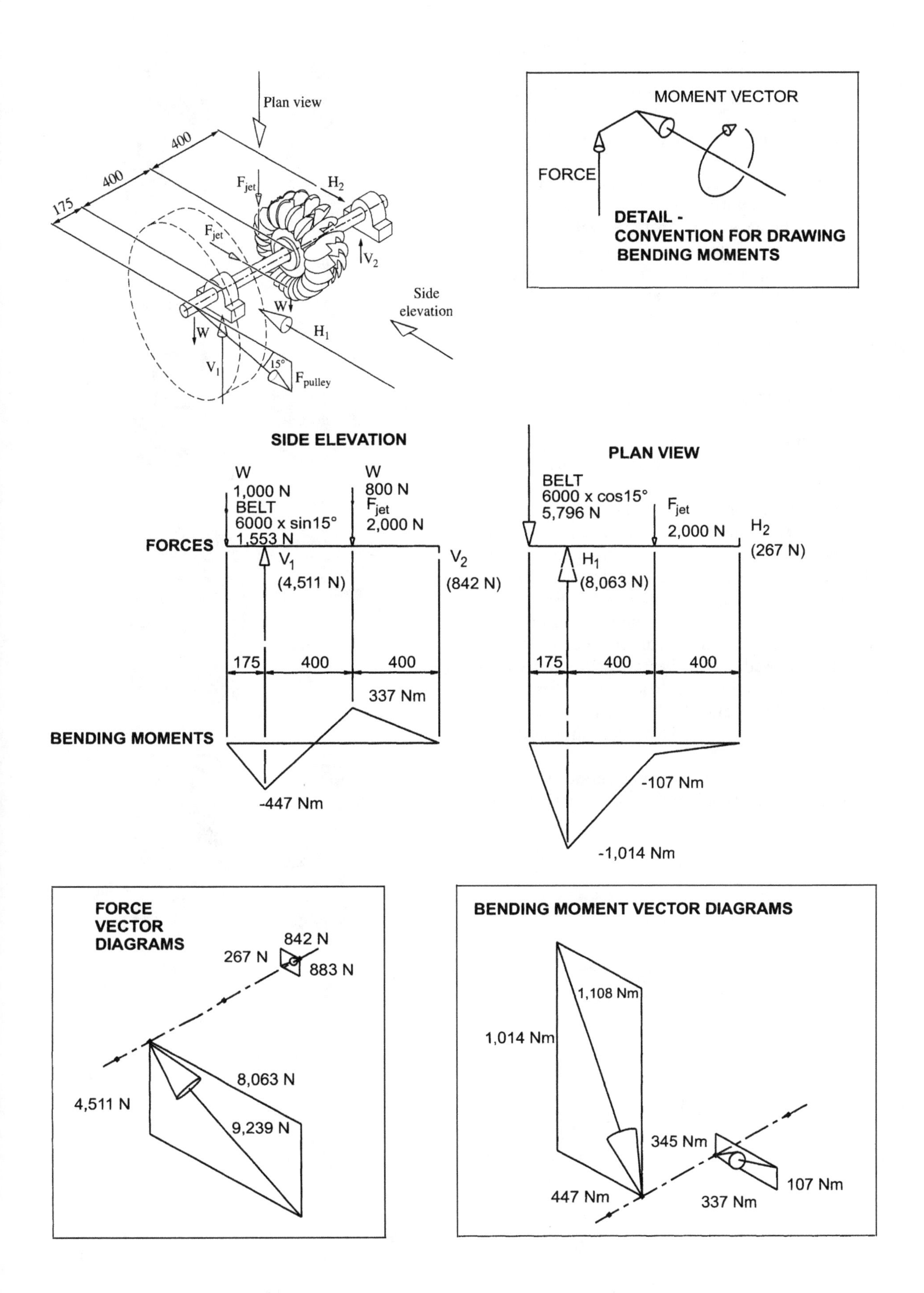

Fig. 4–22: Forces and bending moments on a Pelton shaft

the moments from one side of the shaft (if one takes the moments from the whole shaft, the total will be zero). When the diagrams are drawn, the result is that both the largest bearing force and the greatest bending moment are found to be at bearing 1.

Note that the maximum bending moment may not occur at full power. From Figure 4–23, if a 4-jet turbine is being used, with the nozzles at 90° to each other, when all jets are working there is no net force on the shaft – the force from the opposing jets cancel each other out. The largest force on the runner comes when two jets are working. Be careful to calculate for the worst case.

4.7 Runner shaft bearings

4.7.1 General considerations

This section deals with the selection of the bearings to support the runner shaft. If the bearings are chosen correctly, the turbine can run for years without trouble. If the bearings are wrong, they can wear out very quickly. It is important to calculate the loads carefully, and to pay attention to the engineering detail of the bearing mountings.

Various types of bearings may be chosen, but rolling-element bearings – ball or roller bearings – are usually used. A number of factors determine the choice of bearing type: load, misalignment, cost and availability. Figure 4–24 shows various types of bearings. The first six illustrated are not self-aligning.

If the runner is fixed on the generator shaft, the generator bearings are the runner bearings. The

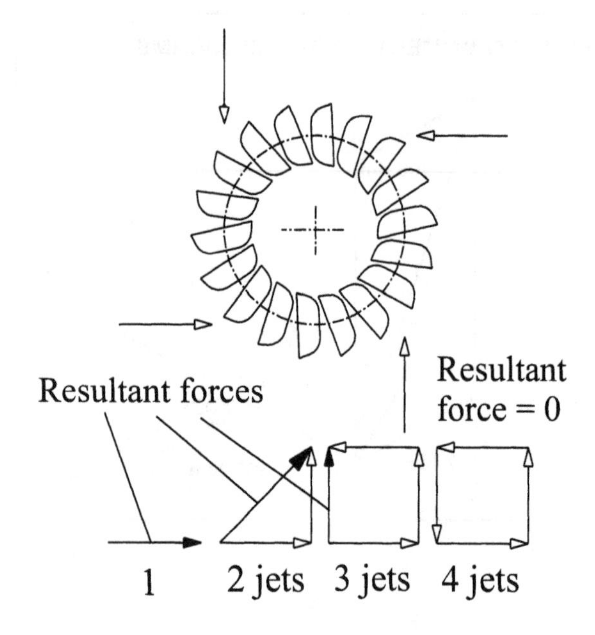

Fig. 4–23: Resultant force from 1, 2, 3 and 4 jets

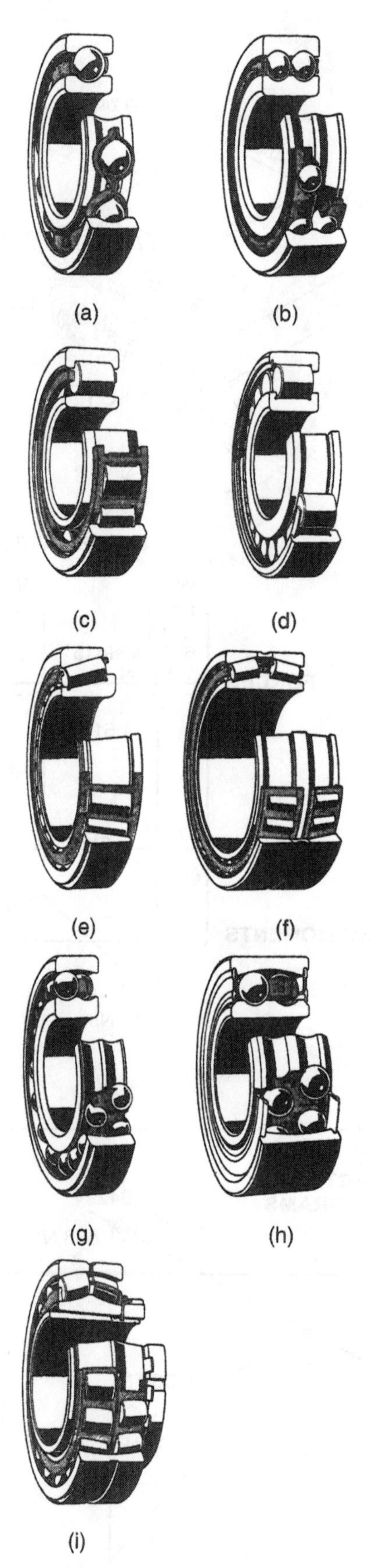

Fig. 4–24: Various types of rolling-element bearings: (a) single-row ball bearing, (b) double-row ball bearing, (c) & (d) cylindrical roller bearings, (e) & (f) single and double-row taper roller bearings, (g) to (i) self aligning bearings: (g) & (h) ball bearings, (i) spherical roller bearing (© SKF).

generator manufacturer should be able to confirm whether the bearings can take the loads. If not, the methods described here can be used to calculate the bearing life.

Load

The bearings must stand the combined loads of the jets on the runner, the weight of the runner and shaft, and any forces from the drive system. The calculation of these forces is shown in Section 4.6. Roller bearings have a higher load capacity for a given size of bearing. If the forces on the bearings are high, roller bearings may be the only possible choice. If the turbine is slow, if it is only used for a few hours a day, and if the loads are small, ball bearings may be adequate.

One advantage of Pelton runners is that there is very little thrust load along the shaft. The bearings only need to be designed for radial loads.

One loading condition that is often forgotten, but which can destroy bearings, can occur during transportation. If the turbine is sent to site along rough roads there can be considerable shock forces in the bearings, leading to brinnelling of the bearing surfaces. The bearings then fail very quickly when running, for no apparent reason. This is avoided by ensuring careful transportation, or by fitting the bearings on site. If the unit has to be transported assembled, over rough roads, then the static load rating (see Section 4.7.3) must have sufficient allowance for the transportation loads.

Alignment

It is usually necessary to use self-aligning bearings. Turbine housings are generally welded together. While great care may be exercised during fabrication to locate the components in the correct places, it is not possible to control dimensions to better than a millimetre or so. Bearing require much better accuracy than this. If two rigidly-mounted bearings are fixed on either side of the casing, they will not line up correctly. To use rigidly-mounted bearings, the casing would need to be rigid, and should be accurately machined after welding. This is often not possible; housings may be too big to fit on any machine tools available to a small workshop. If the housing is not machined after welding, then self-aligning bearings must be used. These allow the bearing unit to swivel to take up the correct line of the shaft.

Cost and availability

As a general rule in choosing bearings, it pays to use bearings that are readily available. Mass-produced automotive bearings are much cheaper than other bearings. Check what sizes are available through car, truck or tractor spare-parts dealers locally, rather than choosing a bearing from a catalogue. If you are asking a bearing manufacturer or stockist for a recommendation, ask for quotations for the next few sizes of bearings above the one they choose; it is possible to find one bearing in a series that is half the price of the other bearings – because it is made in high volumes. It is much simpler to use self-aligning bearings, but they are more expensive than conventional bearings. For small turbines, it may be cheaper to machine the housing and use simple bearings than to use self-aligning bearings.

Take account of the availability. If it takes several months to purchase a non-stock bearing from a dealer, it will take the same length of time to order spares if a bearing fails. It is better to use bearings that are held in stock. If a special size has to be used, then a stock of spare bearings needs to be kept.

Always use good quality bearings. Cheap bearings look just the same when they are new, and save a lot of money in the beginning, but will be much less reliable. It pays to use bearings from a reputable bearing manufacturer. Beware of imitations! Try to ensure that the parts are genuine. In general, reliability is so important in remote micro-hydro sites that it is best to use only good brands.

4.7.2 Bearing arrangement

There are two rules for installing bearings:

- Never have more than two bearing units on one shaft.

This is because the position of a shaft is fixed by two bearings. If a third bearing is then added, it is almost impossible to make it line up exactly with the other two bearings, even with accurate machine tools. The result is that it will try and bend the shaft, all three bearings will be overloaded, and they will quickly wear out. Even in carefully machined automotive gearboxes it is extremely rare to have more than two bearings on a shaft. (There are exceptions, the most common being on the crankshaft of automotive engines, but here plain bearings are used.)

- Only provide axial location for one of the two bearings. The other bearing should be able to float.

This is again to avoid excessive forces being applied to the bearings by the housing. If both bearings are located axially, slight differences in length between the shaft and the housing result in heavy axial loads on the bearing. Changes in length due to expansion can have the same effect. A float of only a millimetre or so in the free bearing is quite adequate. See Figure 4–25.

(There are exceptions to this rule too, notably when using single-row taper roller bearings. These

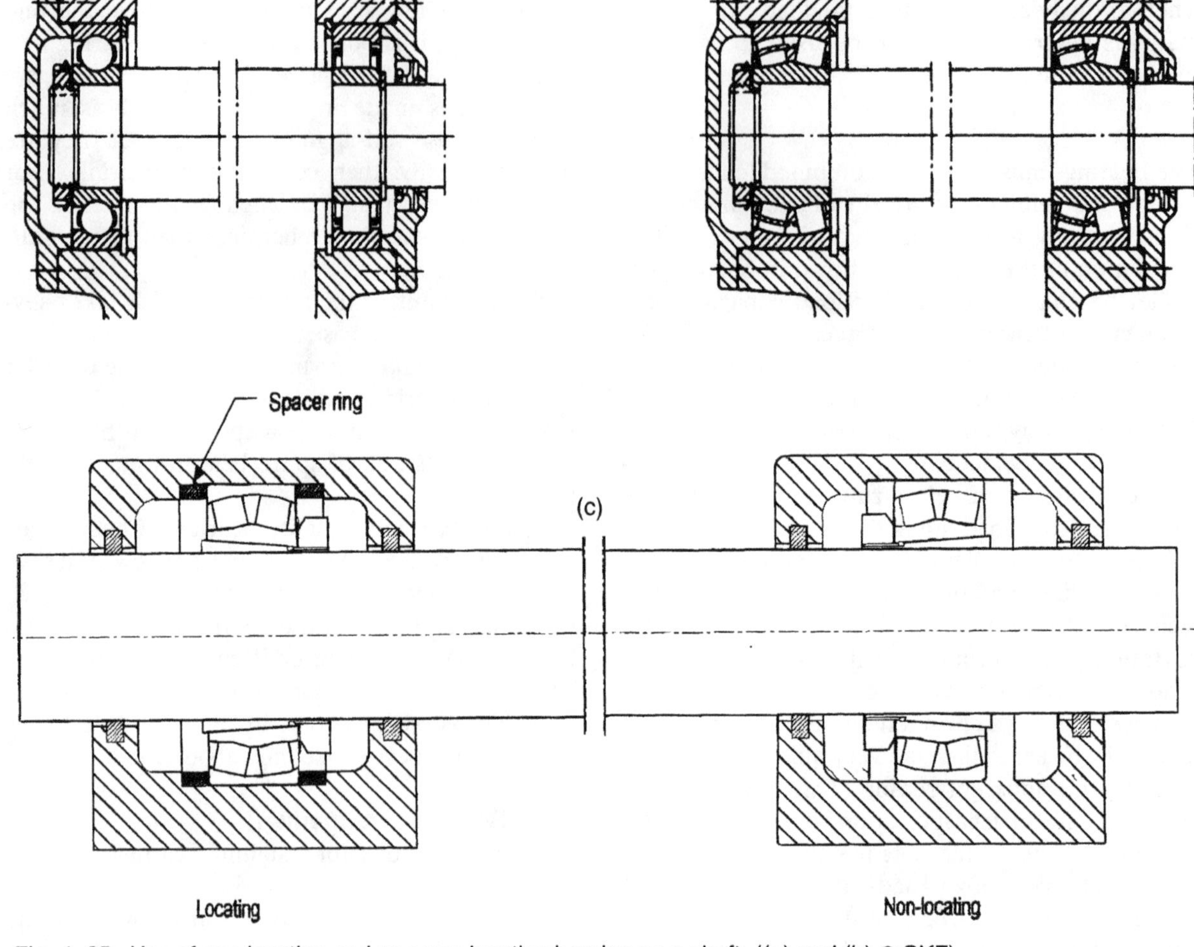

Fig. 4–25: Use of one locating and one non-locating bearing on a shaft. ((a) and (b) © SKF)

bearings require special setting up to give a preload. It is not a simple procedure, and for this reason it is not considered here.)

The inner races of the bearings can be located on the shaft with shoulders (see Figure 4–25(a) and (b)) or with tapered locking sleeves (Figure 4–25(c)). Locking sleeves simplify the arrangement considerably, and they allow adjustment of the runner position so that it can be centred on the jets at site. They do, however, have some disadvantages. Firstly, the very fact that the position of the shaft can be adjusted means that an inexperienced fitter can put the runner in the wrong place. Secondly, if the tapered locking sleeve is over-tightened it reduces the internal clearances in the bearing, and the bearing will run hot and wear out quickly. Thirdly, taper-sleeve bearings are more expensive than plain-bore bearings. Locations for the outer races of the bearings can either be machined into the turbine housing (as in Figure 4–25(a) and (b)), or a proprietary bearing housing can be bolted on (such as that shown in Figure 4–26).

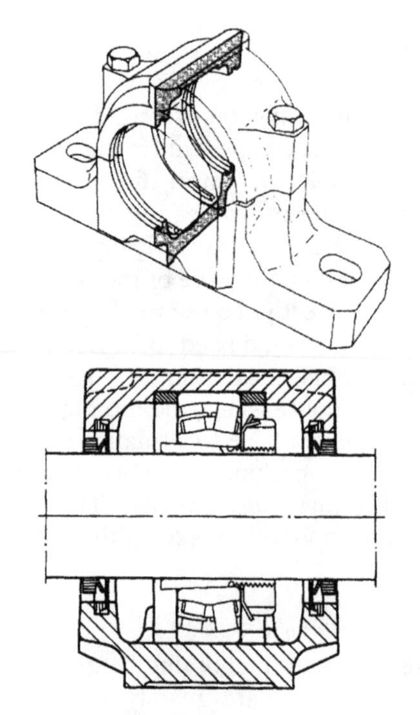
Fig. 4–26: A plummer block bearing housing (© SKF)

Float to make a free or non-locating end can be achieved in a number of ways. In Figure 4–25(a), the right hand bearing is a roller bearing that allows axial slide between the rollers and the inner race. In Figure 4–25(b) and (c), the outer races of the right hand bearings are free to slide axially within their housings. Within proprietary plummer block housings, the locating bearing is held in place axially by spacer rings, which are removed on the non-locating bearing (Figure 4–25(c)).

4.7.3 Selecting the correct bearing size

Normally the shaft in a Pelton turbine has to be quite large to prevent it from bending. The bearings have to be selected to fit on the shaft, and consequently they tend to have higher capacity than is needed. Bearing life is therefore not usually a problem, but it still should be checked.

The following method outlines the general procedure for sizing bearings, and should work for most bearings used for micro-hydro Pelton turbines. The method here is based on the reference SKF (1994), which is in turn based on ISO standards. In the following sections, the following assumptions are made:

- The axial loads are small.
- Shaft diameters are less than 200mm.
- Turbine design speeds are not more than 1500rpm.
- Grease lubrication is used.

If any of these are not true, then a bearing manufacturer's catalogue must be consulted. It is best to check the details of the calculations with the method given in a bearing manufacturers catalogue anyway.

Dynamic load rating

The basic formula for choosing a bearing is:

$$L_{10} = \left(\frac{C}{P}\right)^p \qquad \text{Eq. 4–10}$$

L_{10} – bearing life for 10% failure rate (10^6 revolutions)
C – Basic dynamic load rating (N)
P – Actual dynamic bearing load (N)
p – 3 for ball bearings, 10/3 for roller bearings

The *basic dynamic load rating* C is the radial load at which 10% of this type of bearing will fail after one million revolutions. This figure is quoted in bearing manufacturers' catalogues. The equation predicts the life of a bearing under the load P. Note that the life is given as a probability. Bearings do not all fail at exactly the same time, but follow a statistical distribution. The L_{10} life gives the number of million revolutions at which 10% of bearings of this type will fail.

Bearing manufacturers modify Equation 4–10 by multiplying the life by various factors which modify L_{10} for different operating conditions. Note that some high-quality manufacturers increase the given values of L_{10} because they use improved materials and control the tolerances of the components very carefully. Their bearings exceed the requirements of international standards. This can mean that a bearing from another, lower quality supplier may not be interchangeable with a high-quality bearing. All makes of bearings should reach the life given by Equation 4–10, but if you replace a high-quality bearing in a turbine with a standard-quality one, it is possible to run into problems. Table 4–4 shows the C/P ratios that are required for 90% of bearings to work for 10 years, 24 hours/day without failure at various speeds.

Table 4–4: Ball and roller bearing C/P ratios

Shaft speed n	L_{10h}	L_{10}	*C/P ball bearings*	*C/P roller bearings*
(rpm)	(hours)	(10^6 revs)		
1500	87 660	7889	19.9	14.8
1400	87 660	7363	19.5	14.5
1300	87 660	6837	19.0	14.1
1200	87 660	6312	18.5	13.8
1100	87 660	5786	18.0	13.4
1000	87 660	5260	17.4	13.1
900	87 660	4734	16.8	12.7
800	87 660	4208	16.1	12.2
700	87 660	3682	15.4	11.7
600	87 660	3156	14.7	11.2
500	87 660	2630	13.8	10.6

Dynamic rating: worked example

A turbine is to run at 900rpm, with a load of 10 000N on the highest loaded bearing. For roller bearings, what *C/P* ratio is required?

From Table 4–4 it is seen that 10 years' running is 87 660 hours, equivalent to 4734 million revolutions. To achieve this, the bearing *C/P* must be at least 12.7.

Since P = 10 000N, the bearing must have C = 10 000 × 12.7 = 127 000N minimum.

Note that this is designing so that 9 out of 10 bearings will last 10 years. Statistically, 5 out of 10 bearings will last 50 years or longer! This should give plenty of safety factor.

Static load rating

There is another rating figure quoted in catalogues, C_0, which is the *basic static load rating*. This is the radial load (N) at which a roller will make a small dent in one of the races to a diameter of one ten-thousandth of the roller diameter.

C_0 does not usually have to be considered for turbines as the static loads are small. C_0 is always greater than C.

The static loading may have to include transportation loads on rough roads, as discussed above.

Axial load rating

Axial load can usually be ignored in Peltons. Even if all the flow going into a Pelton bucket went into one half and came out on one side, the sideways force would be about 12% of the jet force, and it is unlikely that the runner will be installed this far out of position. As a rule of thumb, the sideways force is a maximum of 5% of the jet force. The total axial load is then 5% of the jet force multiplied by the number of jets, and is all taken by the locating bearing. This is small enough to be ignored for most types of bearings.

Bearing speeds

Maximum bearing working speeds are listed in the catalogues. These are generally not a problem for micro-hydro turbines, unless very big (for shaft diameters greater than ∅200mm) or heavy-duty bearings are used.

4.7.4 Tolerancing of mating components

The loading of Pelton turbines is fairly straightforward, and this simplifies the mounting requirements. The load is stationary relative to the bearing outer race, and is applied and removed quite gently.

Shaft tolerances

Table 4–5 shows the shaft tolerances required to fit various kinds of bearings for use on a Pelton shaft. The table shows the ISO standard tolerance classes (for more information on tolerances see Section 14) and the actual high and low deviations for machined dimensions. The shaft diameters are shown as 'over' and 'including', so a nominal shaft diameter of 50mm would be found in the fourth row, but one of 50.1mm diameter would be in the fifth row. The third column shows the cylindrical form tolerances recommended. This table is for solid shafts. Hollow shafts require different tolerances.

Bearing manufacturers usually recommend that the bearing seating is ground, but this is rarely possible for micro-hydro and not entirely necessary. R_a is the *Centre Line Average* or CLA measure of surface roughness as defined in ISO 4394 (BS 1134 and ANSI B46.1-1985). It represents the mean distance of the actual surface from the nominal surface position. Other measures of roughness may be encountered, but R_a is by far the most common and generally useful; see Kempe (1989) or Oberg (1992). A turned bearing seating should have a good surface finish, R_a 1.6μm and certainly no worse than to R_a 3.2μm.

Housing tolerances

For typical Pelton runner shaft bearings, the housing for the outside diameter of the bearing should

Table 4–5: Outside diameter tolerances for solid shafts for various types and sizes of bearings

		Type of bearing											
Shaft diameter (mm)		*Ball*			*Cylindrical/ needle roller*			*Spherical roller bearings*			*Tapered adaptor sleeves for all types*		
Over	*Incl.*	*Class*	*Dev.* (μm)	*Form tol.* (μm)	*Class*	*Dev.* (μm)	*Form tol.* (μm)	*Class*	*Dev.* (μm)	*Form tol.* (μm)	*Class*	*Dev.* (μm)	*Form tol.* (μm)
10	18	j5	+5 −3	2.5	k5	+9 +1	2.5	k5	+9 +1	2.5	h9	0 −43	4.0
18	30	k5	+11 +2	3.0	k5	+11 +2	3.0	k5	+11 +2	3.0	h9	0 −52	4.5
30	40	k5	+13 +2	3.5	k5	+13 +2	3.5	k5	+13 +2	3.5	h9	0 −62	5.5
40	50	k5	+13 +2	3.5	m5	+20 +9	3.5	m5	+20 +9	3.5	h9	0 −62	5.5
50	65	k5	+15 +2	4.0	m5	+24 +11	4.0	m5	+24 +11	4.0	h9	0 −74	6.5
65	80	k5	+15 +2	4.0	m5	+24 +11	4.0	m6	+30 +11	6.5	h9	0 −74	6.5
80	100	k5	+18 +3	5.0	m5	+28 +13	4.0	m6	+35 +13	7.5	h9	0 −87	7.5
100	120	m5	+28 +13	5.0	m6	+35 +13	7.5	n6	+45 +23	7.5	h9	0 −87	7.5
120	140	m5	+33 +15	6.0	m6	+40 +15	9.0	n6	+52 +27	9.0	h9	0 −100	9.0
140	180	m6	+40 +15	9.0	n6	+52 +27	9.0	p6	+68 +43	9.0	h9	0 −100	9.0
180	200	m6	+46 +17	10.0	n6	+60 +31	10.0	p6	+79 +50	10.0	h9	0 −100	10.0

be the nominal bearing outside diameter with a tolerance of H7 – H8. This applies to both split and solid housings. The housing can be turned, but should have a good smooth surface – not worse than R_a 3.2μm – and must be cylindrical.

4.7.5 Bearing lubrication and sealing

Choice of lubricant

At first sight it is not obvious why rolling element bearings need lubrication. The components are all smooth and highly polished, and the movement is rolling rather than sliding. However, the motion is not pure rolling. There is sliding between the cage and the rolling elements, and also between the rolling elements and the unloaded portion of the track, where there is a slight reduction in the speed of the balls or rollers. Lubrication is therefore needed to prevent direct contact between the various metal parts. Greases or oils form a thin film between the rollers, the races, and the cages, preventing them from wearing quickly. The lubricant also cools the bearing, prevents dirt and water from entering, and inhibits corrosion.

Oil is the best lubricant, but it is not so easy to use as grease. Being a liquid, oil seeps easily through gaps, and special, high quality seals are required to keep it in the bearing. If splash lubrication can be used, where the bearing sits in a bath of oil, the overall layout may not be too complicated. If a drip feed, or a circulation system, or oil cooling is required, the lubrication system can become quite complex and expensive. In general, oil lubrication is reserved for high speed or high temperature applications. It is not commonly used for micro-hydro, though it is possible.

Grease lubrication is much more simple. Being solid, grease stays roughly where it is put. Cheap and simple seals can be used, including non-contact types. Grease is in fact a type of oil lubrication. It is made up of a soap base. The fibres of soap crystals join together to make a sponge-like matrix that holds oil in its pores. The grease sits around the bearing, and allows small amounts of the oil to go into and out of the bearing (Castrol, 1984).

Any calcium (lime) or lithium-based grease with a consistency of 2 NLGI will normally work. Sodium-based greases should be avoided as they are water-soluble. For standard arrangements, a good quality, general purpose calcium-based grease is recommended. This will have good water resistance, and will also mix with most other greases that customers are likely to buy. It is worth paying a little more for good quality grease of a reputable brand. The extra cost is negligible in terms of the overall operating cost, and cheap grease can lead to very expensive damage. Cheap greases are usually calcium-based, but they degrade quickly and need changing more frequently.

If large bearings are being used at high speeds (shaft diameters >150mm and speeds >1500rpm) it may be necessary to use a lithium-based grease. These greases have better lubrication properties, but their water resistance is not as good. Only use lithium greases if it can be ensured that the operators will use the same grease for relubrication: mixing different types of grease can cause problems. If possible, have operators relubricate the bearings with the same grease supplied initially. Do not use EP (Extreme Pressure), Graphite, Molybdenum or High Temperature greases. These all contain additives which may damage the bearings. Note that the temperature the bearings are allowed to run at is limited by the grease. The bearings themselves can run at temperatures of 200°C or so, but calcium greases can only tolerate 60°C. This has to be the maximum temperature allowed.

Bearing seals

Seals have two functions:

- To keep dirt/water from entering the bearings
- To keep the grease inside the bearing housing.

The simplest way of fitting seals is to buy bearing with integral seals (Figure 4–27(a), (b) and (c)). Bearing units may be 'sealed-for-life' and maintenance free. These are good for small turbines and difficult conditions. However, if it can be guaranteed that maintenance will be done, it is often better to use non-sealed bearings, as regreasing flushes dirt out of bearings, and increases their life. For plummer block housings, bearing manufacturers offer a variety of proprietary seals.

The choice of seal depends on the operating conditions (Table 4–6). There are two main types, contact and non-contact. Contact seals rub against the shaft or a sealing surface, and provide a positive barrier between the bearing cavity and the outside. They are best for dirty wet environments, and are the safest choice for micro-hydro. Non-contact seals make use of small gaps and tortuous paths or labyrinths to prevent grease leaking out and dirt going in. They can be used at any speed, but do not provide such a positive seal.

The simplest type are *felt seals* (Figure 4–27(d) and (e)). These are good for slow speed applications (rubbing speeds up to 4m/s). They can be used at higher speeds but a gap develops between the felt and the shaft. They then become non-contact type. The better the shaft finish, the better the performance, but a fine-turned finish of R_a 3.2μm is adequate.

Lip seals (Figure 4–27(f)) are effective, and can be used at higher speeds. They are very often made of polyurethane, which is wear resistance and retains its elasticity. They may or may not have a

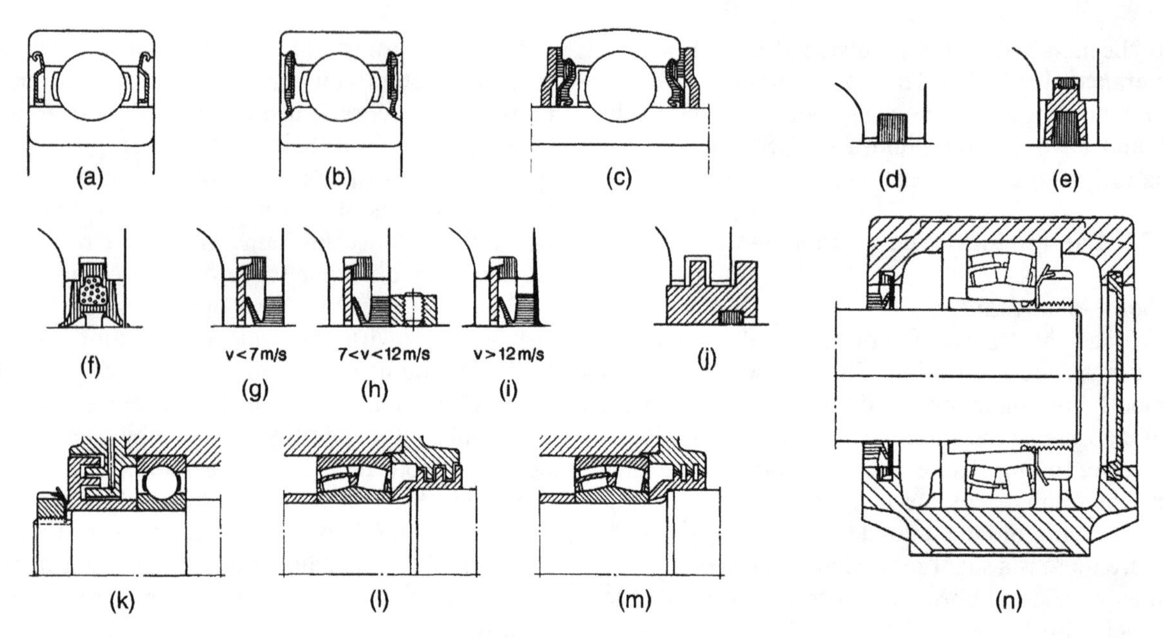

Fig. 4–27: Various types of shaft seals (© SKF)

Table 4–6: Rubbing speed for seals on shafts of various diameters at various speeds

Shaft dia. d (mm)	*Seal rubbing speeds (m/s) at shaft speeds of: 500rpm*	*1000 rpm*	*1500 rpm*	*Type of seal — Contact type: Felt*	*Poly-urethane lip*	*V-ring nitrile rubber*	*Non-contact/ labyrinth*
50	1.3	2.6	3.9	< 4m/s *	< 8m/s	< 12m/s *	
60	1.6	3.1	4.7				
70	1.8	3.7	5.5				
80	2.1	4.2	6.3			Provide axial location over 7m/s (over 12m/s needs full location)	
90	2.4	4.7	7.1				
100	2.6	5.2	7.9				
120	3.1	6.3	9.4				
140	3.7	7.3	11.0				
160	4.2	8.4	12.6			Can be used with full location	
180	4.7	9.4	14.1				
200	5.2	10.5	15.7				

* for the use of felt seals above 4m/s, and V-ring seals above 12m/s, see text.

rubber element with them to increase the pressure on the shaft. These seals need a very good surface finish on the shaft, ideally ground, and R_a 1.6µm at worst.

V-ring seals (Figure 4–27(g), (h) and (i)) are simple but tough and very effective. Made of nitrile rubber, they have a flexible lip that rubs against a mating surface. Above 7m/s rubbing speed they need a shoulder for axial location to prevent them creeping along the shaft. (It is possible to use these seals even above 12m/s rubbing speed, but the whole seal body need supporting to keep it held down on the shaft.) The surface finish of the mating part is not critical. It is often possible to use the smooth surface of a sheet steel washer to seal against.

Examples of *labyrinth* or *non-contacting seals* are shown in Figure 4–27(j), (k), (l) and (m). They use the same sealing principle as the *flinger seals* used to seal the turbine shaft described in Section 4.9. They require a lot of machining, and are expensive to make. They can be used at very high speeds. They are not very effective at peripheral speeds of less than about 3m/s.

If nothing is fitted to one end of the shaft, the simplest seal is to keep the shaft short within the bearing housing and have a blanked end (Figure 4.27(n)).

4.8 Shafts

4.8.1 Shaft design

Designing a Pelton turbine shaft is an iterative process. A rough design has to be sketched out, so that the bearing and drive forces can be calculated. Tentative choices can then be made for the bearings, pulleys, drive belts, and couplings. The shaft diameter is then chosen to fit inside the bearing, but this diameter needs to be checked to ensure that it can withstand the stresses – using the method given in Section 4.8.2. If it turns out that the shaft is too small, the bearing sizes will need to be increased, the layout of everything will need to be modified, and everything must be checked again.

4.8.2 Shaft stress

A Pelton turbine shaft has torsion and bending forces acting on it. The torsion transmits the power out from the runner. The bending moment arises from the jet forces on the runner, the weight of the runner and shaft, the bearing reactions, and from the belt tension if a belt drive is fitted. The axial force on the shaft is usually negligible (which is one of the advantages of a Pelton turbine). The shaft must be designed not to yield, and to withstand fatigue failure from the alternating stresses acting on it.

Diameter required to prevent yield

Draw the bending moment diagram for the shaft, and establish the critical section. If the shaft is of uniform diameter, this will be where the bending stress is maximum. If the shaft has smaller sections away from the maximum bending moment point, the stress at these sections may need to be checked too.

Calculate the minimum shaft diameter required to keep the shaft material from yielding. The relevant equation is derived in Section 10.6.1, combining the torsion and bending stresses using distortion-energy methods:

$$\frac{\sigma_{\text{yield}}}{SF_{\text{y}}} = \frac{32}{\pi . d^3} . \sqrt{M^2 + \frac{3T^2}{4}}$$

$$\therefore \quad d_{\text{y}} = \left[\frac{32.SF_{\text{y}}}{\pi . \sigma_{\text{yield}}} . \left(M^2 + \frac{3T^2}{4} \right)^{\frac{1}{2}} \right]^{\frac{1}{3}} \qquad \textit{Eq. 4–11}$$

d_y – minimum shaft diameter required (m)
σ_{yield} – yield stress for the shaft material (N/m^2)
SF_y – safety factor
M – bending moment (N.m)
T – torsion (N.m)

If you are using a standard grade of steel, the yield stress for it can be looked up in reference books. If you are not sure what steel you have, or cannot trust the supplier, have it tested. If you have no idea what steel is being supplied, and have no test facilities, assume the worst possible figure. Any reasonable mild steel should have a yield stress of at least $200N/mm^2$, though shaft steels are normally much better than this.

The safety factor is a matter of experience. It must allow for loads or the shaft strength being different from the values assumed. When a turbine is started, the forces on the runner rise to their maximum values quite gently. During normal running, the forces are steady, and a generator is at an even, constant load. However, abnormal load conditions should also be considered. A bolt may drop into a piece of mill machinery. A short-circuit can make a generator act as a very strong brake, stopping it almost instantaneously. The forces produced can be huge, enough to snap shafts or break the rotor. However, if there are drive belts or couplings between the machine/generator and the turbine, these will slip or snap first. As a first guess, a safety factor of SF_y = 3–4 could be used.

Diameter required to prevent fatigue

Next calculate the minimum diameter required to withstand the fatigue load on the shaft. The equation is derived in Section 10.6.2:

$$\frac{\sigma_{\text{ec}}}{SF_{\text{f}}} = \frac{32}{\pi . d^3}$$

$$\therefore \quad d_{\text{f}} = \left(\frac{32 M . SF_{\text{f}}}{\pi . \sigma_{\text{ec}}} \right)^{\frac{1}{3}} \qquad \textit{Eq. 4–12}$$

d_f – minimum shaft diameter
σ_{ec} – corrected endurance limit for the shaft material
SF_f – safety factor

The endurance limit, σ_e, for standard materials can be looked up, but needs to be corrected for real shaft conditions: $\sigma_{ec} = 0.7 \times \sigma_e$. Often the endurance limit is not known. A value of:

$$\sigma_{ec} = 0.35 \times \sigma_{ult} \qquad \textit{Eq. 4–13}$$

σ_{ult} – ultimate tensile strength for the shaft material

gives a reasonable, conservative working value. A safety factor of SF_f = 2.0–2.5 should be adequate (but see the notes in Section 10.6.2).

Final shaft diameter

The minimum shaft diameter required is the largest of d_y or d_f. It is normal to round up the diameter to the nearest standard size. A small increase in shaft diameter greatly increases the strength, and it is often worth adding an extra 5mm or so to give that extra bit of safety. Normally, the loads on a turbine are quite predictable, but there can be significant shock loads if a V-belt comes out of its groove and jams, or if a bucket falls off. Making the shaft slightly too big can help it to survive mishaps of this sort.

4.8.3 Attaching the runner to the shaft

There may not be much choice as to how the runner is fitted on the shaft. Standard generators have parallel shafts with keyways, and the hub will need to match. The bore of the hub should be chosen to give a good fit, but one that can be dismantled. If the shaft is made specially for the turbine, there are lots of options (Figure 4–28):

- Parallel fit with keyway (a).
- Taper fit with keyway (b).
- Flanges on the shaft bolted to the hub (d).
- Taper sleeve coupling (c).
- Hub welded to the shaft (f).
- Single-piece hub and shaft.

Think about how the runner is to be serviced. If the runner needs to be changed, is it useful to have to change to whole shaft as well? Usually not, so welding the hub to the shaft or making them out of one piece of metal is not a good idea. Size plays a part too. A good fit between a hub and a parallel shaft can be dismantled easily if the shaft is small. For a big shaft, it may require an enormous force to get the hub off, especially after it has been in place for a while. Taper fits are preferable for larger runners (say, over 300mm PCD).

Using a locking adaptor (as described below) is a more expensive option, but such adaptors are easily fitted and removed. This can be a very significant help. A turbine runner that has been pressed on to a generator shaft and left in the turbine for a number of years can be very difficult to get off. With a locking adaptor, it is always simple to remove it again. Locking adaptors also allow the position of the runner to be adjusted on the shaft, no keyways are required, and the machining is less critical.

Bolted arrangements also do away with keyways, but they take up quite a lot of space. They are easily dismantled, and can be made with simple machinery. They have to be designed properly to give the correct torque capacity, and the bolts need to be tightened correctly.

In designing the shaft, take care not to make sharp changes of diameter or notches, as these will act as stress-raisers and may lead to cracking. Any corners should have a radius of 3–5mm minimum. Special care needs to be taken with welds. Avoid putting welds on the main diameter of the shaft (as in Figure 4–28(e)). Make a special large-diameter sections for welds (Figure 4–28(f)).

Whatever arrangement is chosen, think how it will be dismantled. Leave slots or notches for a gear puller, or provide threaded holes so that the hub can be withdrawn with bolts. Do not wait until you are at a remote site, without the appropriate tools, before you realise that you cannot get the hub off! Standard keys, keyways, taper angles and parallel shaft fits can all be looked up in engineering reference books (such as Kempe (1989) or Oberg (1992). The sections below provide some additional guidance, and some detailed reference information on tolerancing, fits, and keyways are given in the appendices.

Keyed fits

A key transmits torque between a shaft and a hub, but is not the main element of the joint. Both the shaft and the bore of the hub have to be machined to give a good fit. Appropriate fits are discussed in Appendix 14. The runner should fit well on the shaft, without movement. The torque on a runner is only in one direction, and is usually applied and removed quite gently – there are no shock loads – so the design of the key is not critical.

A table of standard metric keys and keyways, and the recommended size of key for a given shaft size, is given in Appendix 13. Parallel keys need some additional means of locking the hub in place on the shaft; Figure 4–28(a) shows the runner located by a bolt on the end of the shaft, which traps the hub between a shoulder and a washer. Taper keys can be used by themselves to locate the hub axially on the shaft, but they are not always reliable. Taper keys alone are adequate for small, lightly-loaded shafts, but additional axial location should be provided for larger shafts.

The tables in Appendix 13 give the appropriate section for the key, but not the length. This is chosen to keep the stress at an acceptable level. For simple calculations (which are adequate in this case, where the torque is only in one direction and there are no shock loads) only two stresses need to be considered: the shear stress in the key, and the bearing pressure on the side of the keyway in the shaft. Stresses are calculated assuming that the torque gives a force F that acts at the diameter of the shaft, d, as in Figure 4.29. The key shears between points 2 and 5, so the shear stress is:

$$\tau = \frac{F}{L.b} = \frac{2T}{L.b.d} \qquad \textit{Eq. 4–14}$$

τ – shear force in the key (N/m²)
F – force on the key (N)

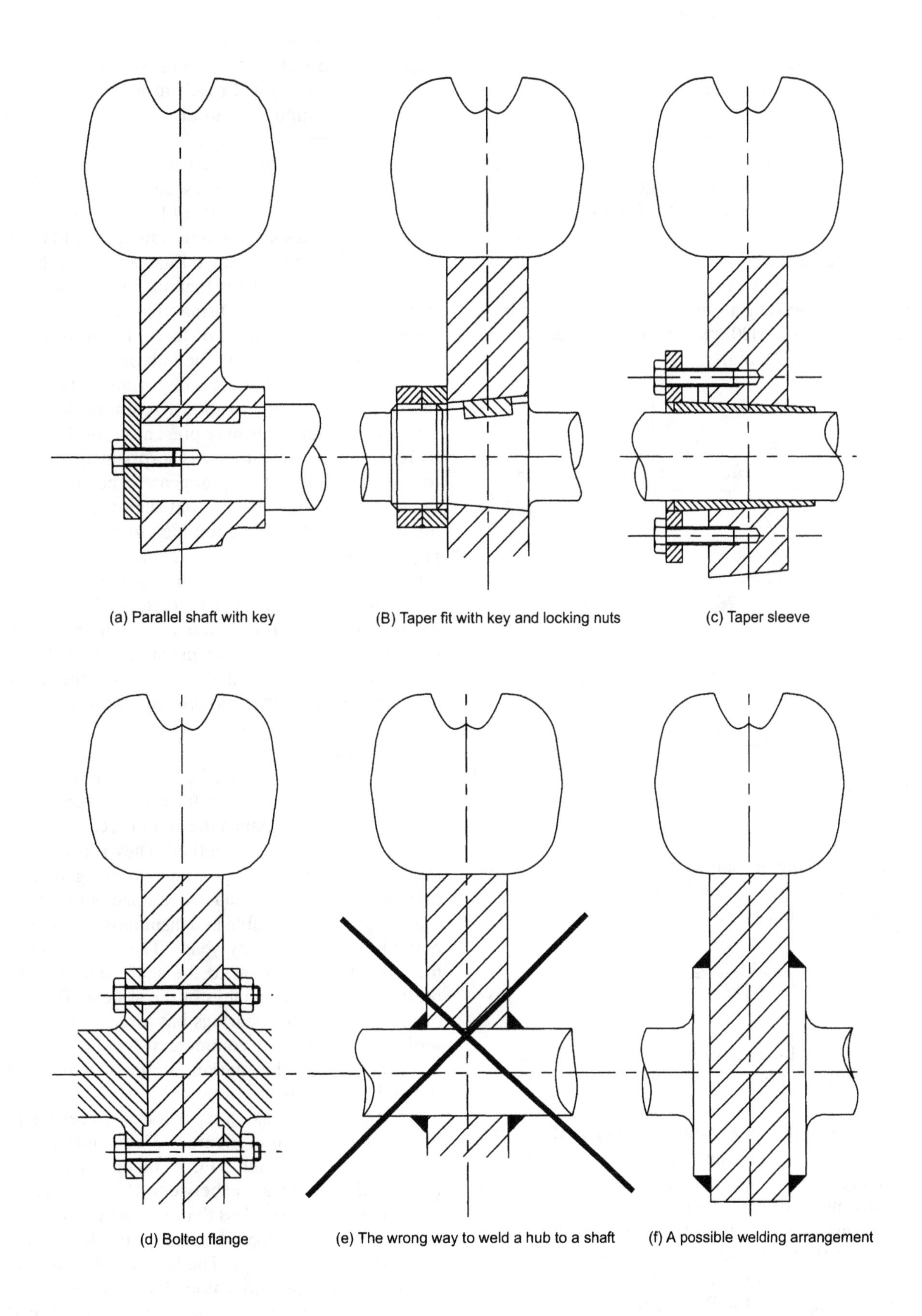

Fig. 4–28: Various runner-shaft attachment methods

T	– torque in shaft (Nm)
L	– length of the key (m)
b	– width of the key (m)
d	– shaft diameter (m)

This stress should compared with the shear strength of the key material, allowing for a safety factor as well. The shear strength of most shaft materials can be estimated as 0.577 of the yield stress; see, for example Shigley & Mischke (1989). BS 4325 specifies bright steel with a minimum tensile stress of 550N/mm² for keys, for example BS 970 070M20. Supposing a safety factor of 2.5, then for a key made of 070M20, τ should be kept below 0.577 × 550/2.5 or 127N/mm².

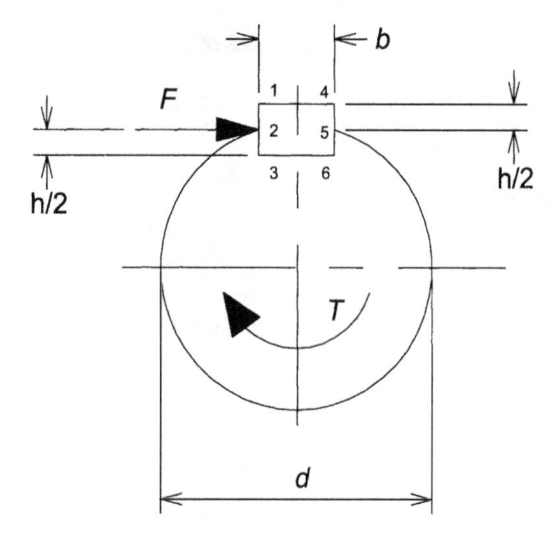

Fig. 4–29: Calculating the stresses on a key and keyway

The crushing or bearing pressure on the side of the keyway is simply the force divided by the area between points 2 and 3:

$$\sigma_s = \frac{F}{L.(h/2)} = \frac{2F}{L.h} = \frac{4T}{L.h.d} \qquad \textit{Eq. 4–15}$$

σ_s	– bearing pressure on side of key (N/m²)
h	– height of the key (m)

This bearing pressure should be compared with the yield strength of the shaft material, again including a safety factor. Shafts should be made of good quality, high strength steel if possible.

The actual stresses are rather higher than those given by these simple formulae because of stress concentrations, but the calculations are adequate for choosing keys for Peltons if reasonable safety factors are used.

Taper fits

Tapers provide a very good, solid means of fitting a hub to a shaft. An inclusive angle of 5–10° (i.e. each side is at 2.5–5° to the shaft axis) is required. American SAE standards use 1.500±0.002" per foot (equivalent to just over 7° inclusive). Too small a taper (less than 1°) will result in a solid assembly that is almost impossible to dismantle. Too large a taper will come apart by itself.

Tapers normally have a key fitted in them too, to transmit the torque and give security. Parallel keys are common, though Woodruff keys are also used. Tapers also need some locking system to hold them together. The taper system in Figure 4–28(b) shows a hub held on by a nut and locking nut. The locking system must be secure. Dismantling, or *breaking* a taper fit can require a lot of force. Leave room in the assembly to hit or pull the hub off.

When making a taper, it is important to be accurate. If the workshop has good machinery and measuring equipment, this may present no problem. If there is only an old lathe, it is surprisingly difficult to repeat a taper. It may appear to be set at the same angle as for the shaft, but then the hub does not lock on to it. It is a good idea to machine both parts at the same time, without changing the angle setting. Whilst the lathe is set up, also make a taper gauge, so that the lathe can be re-set in future. This gauge should be a sturdy shaft, a few hundred millimetres long, with the taper machined on it. If another hub or shaft is needed in the future, the gauge can be put in the lathe, and the angle adjusted to it.

Locking adaptors

Locking adaptors are fitted between the hub and the shaft. They have some form of tightening arrangement which expands the adaptor, causing it to lock the hub and shaft together. They rely on friction to transmit the torque, and do not generally have keys. Their particular advantage is that they are very easy to assemble and dismantle, even after being installed for many years. They can be used for fixing the runner on to the shaft, but are also useful for any other pulleys in the system. The use of taper locking for bearings has already been described. A variety of commercial locking sleeves are available on the market, some of which are shown in Figure 4–30.

It is possible to make a taper locking sleeve, and one design is shown in Figure 4–31. A taper is machined into the runner hub. A taper angle of 2° works well, but any angle between 1° and 5° seems to work. If the angle is less than 1° it is very difficult to withdraw the sleeve. The shaft is machined parallel (without a keyway). The sleeve is tapered on the outside, has two slots along its length to help it clamp, and has a flange on one end. It is machined to a sliding fit on the shaft such as H7-g6 (see Appendix 14). To assemble, the sleeve is fitted on the shaft, and the hub placed over it. Bolts are fitted through the sleeve flange into tapped holes in the hub, and tightened. The action of the taper being

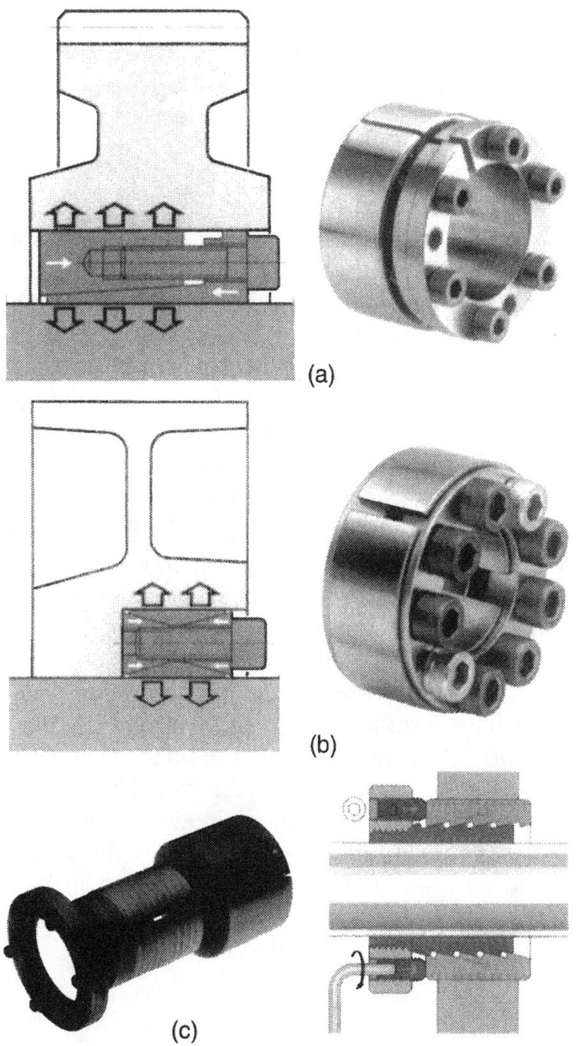

Fig. 4–30: Commercially available locking adaptors (© (a) & (b) Ringspann, (c) SKF)

drawn in creates a very large force between the shaft and the hub, locking them in place. Note the threaded holes in the flange, which are used to withdraw the sleeve from the hub.

Bolted joints

Bolting the runner and shaft assembly together looks a neat idea. It does not require close tolerance machining, and is easily assembled and dismantled. In practice, it often does not fit in very easily into the detailed design. As with all bolted joints, the forces should be carried by friction. Assuming that all the bolts are on a circle at the same radius, the torque capacity of the joint is given by:

$$T = \frac{N_{\text{bolts}} \times F_{\text{friction}} \times r_{\text{pitch}}}{SF} \qquad \textit{Eq. 4–16}$$

T	– torque capacity of the joint (Nm)
N_{bolts}	– number of bolts
F_{friction}	– friction force per bolt (N)
r_{pitch}	– bolt pitch circle radius (m)
SF	– safety factor

The friction force of the bolts can be found from Table 4–2 which assumes that the bolts are tightened to the torque specified, giving the pre-tension quoted. (The frictional force per bolt is then the pre-tension multiplied by a friction coefficient of 0.15, which is a reasonable general figure for steel joints.) The torque can be calculated from the turbine power (see Equation 9–7 in Section 9.4.2). A safety factor of 1.5 is a minimum. Note that each bolt only provides the friction load once, even if it goes right the way through the assembly and holds two flanges in place.

The bolted flanges should be located with a close fitting spigot. Do not rely on the bolts for alignment.

4.8.4 Corrosion protection

Because the turbine shaft and runner is usually assembled on site, it is common not to put any corrosion protection on the shaft. If the shaft is lightly loaded, and the life expectancy of the turbine is not

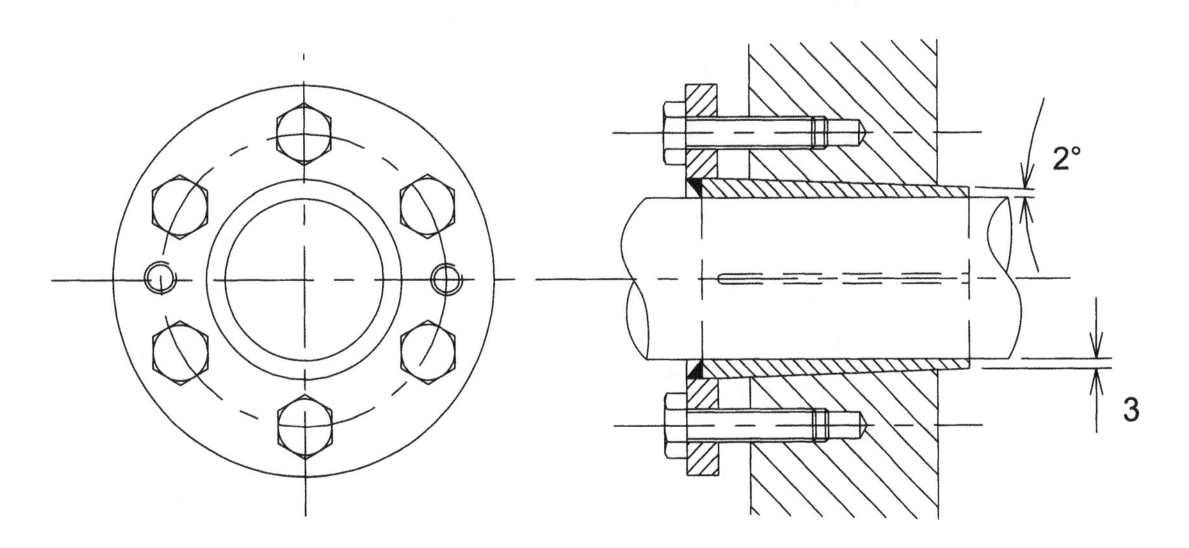

Fig. 4–31: A tapered locking sleeve that can readily be made by a local workshop

great (less that ten years) this may be acceptable. However, corrosion can also lead to corrosion fatigue, as well as making it difficult to dismantle the machine. Corrosion fatigue is seen on large turbines after 20–30 years. It is good practice to paint the parts of the shaft exposed to water after final assembly.

A simple, but adequate, cover is at least two coats of red lead primer. An optional topcoat of a water-resistant paint gives even better protection. The shaft should first be cleaned, removing any rust, grease or dust, and should be dry. Allow adequate time between the coats for the paint to dry. Other paint systems can be used, and the paint manufacturers can be consulted to see which of the available paints are appropriate. Note that red oxide paint is *not the same* as red lead primer. Red oxide is not particularly good for this application. Professional turbine companies use specialist shaft coatings, such as two-pot tar-coal paint systems. These give excellent protection, but are expensive and may not be readily available.

Occasionally, small Pelton turbines are fitted in the pipelines for community water supply systems; if the water has to be brought down from the hills anyway, it makes sense to generate some useful energy from it, rather than just dissipating the head in break-pressure tanks. In this case, the paint system inside the turbine would need to be non-toxic, and red lead paint would not be suitable.

Chrome plating of shafts is not particularly appropriate for hydro applications. If corrosion does manage to penetrate the plating, the surface starts to peel off, causing all sorts of problems with seals.

4.9 Turbine shaft seals

When a Pelton turbine is working, the space around the runner is full of spray, and there is water running down the walls of the casing into the tailrace. The shaft has to emerge from this wet environment, but the water should stay inside. This means there has to be some sort of seal. Pelton turbines run quite fast, and this situation is ideal for non-contact, labyrinth-type seals. The principle is to have a constricted, circuitous path between the rotating and non-rotating parts, so that the water cannot get all the way out. A simple but effective design is shown in Figure 4–32.

The rotating part is clamped on to the shaft by bolts, which tighten a split collar. The fixed part of the seal is bolted to the turbine housing. A small gap 'X' is left between them. Because 'X' is small, water finds it difficult to go down it, but, more importantly, the rotation of the disk on the shaft flings out any water that gets into the gap. The rotation also pumps a small amount of air outwards between the disks, which again adds to the sealing effect. Any water that does make it into the seal housing is drained back into the turbine casing

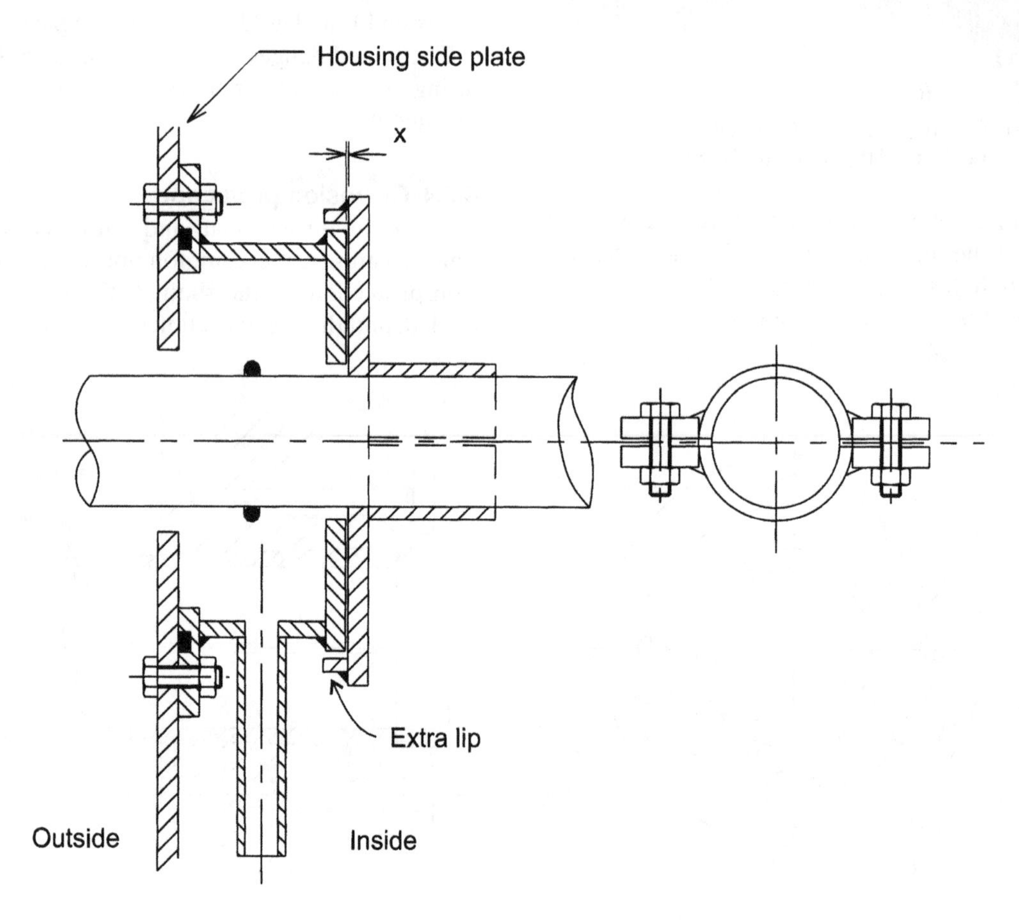

Fig. 4–32: A flinger seal

through the tube at the base. An O-ring can be slipped over the shaft as shown in the figure to make sure that any water that does get onto the shaft is spun off in to the housing. Figure 4–32 also shows an extra lip welded on the rotating disk. This stops water going directly into the gap, giving yet another barrier, but it is not essential. One warning: do not weld the rotating disk to the shaft, because it can lead to fatigue cracking.

These 'flinger' seals have proved to be very effective. Occasionally, spiders webs are found inside them when they are dismantled, so they must be almost completely dry inside. The faster the speed, the better the sealing, but they work even for small diameter disks and low heads. Keep the peripheral speed of the disk above 1.5 m/s; most designs will have speeds much higher than this. The gap 'X' is not too critical, but it should be less than a millimetre. A simple, practical way of setting the gap is to push the rotating disk up against the stationary housing until one can just feel it rubbing.

Figure 4–32 shows an O-ring seal between the flinger housing and the turbine casing. O-rings work well – better than gaskets for this job – because the parts bolt solidly together. However, many casings have multiple parts, and the flinger housing may cross a number of joints in the casing. If the joints are flat and do not have too large a gap, an O-ring can still be used, but a gasket has more flexibility and may be a better choice.

If the runner is mounted on a generator shaft, it may be convenient to bolt the generator front face directly to the housing. If this is the case, try to leave a gap between the two faces so that air can still be drawn into the flinger seal from outside. This is shown in Figure 4–33.

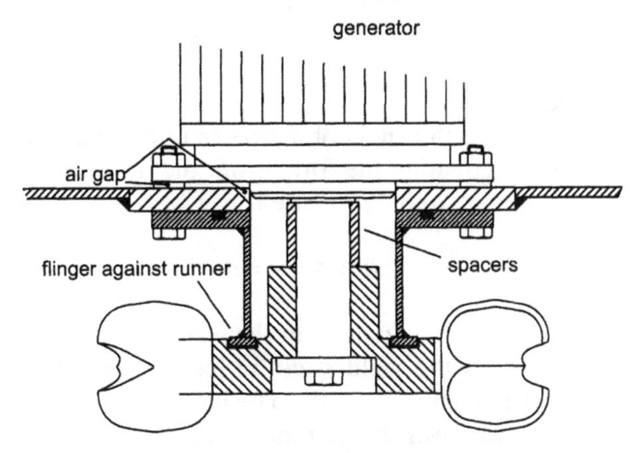

Fig. 4–33: Shaft sealing using the runner as the rotating disk on a Peltric. Note the gap between the generator and the housing to allow air to be drawn into the seal housing

4.10 Manifolds

The word 'manifold' comes from the root 'many', and in the engineering sense is usually used to describe an arrangement of a number of pipes that are joined together, as in the inlet manifold of a car engine. Here the term is used for the pipework that joins the penstock to the nozzle/nozzles, even if it is on a single jet Pelton.

For a single jet turbine, the manifold may be a simple bent pipe that brings the water into the nozzle or spear valve, but for a six-jet turbine it is a large, branching structure. Manifolds are often designed as something of an afterthought, when the rest of the turbine is finished. This is a mistake, and the manifold layout should be considered right from the start of the design. A poor manifold can give major power losses, or take up an enormous amount of room in the powerhouse. It should also be noted that manifolds are situated at the high-pressure end of the system, and stress calculations should be done, particularly for high heads.

4.10.1 Determining the size of the manifold

Water flowing in a pipe loses some of its head because of friction and turbulence. There are additional losses for bends, valves, diameter transitions, and branches. This section gives some general comments on designing pipework to keep the losses small. Appendix 10.7 shows how to calculate losses in detail if required.

Manifolds are generally short compared with the penstock, so the pipe losses are also relatively small. Losses at the manifold bends are proportional to the velocity head of the water, ($V^2/2g$). Since micro-hydro pipe velocities are typically less than 3m/s, the velocity head is 0.46m or less. Most bends, junctions or valves have losses of around 0.1–0.4 times the velocity head, so the total manifold loss is usually less than 1m. Pelton turbines rarely run on heads less than 20–25m, which means that the manifold losses are not too significant.

Problems arise if the manifold is too small. A good guideline is to keep the velocity throughout the manifold similar to the velocity in the penstock. Because penstocks are usually designed to give losses of 5% or less, the flow in them is fairly slow (often around 2-3m/s), and if this speed is maintained in the manifold, few problems will occur. Keeping the velocities in the manifold down to 2–3m/s will make them quite large, and the temptation is to try and cut down the pipe sizes. It is a temptation to be resisted, because small pipework can give very high losses. If the design involves manifold velocities higher than those in the penstock, do the calculations given in Section 10.7.1 to check the losses. It is possible to use higher speeds, and very large power-stations under high heads take the pipe velocities (in large diameter pipes) up to 13m/s or so.

A restricted manifold might give the following problems. A single jet turbine gives very poor performance when the spear valve is fully open, but has the predicted output when the spear valve is half open. Or a four-jet turbine provides the calculated efficiency for one and two jets, but very little extra power for the third and fourth jets. This is because at low flows, the losses in the pipework are small, but as the flow increases so do the losses, reducing the net head at the nozzles. Keeping the pipe size large enough avoids these situations.

Having fixed the pipe size, the bends and junctions must be designed to minimize losses. Sharp corners, and quick changes in pipe diameter all give higher losses. All transitions should be gentle, corners should be radiused where possible, and the flow should be kept as smooth as possible. For guidance, the centreline bend radius of pipes should be about three times the pipe bore, and not less than twice the bore. In this manual, the bend radius of spear valve bodies is shown as twice the bore, because a larger radius considerably increases the overall length, but all other bends are three times the bore.

If possible, keep all the manifold bends – including any bends at the bottom of the penstock – in a single plane. If bends are not in the same plane they can generate a swirling motion in the flow, which can upset the water jets in the turbine. Bends which are more than 30 pipe diameters apart do not cause a problem, because the flow stops rotating in the straight section. If it is necessary to have close, badly out-of-plane bends, make sure there are straightening vanes in the spear valves or just before the nozzles.

4.10.2 Manifold losses

Graphs, tables, and a calculation procedure for estimating the losses in the various elements of a manifold are given in Section 10.7. As has been already stated, it is not usually necessary to go through all these calculations. It is a good idea to go through the procedure at least once to give yourself an understanding of where the losses in a system arise, and their magnitudes. The main points that arise from Section 10.7 are summarised here.

Manifolds may be considered as being made up of bends, joints, and diameter changes, all joined together by pieces of straight pipe of varying lengths. The losses in straight lengths of pipe can be calculated from standard hydraulic fomulae such as the Colebrook-White or Manning equations. Bend, joint and diameter change losses do not lend themselves to easy calculation, but use can be made of graphs of experimental results. These give factors *K* for a particular pipework feature, and the head loss is found by multiplying the relevant *K* by the velocity head in the pipe.

The total loss in the manifold can be estimated by adding together the calculated losses for each component. While this somewhat overestimates the manifold head loss, it gives a good indication of the size of the individual losses, and shows up which parts of the system (if any) are the worst. Losses are strongly influenced by the diameter of the pipework, being roughly proportional to $1/d^5$, where d is the pipe diameter, which means that if the losses are too high, a small increase in pipe diameter can make a big difference. Once the pipe size is correct, good design means avoiding sharp bends, sharp changes in section, or sharp edges on the inside of the pipework. If it looks as if the water could flow smoothly through the manifold, it will probably be acceptable.

4.10.3 Manifold stresses

The manifold is at the bottom of the penstock, and can be under high pressure. When using high heads and large pipes, a burst is very dangerous, and can cause major damage to the plant. The thickness and strength of the manifold material has to be checked to see if it can withstand the predicted surge pressure. If ready-made piping is being used, there should be specifications or standards specifying the maximum and working pressures. If the pipe is manufactured, calculations need to be done. This section gives the standard formula for pipes, and a well-proven formula for mitred joints. There is then a brief discussion of calculations for bifurcations.

Manifolds in large powerhouses are designed for 'leakage before rupture', that is, any defects or fatigue cracks should propagate right through the material before they lead to catastrophic failure. This means the plate thickness must not be too thick, and the material not too brittle. The calculations involved are sophisticated, but are beyond what is necessary for micro-hydro. Nevertheless, the designer of any turbine should consider what would happen if there was a failure. Pressure testing a manifold before it is installed is good practice.

Straight pipes

The required thickness of a pipe can be calculated simply by equating the force from the hoop stress in the pipe, σ, to the pressure force:

$$2t \times \sigma = p \times d \qquad \text{Eq. 4–17}$$

The pressure, p, comes from the water. The allowable stress is the yield stress reduced by a safety factor. The thickness needs to be increased to allow for corrosion over the lifetime of the plant. Putting all these together:

$$t = \frac{H_{\text{total}} \cdot \rho \cdot g \cdot d}{2\left(\sigma_{\text{yield}} / SF\right)} + t_{\text{corrosion}} = \frac{H_{\text{total}} \cdot \rho \cdot g \cdot d \cdot SF}{2\,\sigma_{\text{yield}}} + t_{\text{corrosion}} \qquad \text{Eq. 4–18}$$

t – pipe thickness (m)
H_{total} – total head (m)

ρ	–	density of water (1000kg/m^3)
g	–	gravity (9.8 m/s^2)
d	–	pipe diameter (m)
σ_{yield}	–	yield stress (N/m^2)
SF	–	safety factor
$t_{corrosion}$	–	corrosion allowance (m)

The total head is the gross head plus the surge head, as calculated in Section 4.3.5. A safety factor of at least 2 should be applied, using the minimum value of the yield stress expected. A set of recommended corrosion allowances for micro-hydro pipework is given in Table 4–7.

Table 4–7: Recommended corrosion allowances for penstocks and manifolds. Based on Waltham (1994)

Internal finish	*Corrosion allowance* $t_{corrosion}$ (mm)			
Scheme size	<8kW	8–20kW	20–50kW	>50kW
Galvanized	0.5	0.7	0.9	1.0
Bitumen, red lead or other recommended paint systems	0.8	1.1	1.4	1.6
Standard paint/red oxide	1.0	1.4	1.8	2.0
Unpainted	1.5	1.8	2.2	Needs protection
HDPE, uPVC	0	0	0	0

Mitred joints

Mitred joints increase the stress in the pipe, and a factor is added into Equation 4–18 to allow for this. The maximum stress is on the inside of the joint, and the thickness is given by:

$$t = \frac{H_{total}.\rho.g.d.SF}{2\,\sigma_{yield}}.\left[1 + 3\cdot 8\tan(\theta/2)\right] + t_{corrosion}$$

Eq. 4–19

θ – bend angle – see Figure 10–15

This formula has been proven by finite element analysis and strain gauge measurements, and is valid for any mitred bend without reinforcement, even if the bend is set in concrete (Brekke, 1994).

Bifurcations

Bifurcations do introduce extra stresses into the manifold, but unfortunately the effect is difficult to calculate. For most micro-hydro schemes, bifurcations are unlikely to give problems, using the same plate thickness as for rest of the manifold. There are potential problems with large pipes at higher heads. If the thickness of the manifold is near the minimum allowable for the surge head, some improved design should be considered.

The problem with bifurcations is illustrated in Figure 4–34(a). This is a simple 60° junction with the branch pipe welded on to the side of the main pipe. Section A-A shows that in the angle between the two pipes there are V-notches. Pressure in a pipe is normally carried by tensile hoop stress that runs around the circumference, but in these V-sections a hoop stress would act with the pressure rather than opposing it. The pressure stresses have to be resolved partially into bending at right angles to the section A-A, partially into tensile stresses at an angle in the surrounding material.

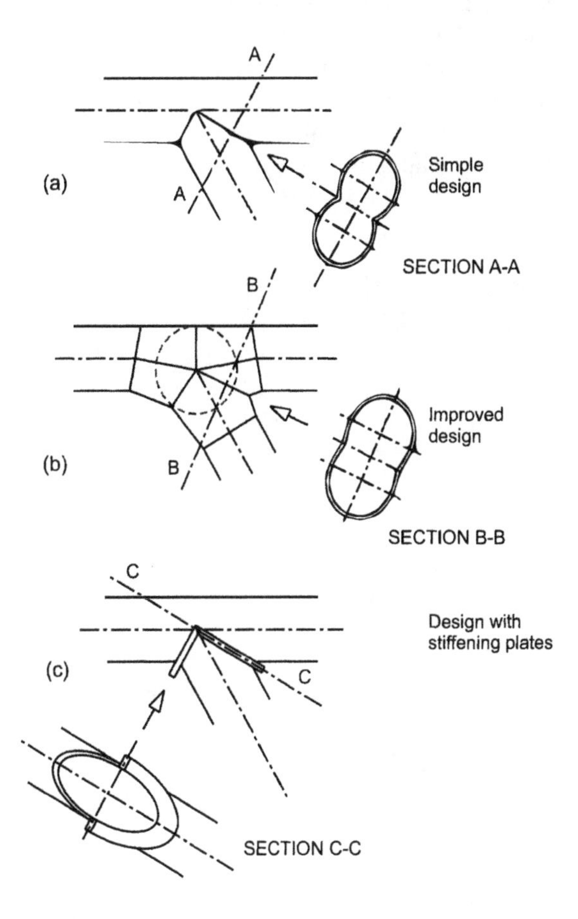

Fig. 4–34: Bifurication designs: (a) Simple. (b) Improved. (c) With stiffening plates

The best way to improve the situation is to make the joint more bulbous, as shown in Figure 4–34(b). The design philosophy is to make all the surfaces tangential to an imaginary sphere at the centre of the joint. The sphere needs to be a little larger than the largest pipe diameter to achieve the best layout. The various branches then take a tapered shape, and this greatly reduces the V-shapes between them, as in section B-B. The spherical shape means that when the section is weak in one direction, it can take tensile stress at right angles. The tapered branches also make these bifurcations good hydraulically. This type of joint is standard for very large, high-head hydropower plants, where manifold stresses can be the limiting factor on the head that can be used. Their disadvantage is that they are more difficult to manufacture. It is often not necessary to go all the way to making a spherical-based junction. Almost any bifurcation incorporating a taper to improve the flow, such as that illustrated in Figure 10–19, improve the stresses in the joint.

Another way of improving bifurcation strength is shown in Figure 4–34. This is a simple joint with reinforcement plates between the pipes. These are thick, horseshoe-shaped plates that take the pressure stresses in bending. To form a good joint, the pipes need to be welded to the plates, rather than welding them on to the outside of a finished joint. This sort of reinforcement is dealt with in detail in reference (Bier, 1966). The calculation method given in the paper results in enormously thick reinforcement plates because it takes all the pressure stress into the reinforcement, ignoring the support from the surrounding plates. Reinforced bifurcations are easier to produce than spherical joints, but are heavier and not as neat.

In summary, when the manifold pipe has a large safety factor, no special attention needs to be given to pipe joint stress. If the pipe is near its stress limit, simple junctions are not adequate. Increasing the plate thickness at the joint is one solution, but when the strength is marginal, a tapered, spherical or reinforced joint should be used.

4.11 Shut-off valves

If a turbine has nozzles, not spear valves, some way is needed of turning the jets on and off. As stated previously, on small, simple schemes, it may be possible to stop the flow at the entrance to the penstock, but this is not very convenient. It is more normal to fit shut-off valves at some point in the manifold. The three main types used for hydropower are shown in Figure 4–35.

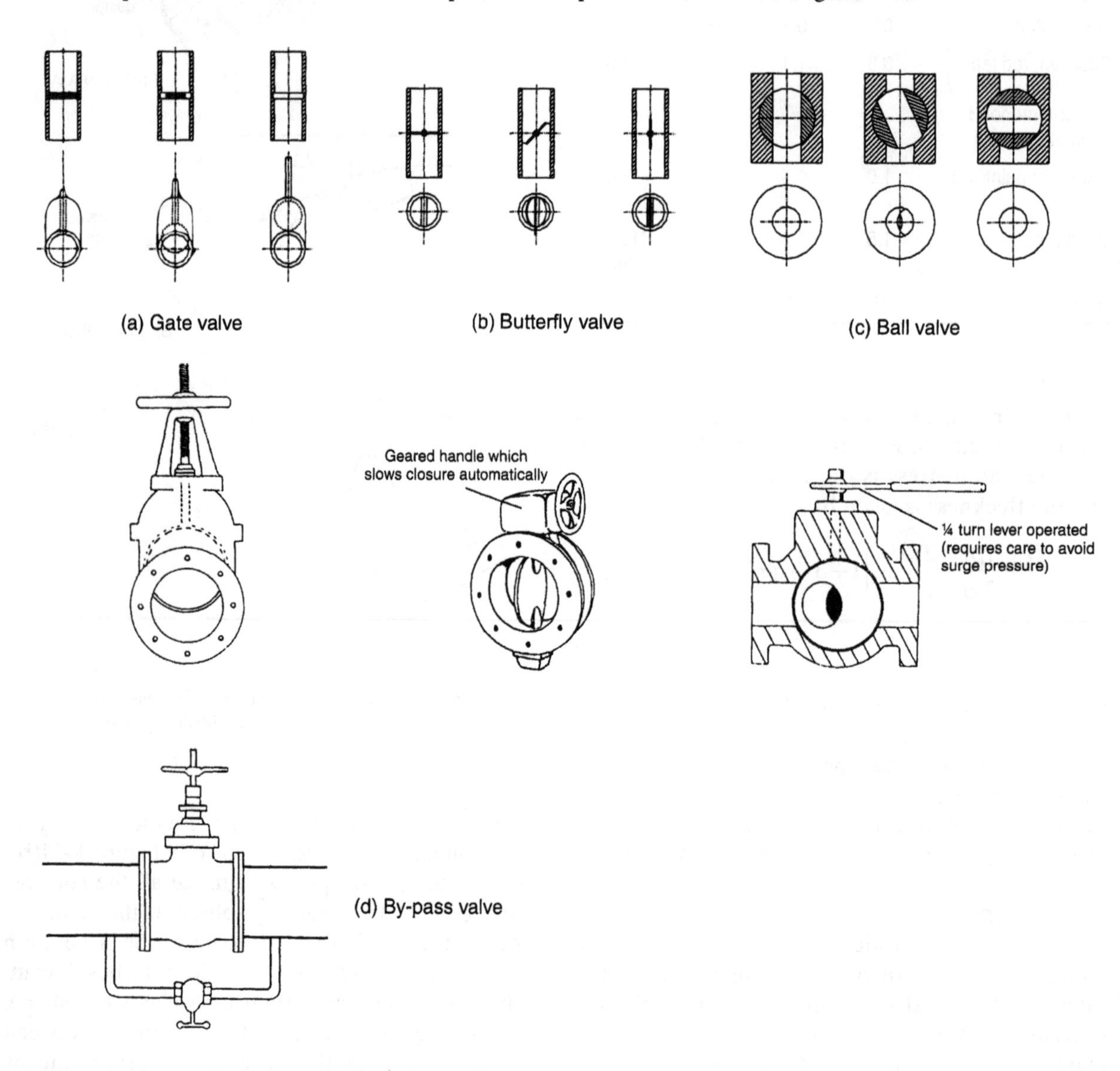

Fig. 4–35: Shut-off valves

Shut-off valves are also used with spear valves, though at first sight this seems unnecessary. There are, however, a number of advantages in fitting shut-off valves as well. Firstly, they make it possible to work on the turbine without having to drain the penstock. Secondly, they provide an extra way of shutting down the turbine in an emergency. Thirdly, though well-made spear valves should not leak at all, in practice, micro-hydro spear valves often do leak, and the back-up valves are essential to stop the runner or to work inside the casing.

The three types of valve are gate valves, butterfly valves and ball valves. The one feature in common for these valves is that they present a full-bore, nearly unobstructed pipe to the flow when fully open. Other sorts of valves, such as globe valves, or angle valves, take the flow round bends and introduce a considerable amount of turbulence. They are not suitable for manifolds. Note too that, unlike spear valves, shut-off valves are not used to control partial flow; they are either used open or shut. When partially open they produce a lot of turbulence and give large head losses.

Valves are usually specified for two different pressures. The higher pressure is often called the 'body' pressure, and is the maximum that can safely be withstood by the valve when it is open. The lower figure, often called the 'seat' pressure, is the maximum when the valve is fully closed with pressure on one side only. If the valve is controlling the whole flow into the turbine, then the body pressure needs to be more than the gross head plus the worst case surge head. The seat pressure only needs to be the gross head plus any surge due to closing the shut-off valve itself (which will be very small if the valve can only be shut slowly). However, if there are multiple jets and the shut-off valve is on one of the branches, the situation is quite different. It must be able to take the gross plus surge head from a jet on another branch, and this may occur when the shut-off valve is closed. Both the seat and the body pressures must cope with the high surge pressure.

4.11.1 Gate valves

Gate valves – also called *sluice* valves – are commonly used on water systems, and are consequently readily available. They work by sliding a circular gate across the pipe. There is usually a tapered metal seat running around the inside of the bore for the gate to sit in when the valve is fully closed. Some valves have a flexible seal in the seat. The gate is fixed to a threaded rod or *spindle* which comes out of the top of the valve housing, and this is moved up and down by turning a handwheel.

When a gate valve is shut against a high head, the pressure pushes the gate against its seat, and a large force is required to open it. While the valve may be specified to withstand the pressure, it is quite possible to break the spindle in trying to open it under pressure. If possible, find out from the manufacturer the maximum head at which the valve can be repeatedly opened. As a rough guide, the valve should not be operated if the pressure force (pressure times area) is more than 2000kg. If the operating force is too high, a gate valve can still be used if a by-pass valve can be fitted, as in the detail in Figure 4–35(d). The by-pass valve allows the pressure to be equalized on both sides of the gate, so that the valve can be opened easily. This system only works if the downstream pipe can be closed, which means that there must be a spear valve downstream.

The loss coefficient for a fully open gate valve is typically K_v = 0.1–0.2. As usual, the head loss across the valve is calculated by multiplying this coefficient by the velocity head, as in Equation 10–89. Gate valves have slightly higher K_v than ball valves because of the recess for the gate seat.

4.11.2 Butterfly valves

A *butterfly* valve has a disk with a central pivot which can be turned so that it either blocks the pipe, or else lies along the axis of the pipe. It has some sealing arrangement on the disk or bore to give a watertight fit when the valve is closed. The disk, or 'butterfly', is fixed to a shaft or *spindle* which can be turned either directly with a lever or indirectly through a gearbox.

An advantage of a butterfly valve is that any pressure force acts equally on either side of the spindle, so the valve can be turned even under pressure. They can still be difficult to open from the fully shut position because of large friction forces. Nevertheless, it is easier to open than a gate valve. The light operating forces mean it is possible to use direct lever operation even for quite high heads, but this is not always advisable. A quarter-turn lever can be shut very quickly, giving a high pressure surge in the penstock. It is better to operate the valve through a gearbox, as this slows down the closing. Such gearboxes are a standard option for butterfly valves, though they are more expensive than simple levers.

The head loss coefficient for a commercial butterfly valve is typically K_v = 0.3–0.5 for full bore valves. The smaller figure is for a valve where the thickest part of the disk in the flow is less than 0.15 times the bore, the larger figure for 0.2 times the bore.

4.11.3 Ball valves

A ball valve comprises a sphere with a hole in the middle. This is fitted into the body with a seal to prevent leakage around the ball. When fully open, the valve presents a plain bore to the flow, the same diameter as a piece of the pipe. The ball is fixed on a spindle, and can either be turned with a lever or a gearbox. The force required to open a ball valve against pressure is somewhere between the forces for a gate and a butterfly valve. A gearbox is again preferable to a lever as it prevents accidentally fast closure and high surge – as described in the previous section.

Ball valves are very good, but tend to be expensive. They are standard equipment for large powerstations. The loss coefficient is $K_v = 0$ for a fully-open, full-bore valve.

4.12 Turbine housing

A Pelton turbine casing has a number of functions:

- To hold the runner and the nozzles in the correct position relative to each other.
- To keep the water inside.
- To catch the water coming off the runner and take it to the tailrace without letting it fall back onto the runner or jets.
- To stop buckets or pieces of buckets from flying out if they come loose.
- To take any stresses imposed by the manifold, the runner, or the drive system.

4.12.1 Controlling flow within the casing

Basic dimensions for horizontal-axis and vertical-axis housings are given in Figure 4–36. These are good minimum dimensions, but they can be varied. The performance of the turbine will not be affected if larger dimensions are used, and it may even improve marginally. Care should be taken when reducing the dimensions.

Exhaust water from a Pelton comes sideways out of the wheel. In a horizontal axis machine, this water hits the side walls and spreads out on them. Some water goes downwards into the turbine pit, but some goes upwards and must then fall back to the tailrace. This water must not splash back on to the runner, so the casing has to be quite wide. In vertical-axis Peltons, the exhaust water from the top of the runner is thrown up on to the top of the casing, spreads out to the edges, and then runs down the outside walls. The water from the lower side of the runner falls straight into the turbine pit. Thus the flow paths inside a vertical turbine are better defined, and the top surface can safely be nearer to the wheel.

Though the majority of the exhaust flow goes sideways, some drops of water are flung radially outwards, at high speed. In order to keep these droplets from splashing back, the dimensions R_h have to be kept to a minimum size. In vertical turbines, droplets hitting the walls tend to join with the general flow down the sides, and fall away into the tailrace. In horizontal machines, there is no fixed flow on the end and top plates, and they have to be kept further away.

It is recommended that, in a horizontal-axis machine, the distance from the centreline of the runner to the walls, W_h, is at least the same as the pitch circle diameter, D. It is not always possible to achieve this, especially when direct mounting the runner on a short turbine shaft, but W_h should be kept as large as possible. In theory, the minimum width is determined by the flow. The casing need to be wider for two full-sized jets than for a single, small jet. Some major manufacturer do allow W_h to be as small as $D/2$, so this can obviously be made to work. For vertical-axis machines, the top only needs to be set about $D/2$ from the centreline, dimension H_h. In small Peltric sets, where the runner is on the shaft of a standard induction motor, H_h sometimes has to be a little smaller still.

Consideration must be given to the flow paths. The inside surfaces should be smooth, allowing water to flow down to the tailrace. Be careful not to put strengthening members or flanges on the walls that block the flow or direct it back on to the runner. For vertical-axis machines, it is beneficial to put a radius in the top corners to aid the flow – see the dotted variants in Figure 4–36.

Some horizontal-axis machines have very narrow top casings, not much wider than the runner. A diagram is shown in Figure 4–37. A 'wiper blade' is fitted around the runner just before it goes into this top section. The blade fits very tightly around the runner, leaving a gap of only 0.5–1mm. The top casing then has a clearance of 10–20mm to the runner. The wiper catches a large percentage of the water droplets, spray and air being carried around with the buckets, and lets the water fall back into the tailrace. Manufacturers claim that there is a noticeable increase in efficiency from this system, because the windage losses due to entrained water and air turbulence in the upper part of the casing are much reduced. Obviously, the runner bearings need to be mounted rigidly and accurately for this arrangement, because any movement would result in the buckets hitting the wiper and being damaged. The added complication of this design mean it is not commonly used for micro-hydro.

Another variant is shown in Figure 4–38. The lower jet is pushed round so that it points slightly upwards, and the angle between the jets is less than 90°. The combined effect is to bring the upper jet round so that its discharge is into the lower part of the casing, and flows more easily into the turbine pit. This arrangement gives good efficiency, but also has disadvantages. Firstly, the spear valve, pipework, and manifold valves for the lower jet are below floor level, which complicates the powerhouse layout. Secondly, the variety of angles on the casing make it harder to manufacture; it is much easier to produce a casing where all the jets are fixed at right-angles. This sort of casing might be considered for larger schemes, say over 100kW, but seems an unnecessary complication for small micro-hydro plants.

The exhaust water falls from the runner and the casing into a pit, which is the start of the tailrace – the channel that leads the water away from the

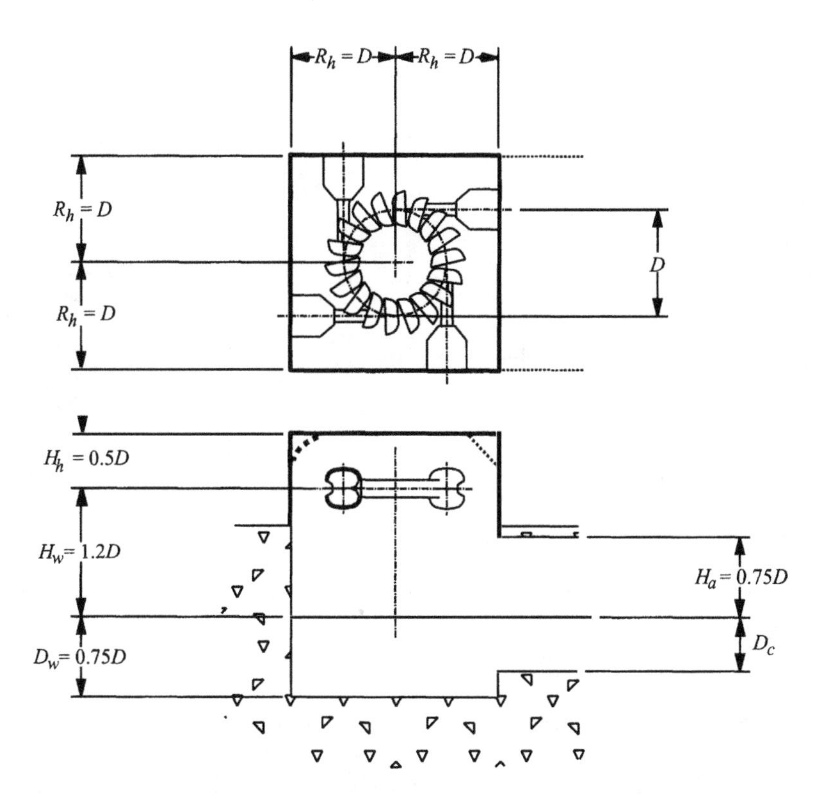

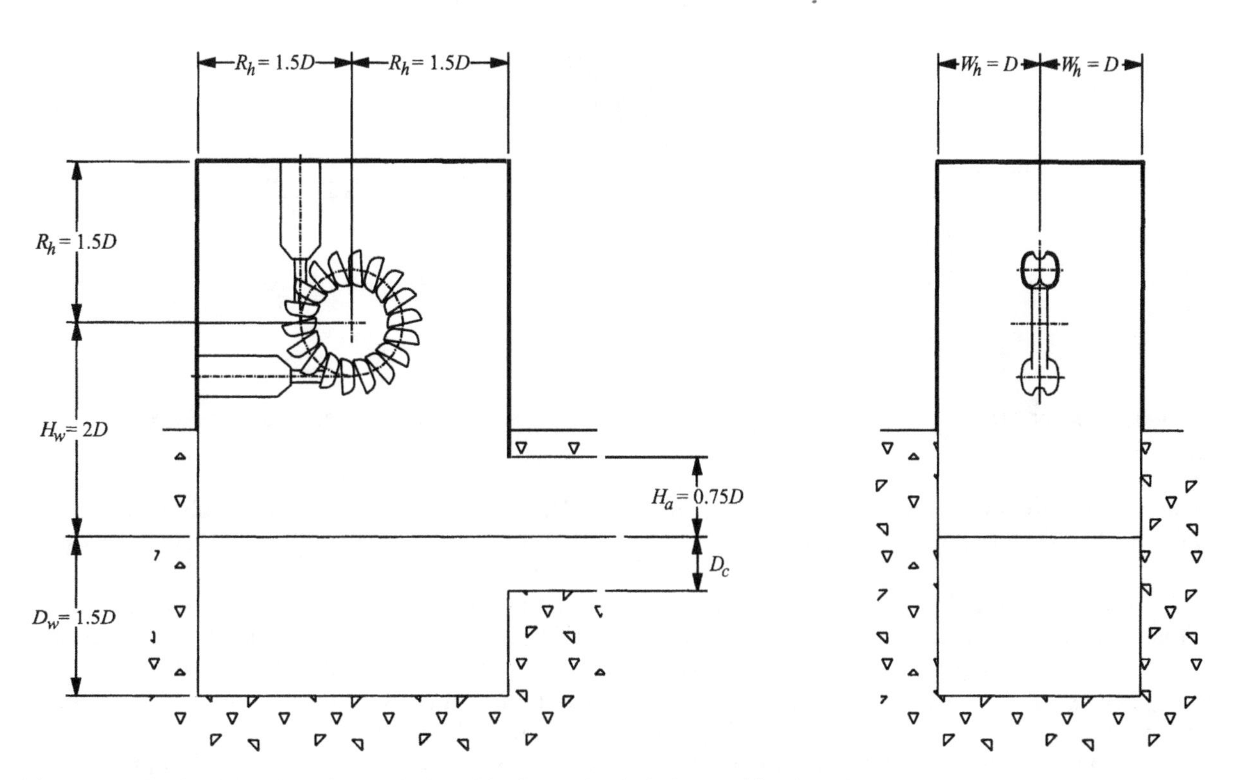

Fig. 4–36: Basic dimensions for vertical and horizontal-axis Pelton turbine housings

powerhouse back to the stream. The level of water in the pit depends upon the flow in the tailrace, which is discussed in Section 4.13.3. The water in the pit is actually a frothy, turbulent mass of bubbles, and so rises rather higher than the calculated depth. There is a very marked drop in power if the water in the pit starts to touch the runner, and it is important to leave an adequate gap. At maximum flow, it is recommended that the minimum distance H_w between the runner and the nominal water surface is $1.2D$ for vertical-axis turbines, and $2D$ for horizontal-axis turbines.

Note that the clearance between the runner and the water affects the maximum head available from the turbine. As far as the turbine is concerned, the head H_w is 'dead' – it produces no useful power. So

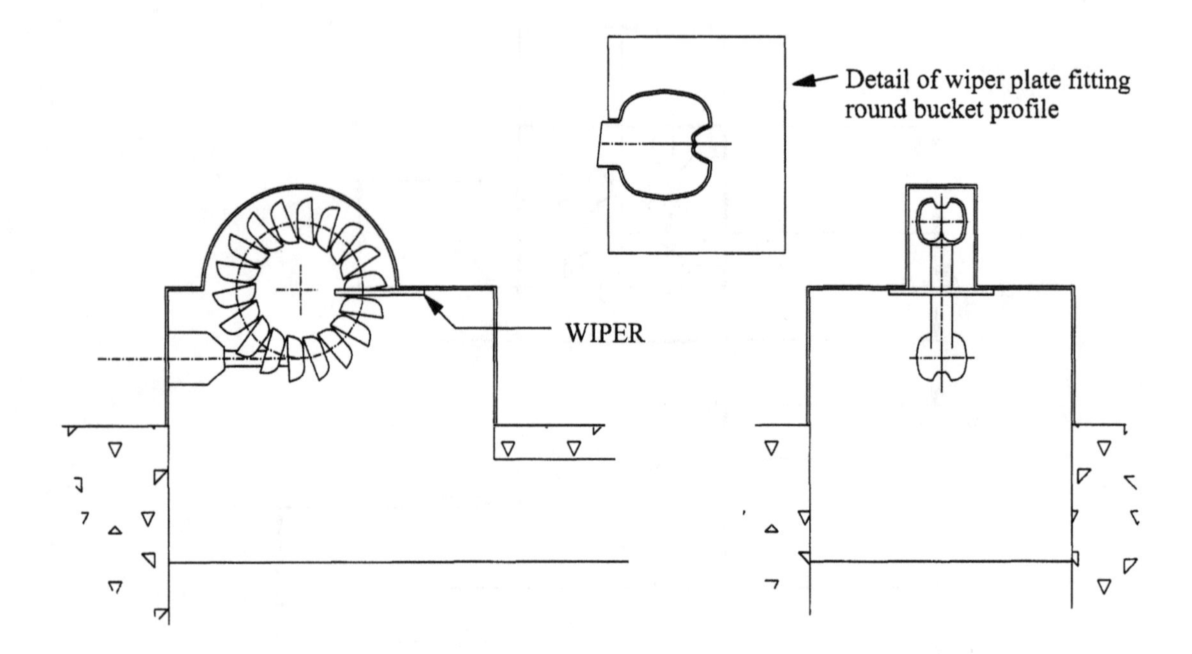

Fig. 4–37: Horizontal shaft casing with a narrow top casing to reduce windage losses

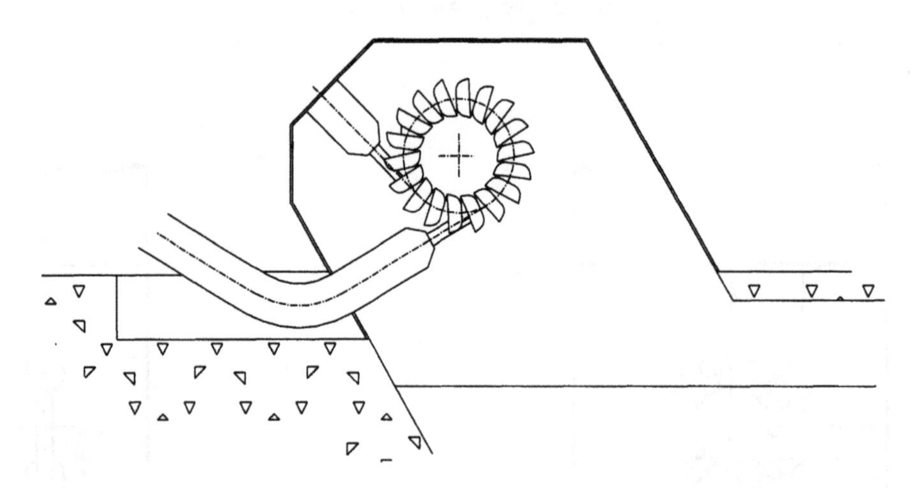

Fig. 4–38: Horizontal shaft casing with the nozzles angled to help the exhaust water flow into the tailrace

while it is important to keep the runner clear of the water in the pit, the clearance should not be too large.

The turbine pit should be deeper than the tailrace so that a pool of water always sits there. This pool helps to dissipate energy remaining in the exhaust water, begins to settle out the entrained air, and prevents erosion of the lining of the pit. Recommended dimensions for D_w are given in Figure 4–36.

4.12.2 Casing construction

The main forces on the casing come from the manifold (see Section 4.13.2) and the bearings (Section 4.6), and the structure must be able to carry these forces into the baseframe. The plates holding the nozzles need to be stiff enough to prevent the jets being moved by the manifold forces. The bearing mountings should also be rigid, especially if the bearings are not self-aligning (Section 4.7.1).

The thickness of the casing walls needs to increases with size, head, and speed. For horizontal-axis machines, the side plates tend to be thicker than the rest to reduce vibration and noise from the water hitting them. The top and end plates around the circumference of the wheel will be thinner. Small units of a few kilowatts might be made of 2–3mm steel plate. Turbines of 50–100kW would typically have 6–8mm sideplates and 5mm endplates. For reference, a 2–3MW turbine might have 20mm side walls and 12–15mm endplates.

The natural shape for the casing is a large box, and this is quite a strong structure. Care has to be taken not to create a large sound box which amplifies the turbine noise to produce a deafening roar in the powerhouse. It is not a problem with small turbines of a few kilowatts, but can be severe for units of 50–100kW or more. Avoid large, thin, unsupported plates. Test the casing in the workshop by tapping it to check that it does not resonate like a drum. If any plate does give a resounding 'clang' when it is hit, make it thicker or stiffen it with gussets on the outside.

Any housing plate that is opposite a jet should be reinforced. During overspeed, part of the jet goes straight through the runner and will hit the casing on the other side. This can damage thin plates, as well as being very noisy. Also, in normal operation some droplets are thrown off the runner on to this area and can cause pitting.

As mentioned in the previous section, the inside of the casing should be smooth and without obstructions on the walls. The idea is to provide a smooth flow path for the exhaust water into the pit. Be careful not to make ledges or barriers when fitting stiffening gussets, flanges between sections, or access panels.

Very small Peltric units can have one-piece casings. These turbines are light, and servicing can be carried out by disconnecting the penstock and lifting out the whole housing/manifold unit – see Figure 4–39. Even then, it may be more convenient to provide access to the inside of the turbine when it is fixed in place.

Various other casings are shown in Figure 4–39. When deciding how to split the casing, bear the following points in mind.

- For all but the smallest turbines, it is best to leave the pipework in place for servicing, so the manifold and nozzles should be on fixed sections of the casing. With a large, heavy manifold, it can be quite difficult to fit sections back in once they have been removed.
- Make sure that the access panels allow room for removing and refitting the runner, for adjusting the flinger seals and deflectors, and for unblocking the nozzles.
- Pay particular attention to how the shaft seal housing goes over any joints between the different sections.
- It is easy to drop tools, bolts and components into the turbine pit when working on a Pelton. Is it going to be possible to retrieve them?
- For larger units, is it possible to enter the tailrace to service the turbine? It may not be a pleasant job, but it is a very easy way of getting access to the turbine without dismantling it.

If a bucket breaks off from the runner, it can be travelling at a very high speed, particularly if it comes off at runaway. The side and end plates need to be strong enough to contain such shrapnel, which might otherwise cause severe injury to someone standing nearby. There is no easy formula for calculating the size of plate required, but experience shows that buckets do not puncture the thicknesses of plate generally used. However, care should be taken not to put plastic, glass, or very thin metal inspection panels around runner.

The housing should be painted to protect it from both the inside and outside environments. Any reasonable waterproof paint system should do. The internal paint is likely to be blasted off by the exhaust water over time, and will need touching up. Good protection is provided by coating the casing, inside and out, with a minimum of two coats of a red lead primer, and then covering this with an enamel topcoat on the outside. The inside can be left with the red lead finish, which can easily be repainted or touched up. As with all paints, the surfaces should be prepared first to remove oil, grease, dirt, dust or water, and should be thoroughly dry.

4.12.3 Open turbines

It is possible to have a Pelton without a casing. The author has seen cheap installations where the runner is left in the open, and the water sprays in all directions. For generator-mounted runners, the generator can be put in a simple box to keep the water off it. Another installation had the turbine on the outside of a building, with mill machinery inside; the shaft went through the wall. Another idea for small Peltons is to turn a truck or tractor tyre inside out and use it as a partial cover. These designs are a little idiosyncratic, and the operators get rather wet turning them on and off, but they have a certain attraction.

4.13 Baseframes, foundations and tailrace

4.13.1 Baseframes

There are forces on a turbine from the manifold, and these can be quite considerable (see Section 4.13.2). There is also vibration from the jets and the rotation of the runner. These effects mean the turbine needs to be fixed to something firm. It is usual to bolt it down to a concrete base.

A small Peltric set, where the generator is integral with the turbine and sits on the top of it, will normally be bolted directly to the floor. In other layouts, the driven machinery needs to be considered as well. For a Pelton with a runner directly mounted on a generator shaft, as in Figure 4–19(a), the turbine and generator should be mounted on a single frame, with the frame then fixed to the floor. This frame needs to be strong and rigid to hold the generator and turbine in the correct positions.

Even when there is a belt drive or flexible coupling between the turbine and the driven machinery, it is still a good idea to link them with a baseframe. Belt drives in particular put a lot of force on the machines, and can give considerable vibration. While it is possible to fix a turbine and a generator

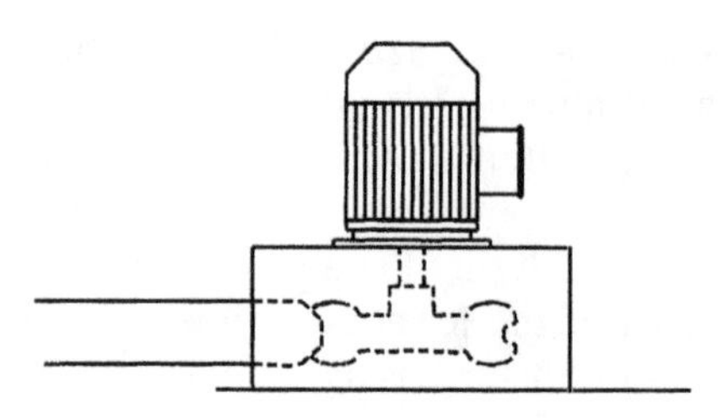

(a) Single piece Peltric casing; the whole unit and manifold have to be lifted out for servicing.

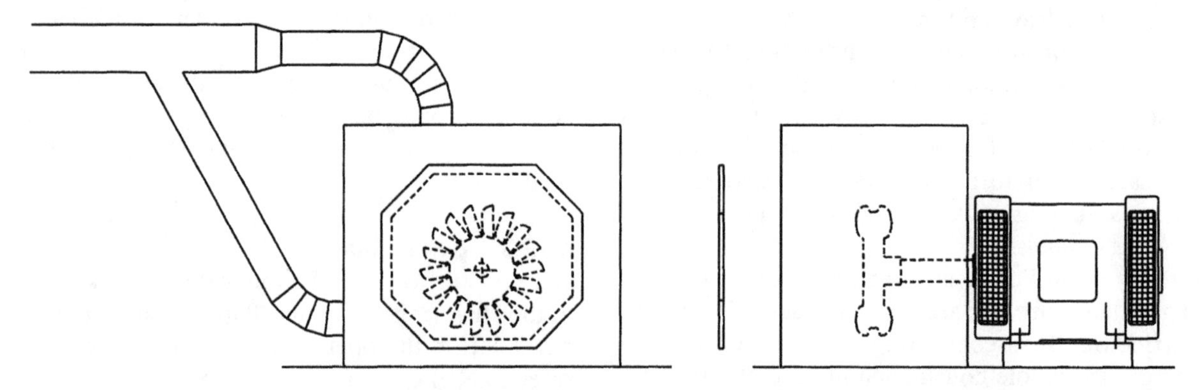

(b) Casing for a two-jet unit with the runner mounted on the generator shaft; a large access panel is provided for servicing and removing the runner.

(c) A single-jet Pelton with separate shaft bearings; the whole top half is removed for servicing.

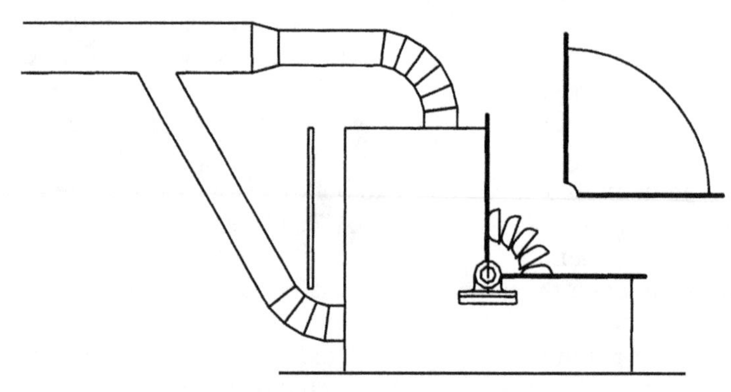

(d) A two-jet unit with separate shaft bearings; there are two removable sections.

Fig. 4–39: Various casing arrangements

separately to a concrete floor, and have a drive between them, it has to be done carefully. It is not too critical for small machines of less than, say, 10kW, but for larger units the floor should be reinforced, and ideally the reinforcement should tie into the mountings. It is often easier – and more accurate – to make a baseframe that connects both pieces of machinery. The author has seen the separate concrete mounting of a 16kW generator crack up from the forces of a V-belt drive. Baseframes should be made of substantial steel sections, so that they are quite rigid. They should not deflect noticeable under the weight of the machinery or the forces of a drive system.

Figure 4–40(a) shows a joint between a baseframe and a turbine housing. The frame is made out of channel section, and is set half way into the concrete of the powerhouse floor. The turbine outer plate has a flange welded to it, which bolts to the top flange of the channel. There is a gasket between the flange and the baseframe to prevent leakage. Note also that the sideplate overlaps the baseframe. This covers the gasket joint and gives added protection against leakage. An even better overlap is shown as a dotted outline. This stops spray hitting the joint between the concrete and the baseframe and eating away the concrete.

The baseframe should be fixed to a concrete base with anchor bolts. Design suggestions for these are given in the following sections. Use too many bolts rather than too few. Extra bolts give extra rigidity to the frame. If long, flexible sections of frame are left between the anchor bolts, the vibration of the machinery can cause gaps to develop between the metal and the concrete, leading to leaks from the turbine. The inside of the casing is usually slightly pressurized, and it is amazing how water finds its way through any cracks that do appear. When fixing the anchor bolts to the frame, make sure that they bolt down on a solid section, not a flexible one. If the anchor bolts are clamping a flexible piece of frame, or a piece of gasket, they will not carry the load as well as in a solid connection. Figure 4–40(b) shows a tube welded between the flanges of the channel to make it solid for an anchor bolt. Note that the anchor bolt does not clamp the gasket.

4.13.2 Foundations, floors and anchor bolts

A beautifully designed turbine fixed to a weak floor will still give problems. The powerhouse floor, turbine pits and tailrace channel need to be excavated carefully, and the foundations prepared well. Repairing machine foundations that are cracking or breaking up can be a huge, expensive job. The technique for installing the turbine is dealt with in more detail in Section 6.1.

For turbines up to about 50kW a simple cast concrete floor or plinth is usually adequate. The concrete should be at least 150mm thick, and laid on a well-compacted layer of stones or rubble. For larger installations, the concrete should be reinforced with steel bars. A grid of steel should be laid to spread the load of the machinery into the floor, and to strengthen the sides of the turbine pit.

The main connection between the concrete and the machinery is the anchor bolts. These can be made in lots of ways, a number of which are shown in Figure 4–41. Type (a) is best for fitting in holes drilled after the foundation has been made, the other three are for casting into the foundation. The umbrella-handle shape (d) is particularly useful for bolts around the turbine pit, with the bend taken into the concrete away from the pit.

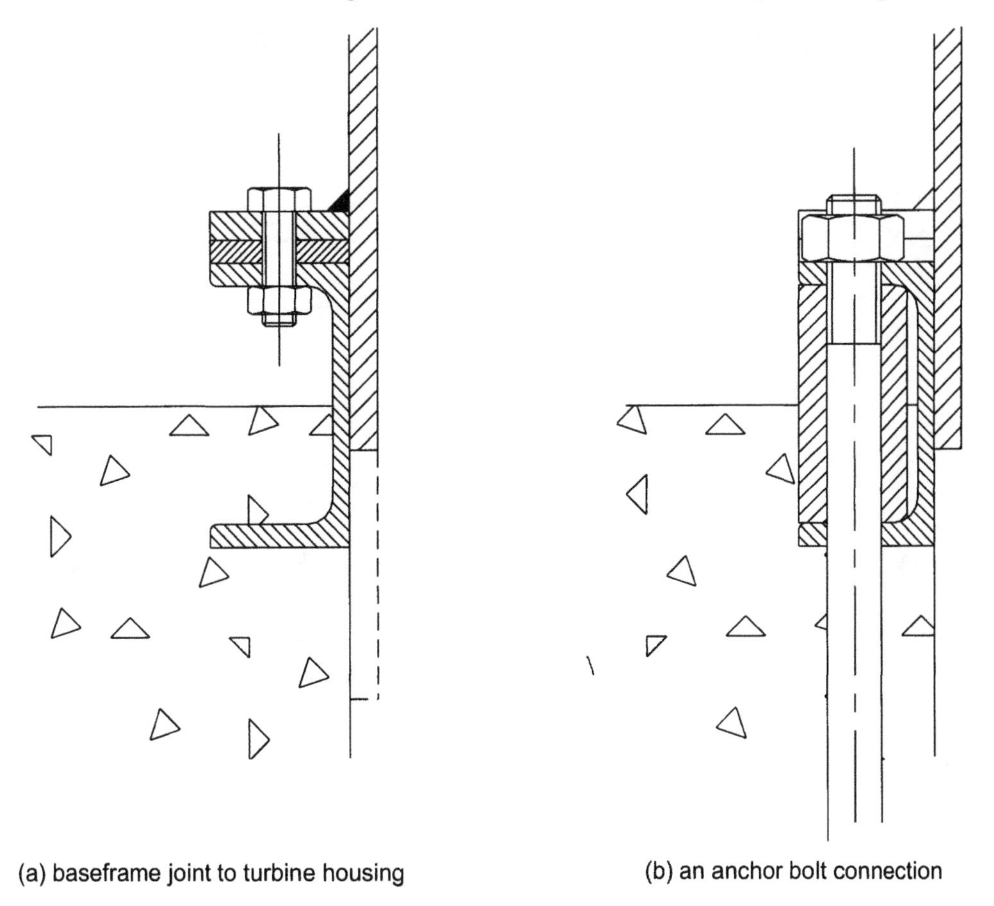

(a) baseframe joint to turbine housing

(b) an anchor bolt connection

Fig. 4–40: Detail of (a) baseframe joint to turbine housing, and (b) an anchor bolt connection

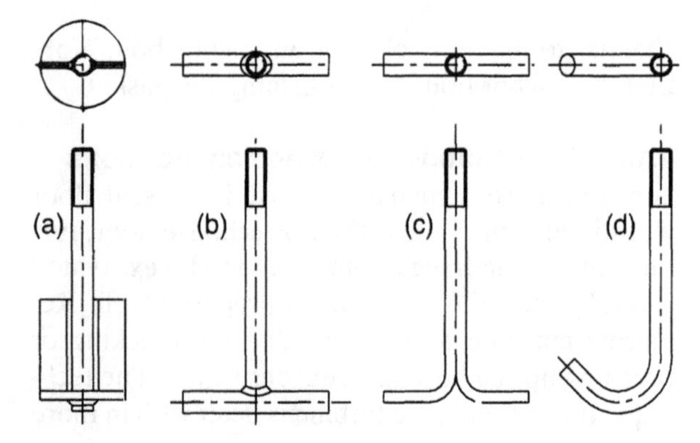

Fig. 4–41: Various anchor bolts

Anchor bolt loads

Accurately calculating the loads in the anchor bolts is very difficult, because the structure is redundant, and the distribution of the loads depends on the relative stiffness of the components. The method below calculates the loads for an ideal case, which gives an indication of the magnitudes involved. This allows an informed judgement to be made on the size of bolt required. The background to these calculations is discussed in more detail in Section 10.8.

Consider the example in Figure 4–42. This shows a turbine and manifold as a free body, detached from their surroundings. The manifold is connected to the penstock through an expansion joint. Ignoring the friction in the joint, the force pushing down along the manifold is the pressure force, P. In this example, the pipe diameter is Ø100mm, and the maximum pressure (gross head plus surge) is 65m. The force P is therefore:

$$P = pressure \times area = H.\rho.g.A$$

$$= 65 \times 1000 \times 9.8 \times \frac{\pi}{4} \times 0.1^2 = 5003N$$

This force acts horizontally, and if we assume that it is shared equally by all the bolts, the shear force in each is 5003/12 = 417 N – not very high.

The force P also applies a bending moment to the turbine, which tends to pivot it around its right hand edge, along the line B-B. Using the method

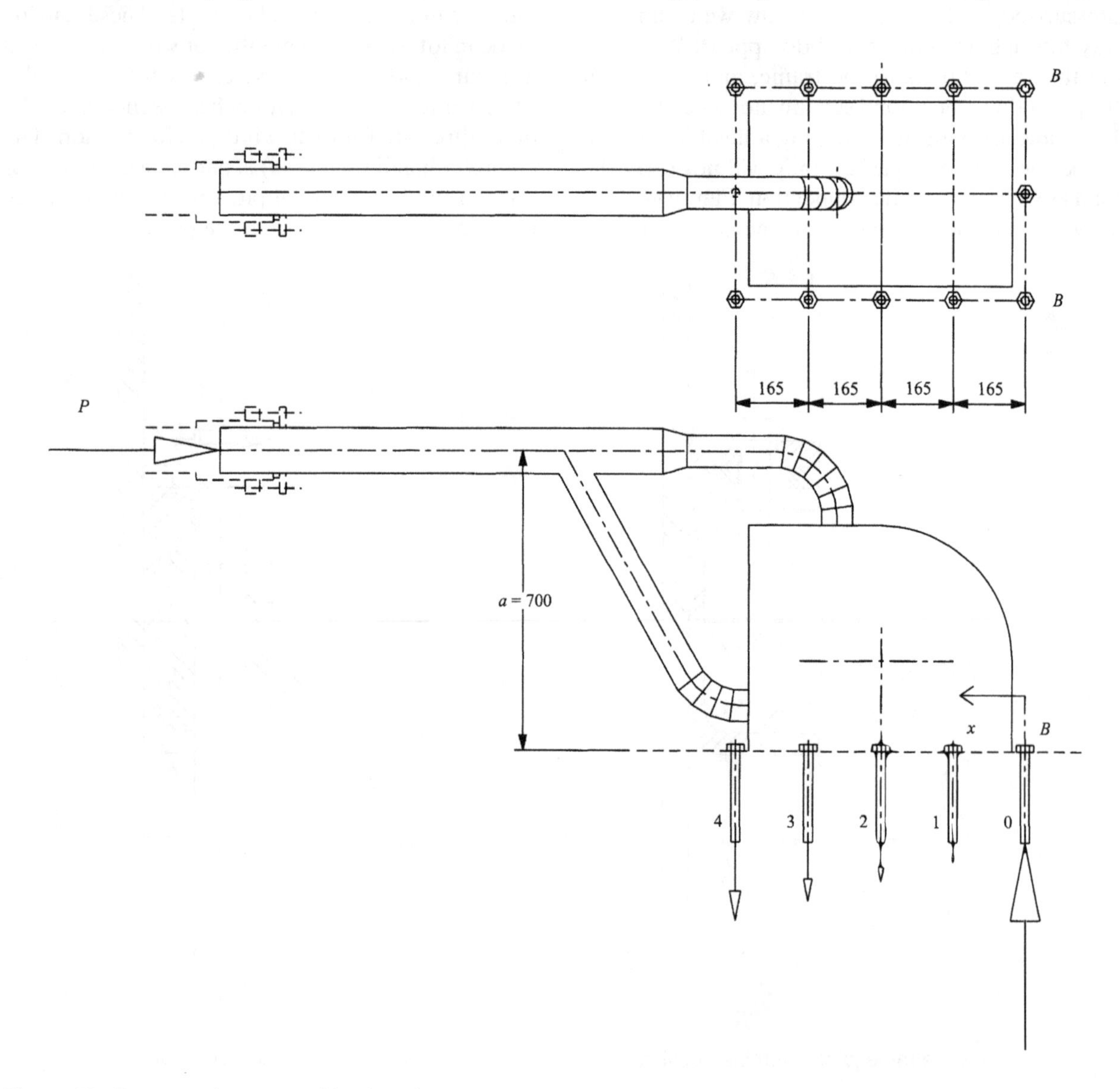

Fig. 4–42: Forces acting on a turbine housing

from Section 10.8, the force in a bolt a distance x_i from the pivot line is given by:

$$F_i = \frac{P.a.x_i}{\sum n_i.x_i^2} \qquad \textit{Eq. 4–20}$$

- x_i – distance from pivot line to point i (m)
- F_i – force on each bolt at x_i (N)
- n_i – number of bolts at position x_i
- P – pressure force on manifold (N)
- a – moment arm of P about pivot line of housing (m)

At position 4, there are 3 bolts at a distance $x_4 = 4 \times 0.165 = 0.66$m. The extra tensile force in one of these bolts due to the force P is:

$$F_4 = \frac{P.a.x_4}{\sum n_i.x_i^2}$$

$$= \frac{5003 \times 0 \cdot 70 \times 0 \cdot 66}{\left(2 \times 0 \cdot 165^2 + 2 \times 0 \cdot 330^2 + 2 \times 0 \cdot 495^2 + 3 \times 0 \cdot 660^2\right)}$$

$$= 1117N$$

Similarly, F_1 = 279N, F_2 = 559N, F_3 = 838N. As would be expected, the bolts at position 4 are the most heavily loaded.

However, as described in Section 10.8, the real forces will be slightly different. In deriving Equation 4–26 it was assumed that the turbine was infinitely stiff. In practice, there will be stiffer paths to some bolts than to others, and the bolts near these stiff parts will carry more of the load than Equation 4–26 indicates.

It was also assumed that the load F_i is carried entirely by the bolt. This is also not true. F_i is shared between the bolt and the parts being clamped in proportion to their relative stiffness. Anchor bolts are usually long, so their stiffness is low, but Young's Modulus for concrete is much less than for steel (10–17kN/mm^2 instead of 210kN/mm^2). As a *very* rough approximation, for an anchor bolt about 10 diameters long clamping a rigid part of the turbine frame to a concrete floor, the load will be shared equally between the concrete and the bolt. This means that the bolts in position 4 will carry about 838/2 = 419N from the pressure force. However, if there is flexibility in the part of the turbine frame being clamped, or a gasket, then the bolt will carry *all* the load. This load will be in addition to the preload. In view of the extensive approximations used to derive these results, the anchor bolts need to be generously planned for the loads calculated.

4.13.3 Tailrace

A tailrace channel has to be cut from the turbine pit to the outside of the powerhouse, so that the water can be taken back to its stream. In most instances, the turbine is installed on the powerhouse floor, so the tailrace has to be underground till it gets outside the powerhouse. This first section is usually either a rectangular concrete or masonry channel, or a section of pipe. Whichever is used, it must be able to take the full design flow.

The Manning formula can be used to determine the required dimensions of the channel and its slope:

$$V = \frac{r^{2/3}.S^{1/2}}{n}$$

$$r = \frac{A}{P} \qquad \textit{Eq. 4–21}$$

- V – average water velocity in channel (m/s)
- r – hydraulic radius (m)
- S – slope of channel = h/l
- n – roughness coefficient (Table 4–8)
- A – cross-sectional area of water (m^2)
- P – wetted perimeter of channel (m)

For the rectangular channel in Figure 4–43 the hydraulic radius is

$$r = \frac{A}{P} = \frac{b.c}{b+2c} \qquad \textit{Eq. 4–22}$$

The result for a partially full pipe is rather more unwieldy:

$$r = \frac{A}{P} = \frac{\left[\frac{d^2}{4}.(\theta - \frac{1}{2}\sin 2\theta)\right]}{d.\theta} = \frac{d.(2\theta - \sin 2\theta)}{8\theta} \qquad \textit{Eq. 4–23}$$

- θ – angle from base of pipe to edge of water – see Figure 4–43 (radians) – note units!

As a rough guide, the flow velocity in concrete or masonry should be kept to 1.5m/s for very shallow

Table 4–8: Values of the Manning roughness coefficient (n)

Type of pipe/channel surface	*Roughness coefficient n (s/m$^{1/3}$)*
Smooth concrete or plaster	0.010
Unfinished formwork concrete	0.015
Rough concrete, irregular masonry	0.020
Welded steel, light rust	0.012
Welded steel, brush enamel or bitumen paint	0.016
Plastic pipes with smooth joints	0.010
Plastic pipes with internal beads at the joints	0.014

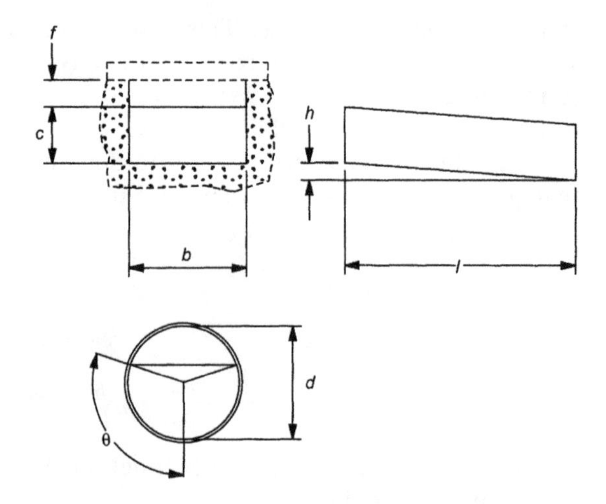

Fig. 4–43: Flow in channels and partially filled pipes

flows (less than 300mm deep) and 2m/s for deeper flows. The value 2m/s is a reasonable limit for open flow in pipes too. These equations are not very easy to use, as they often require an iterative or numerical solution to get the result wanted. An illustration of this is given in the examples below.

A good 'freeboard' (f in Figure 4–43) should be allowed in the channel. The water coming from the turbine is frothy, with lots of bubbles in it, and so takes up more space than the calculations show. The flow should not touch the top of the channel or pipe, or it may back up inside the casing. If possible, f should be from $0.4c$ to $0.5c$ where the channel leaves the turbine pit. If the level in the turbine rises to the point where it touches the runner, there will be a marked loss of power.

Example 1

The maximum flow from a turbine is 25l/s. The tailrace is to be a rectangular roughly finished concrete channel the same internal width as the turbine housing, which is 400mm wide. The tailrace discharges into an existing channel 23m away whose surface is 0.25m below the design water level in the turbine pit. What depth should the channel be?

Solution to Example 1

The maximum slope of the tailrace is fixed by the need to get the water into the existing channel nearby, and is 0.25/23 = 0.011, or 1 in 1/0.011 = 92. Knowing the flow, the width, and the roughness coefficient (n = 0.02) there is only one possible solution to the equations, but it can only be calculated numerically. To tackle the problem manually, first assume a velocity. Try the maximum allowable for concrete of 2m/s. Knowing the flow is 25l/s, the depth of the channel can be calculated:

$$Q = V.A = V.b.c$$

$$\therefore\ c = \frac{Q}{V.b} = \frac{0{\cdot}025}{2 \times 0{\cdot}4} = 0{\cdot}031\,\text{m}$$

This is only 31mm deep, which is well below the 300mm limit given for 2m/s velocity above. Try V = 1.5m/s, and the depth comes out to c = 42mm.

$$r = \frac{b.c}{b + 2c} = \frac{0.4 \times 0.042}{0.4 + 2 \times 0.042} = 0.035$$

Rearranging Equation 4–21 to make S the subject:

$$S = \frac{n^2.V^2}{r^{\frac{4}{3}}} = \frac{0{\cdot}02^2 \times 1{\cdot}5^2}{0{\cdot}035^{\frac{4}{3}}} = 0{\cdot}079$$

A slope of 0.079 is equivalent to 1 in 13, which is quite steep for a channel, and too steep for this situation. So, choose a lower velocity, and go through the process again. A few results are shown in Table 4–9.

Table 4–9: Water depth and slope for various trial values of V

V	*c*	*1/S*
(m/s)	(mm)	(1 in –)
1.5	42	13
1.0	63	43
0.8	78	84
0.7	89	124

The target is a slope of 1 in 92, so the velocity will be between 0.7m/s and 0.8m/s. Extrapolating from the values above, the water depth will be around 80mm. Allowing a 50% freeboard, the depth of the channel will be a minimum of 120mm. If computing power is available to find a numeric solution, or if you continue the iterations till you get a slope of 1 in 92, the water depth is actually 81mm, with a velocity of 0.78m/s.

Example 2

For the same scheme, there is some ∅202 mm bore HDPE pipe left over from the headrace construction. Could this be used instead?

Solutions to Example 2

The problem is even harder to solve for a circular pipe. The best approach is to try values of θ, check to see that the flow velocity is acceptable, then see what slope is produced. Assuming $n = 0.014$ for the pipe, the trials are shown in Table 4–10.

Table 4–10: Water velocity and slope for various trial values of θ

θ	*V*	*1/S* (1 in x)
(deg)	(m/s)	(x)
90	1.56	39
100	1.28	66
105	1.18	82
108	1.12	91
109	1.11	95

So the θ we require is about 109°. A numeric solution gives the more accurate result θ = 108.2°, V = 1.12m/s.

The water depth for θ = 109° is:

$$h = \frac{d}{2}(1 - \cos\theta) = \frac{0{\cdot}202}{2}(1 - \cos 109^\circ) = 0{\cdot}134 \text{ m}$$

Now, in a 202mm pipe, a depth of 134mm is quite close to the top; about 70% of pipe is filled, so there is not really have enough room for safety. Perhaps two pipes in parallel could be used, which would definitely work. If one does the calculations to prove it, the angle θ comes out at 82°, V = 0.95m/s, and the pipes are only 40% full.

4.14 Associated components

The influence of the turbine on the design of the penstock has already been discussed in Section 4.3.5. This section details other parts of the whole system whose design is affected by the turbine specification.

4.14.1 Desilting basin

Most water sources have silt in suspension in the water, and desilting basins are used to settle this out and clean the water. A muddy river in a monsoon can have so much silt in that it blocks the headrace channels and forebay tank, and any branches of the turbine manifold that are not being used become completely full of silt. Most sites are not so bad, but silt can still cause erosion of the turbine.

The amount of erosion is dependent both on the particles and the head. Quartz or sand particles are particularly abrasive, and the effect is worse for higher heads (it is approximately proportional to the cube of the jet velocity). Clay particles cause very little damage. Abrasion affects nozzles, spear valves, and the runner. Large particles tend to cause wear on the runner, smaller particles erode the spear and nozzle. Experimental evidence for high heads shows that if 90% of the silt is less than 60μm, damage is concentrated in the spear valve, with minimal damage to the buckets (Brekke, 1994). For the heads encountered in micro-hydro, erosion is less of a problem. As a general guidelines for micro-hydro, with heads of less than 200m, settling basins should be designed to remove particles larger than 0.3mm.

4.14.2 Trashrack

The trashrack stops leaves, twigs and other debris from entering the penstock, and it also serves as a final barrier to stones or pieces of concrete. A very large stone coming out of a nozzle could fracture a bucket, and the maximum gap between the trashrack bars should not be more than 8–10mm to keep the pebble size down.

The bar spacing may need to be smaller than this so that pebbles do not get caught in the nozzles. For simple nozzles, the gap between the bars should always be less than the size of the smallest nozzle.

For spear valves, the maximum diameter pebble that can pass through is determined by the gap between the spear and the edge of the nozzle. For the design recommended in this book, the spear angle is 60° and the maximum travel is 0.6 × d_{noz}. This gives a maximum pebble size of:

$$0{\cdot}6 d_{noz} . \sin\frac{60^\circ}{2} = 0{\cdot}3 d_{noz} \qquad \textit{Eq. 4–24}$$

(see Section 10.2)

To be safe, the maximum bar spacing should be about $d_{noz}/4$.

4.14.3 Generators

As mentioned in Section 2.3.7, normal stock generators may not be able to withstand the overspeed of a Pelton turbine. A 1500rpm rated generator will be driven at 2700rpm at full overspeed, and standard generators will not tolerate this. There are two possible solutions. The first is to fit an overspeed limiting device, as discussed in Section 4.4, that ensures that the runner can never overspeed by more than a certain amount. The second is to specify a generator that can cope with the higher speed. For synchronous generators this can involve bonding the generator rotor windings into their slots with special resins so that they cannot be flung out, and sizing the bearings so that they can withstand operation at high speed. Manufacturers will usually put a time limit of, say, 10 minutes or 20 minutes on the generator running at high speed, this being determined by the bearings. The best protection is to have both overspeed limiting devices, and a generator that is not damaged by high speeds.

4.14.4 Governors, load controllers and flywheels

Governing is the control of the speed and power output of a turbine. Suppose a hydroelectric scheme is producing electricity for a village. The plant can produce 25kW, but when it is started up, only 10kW of load is turned on in the village. The operator would only have to open the spear valves part way to meet the demand. By adjusting the flow, the operator could set the frequency to the correct value, and the system would run perfectly well. Until, that is, the load changed. If more villagers switch their lights on, the load will increase and turbine will slow down, the generator will cut back the voltage, and all the lights will go dim. If load is switched off, the turbine will speed up. If safety circuits are fitted, the system should shut down before the speed gets too high. If not, the voltage will rise, and the lights will glow brighter

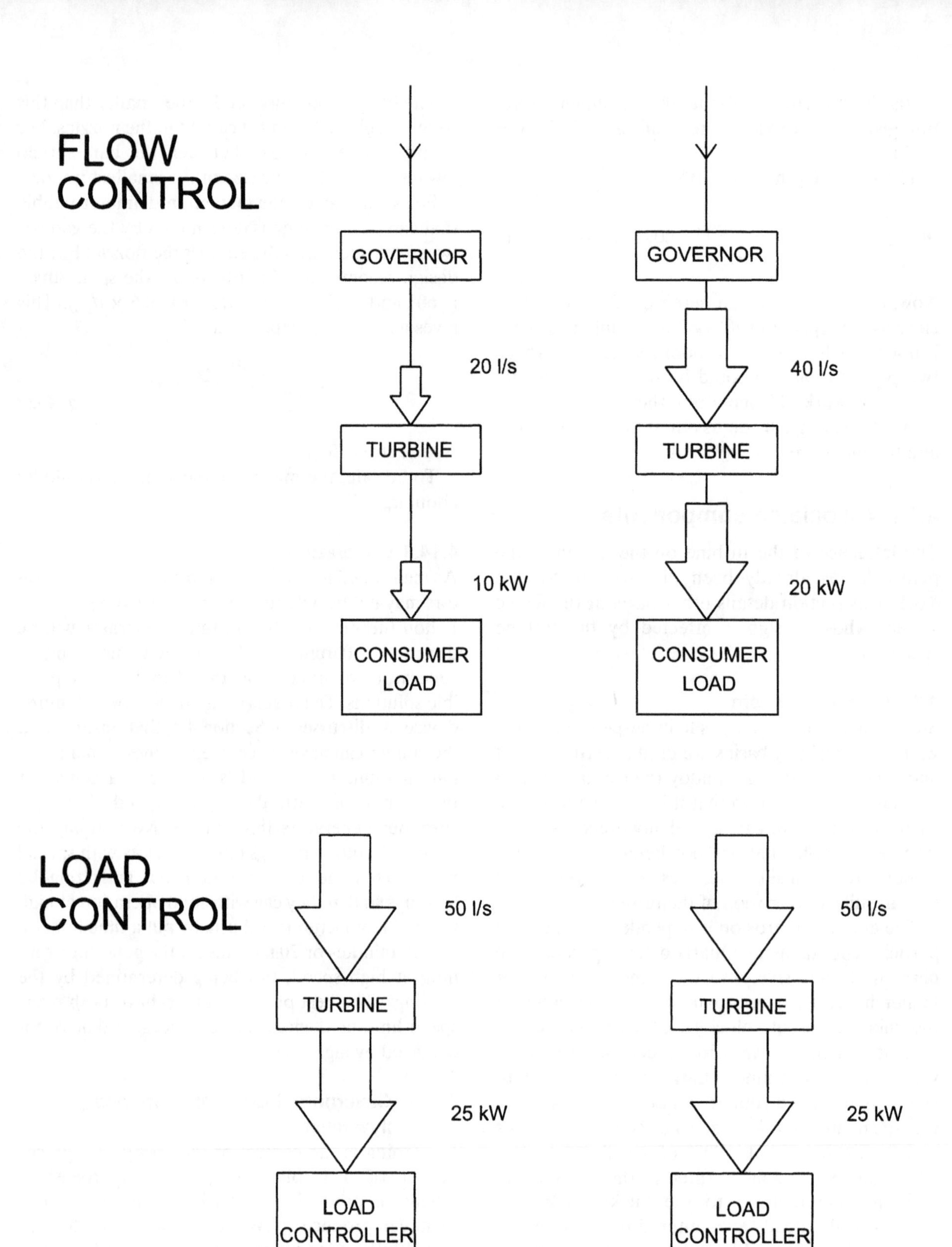

Fig. 4–44: Principals of governing: flow control and load control

before burning out. Non of these situations is very good.

One solution is to sit the operator in the powerhouse, with instructions to adjust the flow as the load changes. As evening falls and the lights go on, the operator would need watch the frequency and open the valve to keep the speed constant. Provided the load changes slowly, and the operator is reliable, this system is quite adequate. If the operator goes to sleep, the system can go wrong.

Instead of manual governing, an automatic system can be fitted. There are two main types: those which control the flow, and those which control the load. These are illustrated in Figure 4–44. A *flow* control governor senses the turbine speed, and adjusts the flow accordingly. The turbine only generates as much power as is needed. As more load is switched on, the speed starts to drop, and the governor opens up the valves to increase the flow. If the load decreases, the governor cuts back the flow. *Load* controllers also sense the speed, but they control a dummy or ballast load. The turbine is set to produce too much power, and some of this is wasted. Suppose the village is consuming only 10kW. The turbine is set to 25kW, and the load controller diverts 15kW into the ballast load. If the village load increases to, say, 20kW, the total load on the turbine would be 35kW, and the turbine would start to slow down. The load controller reacts by reducing the power to the ballast to 5kW, restoring the equilibrium.

Flow control governors have the advantage that they only use the water they require. This is particularly important for storage schemes, where the water not being used is kept in a reservoir to meet peak demand at other times. The disadvantage of flow control is mechanical complexity. The governor has to be able to control valves accurately against quite large forces, and most systems use pressurized hydraulic oil for this. A hydraulic system introduces a lot of components – tanks, filters, valves, actuators, hoses – which are expensive and require careful adjustment and maintenance. All flow control systems require careful matching of the system response time to the changes in the load. The response characteristics of all the components – penstock, spear valves, turbine, generator and load – have to be known, and the calculations are complex. It is often necessary to introduce large, heavy flywheels into the driveline in order to increase the time characteristic. Another disadvantage with flow control is that it induces surge in the penstock when the flow is reduced (unless it operates using deflectors – see Section 4.4.2). Limiting this surge, but still giving adequate response time to load changes, requires careful balancing and compromise between the various parts of the plant. For all these reasons, flow control governors are rarely used for micro-hydro, and they are not covered in this manual. If the reader requires further information on such governors for micro-hydro, details can be found in Fischer et al (1990).

Fig. 4–45: A 50kW Electronic Load Controller (Ghandruk, Nepal)

Load control systems are more appropriate to micro-hydro, and have been widely used. The controller is usually electronic, leading to the name Electronic Load Controller, or ELC. The circuit senses the speed from the frequency, and adjusts the load by solid state switching, usually using thyristors. ELCs can react very quickly to changes in the load, so that there is no need for flywheels. The quick response also means that they can cope with switched loads which constitute a high proportion of the total load (such as an electrically driven sawmill in a small village scheme), something which flow control governors find difficult. Because ELCs work on the output side of the turbine, they have no effect whatsoever on the flow and surge pressures, so the penstock can be designed completely independently of the governing system. While the electronics can be complicated, the ELC can be treated as a 'black box' which is replaced as a unit if it goes wrong, and there are no complicated adjustments or maintenance for the operators to do. ELCs are relatively cheap, being much cheaper than flow control governors for most micro-hydro applications. It is possible to make very large power ELCs, but above 100–200kW the ballast load tanks become very large and unwieldy, and governors start to become economic. One of the few disadvantages with ELCs is the electrical effect of the thyristor switching. This generates high-frequency harmonics in the load, which causes additional heating in the generator. Generators have to be oversized to cope with a thyristor load. There are ways of reducing this effect – by switching in step loads in addition to the thyristor load – but some over-rating is still necessary.

For loads over 100kW, a new breed of hybrid systems is becoming available, which mixes flow control and load control. These systems use electronic load control to respond to quick changes in the load, but then slowly operate the turbine valves to cope with larger, longer-term variations. These systems have the advantage that they have the fast response of ELCs, do not require the control calculations and tuning of a flow control system, and have the simplicity and flexibility of electronic control. They still require mechanical actuators on the valves, and this often means a hydraulic oil system is still needed.

5
MANUFACTURE

5.1 Buckets

Practically the only way of making Pelton buckets is casting, and by far the most common casting process is sand moulding. This is described in some detail; it is possible to make single buckets, segments comprising a number of buckets, and whole runners using this method. A more complicated process, but one that is used, is lost-wax or investment casting. This is can be used for single buckets, but is more common for complete runners.

5.1.1 Pattern making

The first stage in making a casting is to make a full-sized model of the bucket, called a *pattern*. This pattern has to be the correct shape, and made to exact dimensions, because all the castings will be copies of this master model.

Pattern material

The pattern is usually made by hand, so it has to be made from a workable, easily-shaped material. It must also be stable and durable, keeping its shape over time. However, for making the casting, the pattern has to be tough and strong, or else it may break as the sand is compacted around it. For this reason, patterns are often made in two stages. The first pattern is made from a soft, readily-worked modelling material, and this is used to cast a metal working pattern. The metal pattern is used to make the final castings.

Patterns were traditionally made out of wood, and wood is still widely used by foundries for large patterns. Wood is easily cut and shaped, pieces can be readily glued together, and modern synthetic fillers can be used to add material and fill holes. Its main disadvantage is that it can crack and warp, changing the shape of the pattern. It is still obviously possible to use wood for a Pelton pattern, but for a bucket, a much better pattern material is epoxy resin. This is supplied in two parts, a resin and a hardener, which are mixed together in the required quantity just before use. It may be sold as a proprietary filling or modelling material, but is also widely available as a two-tube epoxy glue. Dentists also use an epoxy resin, and this may be available when other resins are not. Dental resin works very well for patterns, giving an excellent smooth finish.

A neutral *filler* needs to be added to the resin mix, and talcum powder can be used for this. Mix the resin and the hardener together in the proportions given on the pack, and then stir in an equal volume of talcum powder. Mix well, making sure that there are no lumps of powder and that the hardener is well distributed in the resin. Some brands of epoxy resin set very quickly, so be careful not to mix more than you can use before it starts to harden. When hard, epoxy can be cut easily with a fine-tooth (24 tpi) hacksaw or jig-saw. It can be shaped with files and emery cloth or sandpaper. Special shaping tools, made out of a mesh with sharpened edges, work well too.

Working patterns for production are normally made of aluminium, though brass, bronze or other metals could be used. These are cast from the epoxy pattern, using the same casting process as that used for production of buckets. This process is described in Section 5.1.2.

Pattern dimensions: shrinkage and machining allowances

The pattern is used to make the mould into which the molten metal is poured. When the metal sets, it is exactly the same shape and size as the pattern, but it is still very hot. As the solid casting cools to room temperature, it contracts, and so ends up smaller than the pattern. To counteract this, the pattern needs to be larger than the required bucket. The 'shrinkage allowance' depends on the material used for the casting. Approximate multiplying factors for various materials commonly used for buckets are given in Table 5.1. Note that these are typical figures, and varying compositions may have slightly different factors.

Table 5.1: Shrinkage factors for various metals

Metal	*Factor*
Aluminium	1.013
Brass	1.015
Bronze	1.017
Grey cast iron	1.009
Plain carbon and low allow steels	1.017
High alloy steels	1.023

In order to obtain a pattern dimension, multiply the required bucket dimension by the shrinkage factor for the metal being used.

Example

A pattern is required for making grey cast iron buckets. The PCD of the turbine is 250mm, and the buckets are to have a width of 95mm. An initial pattern is to be made from epoxy resin, and this will be used to make aluminium working patterns. What should the PCD and the width of the epoxy pattern be?

Solution

From Table 5.1, the shrinkage factors for aluminium and grey cast iron are 1.013 and 1.009 respectively. All dimensions on the epoxy pattern should be 1.013 × 1.009 = 1.022 times the bucket dimensions. The pattern width should be made as 1.022 × 95 = 97mm, and the pattern PCD will be 1.022 × 250 = 256mm.

Note that the aluminium pattern will be 1.009 (the shrinkage factor for cast iron) times larger than the bucket, so its width will be 1.009 × 95 = 96mm.

Castings that are to be machined generally have a few extra millimetres of material, a *machining allowance*, to make sure the surface is clean after machining. This is generally 3mm or so. With the design used here it is not normally necessary to add a machining allowance to the back of the bucket; this can usually be skimmed clean, removing very little material. The front of the stem may require a little extra, and so the pattern shown in Figure 5–6 has a 1% PCD allowance added to the front of the stem.

Making the pattern

The pattern discussed here makes the bucket drawn in Figure 4–1. The base and sides are made out of flat slabs of epoxy, the splitter ridge is added as a separate piece, and the curves are filled in later. The pattern described here has a stem suitable for machining. If the bucket is to be welded to the hub, the pattern needs to be modified accordingly. In theory, the sides of the stem should have a 1–2° draft angle on them to help the pattern come out of the mould without pulling away bits of sand. The rounded shape and sloped sides of the bucket have ready-made draft angles in them, so it is only the stem that may need additional draft angles. In practice, for small patterns, straight sides can be used. For large buckets, say over 500mm PCD, some draft angle may be needed on the sides of the stem. The split line for the two halves of the moulding box is always along the top face of the cups, so the sides of the stem need to slope inwards from this face – see Figure 5–1.

The first step is to decide on the runner PCD, and multiply this by the appropriate shrinkage factors (see above). This gives a pattern PCD, which is used to calculate all the dimensions. In the example above, the pattern PCD for a 250mm cast iron runner was 256mm. Various gauges and templates are

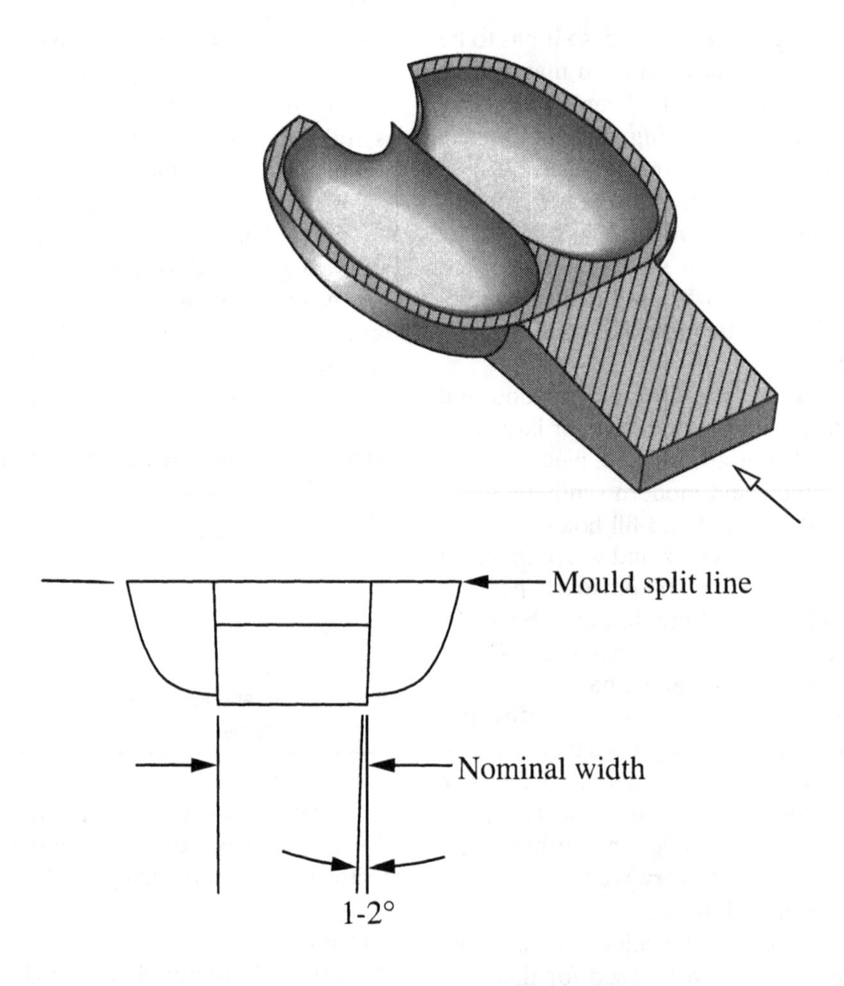

Fig. 5–1: Draft angles for casting bucket stems. The split surface for the mould is shown shaded

required to make the pattern, and these are shown in Figure 5–2. Note that in Figures 5–2 to 5–6, all dimensions are given as a percentage of the pattern PCD. It is best to cut these pieces out of a sheet of a plastic, such as nylon, which does not bond to epoxy. The thickness gauge is used to measure the thickness of the slabs of epoxy before they are fitted together. The angle gauge is used to position the side and end slabs on the base of the pattern at the correct angle. The inside and outside radius templates are used to form and check the radii on the inside and outside of the pattern.

Ten flat slab components are required to make the base and the sides of the pattern. These are shown in Figure 5–3. They are all 1.9% PCD thick. They can be made by gluing strips of plastic 1.9% PCD high to a flat plastic base to make moulds approximately the right size, and filling these with epoxy. Make a separate mould for the base plate. The end and edge pieces can all be made from a slab approximately 60% × 25% PCD, and the stem pieces from a slab approximately 45% × 18% PCD, as shown in Figure 5–4. Leave the slabs to set for about 18 hours before removing them from the moulds. Sand both sides of the slabs till they are smooth and flat, checking the thickness with the gauge.

Mark the pieces out accurately, including the centrelines and the shaded areas on the base plate (which are used to align the parts later). It saves a lot of time later if the marking out is done carefully. Cut out all the pieces, then file and sand the edges to make them smooth and straight. Using the angle gauge, sand or file the upper and lower edges of all the end and side pieces, so that they stand at the required angle on a flat board. Be careful to put the angle on the correct way round on the side pieces!

The splitter ridge is not made as a flat slab, but is *cast* out of epoxy using a *triangular section* box. The

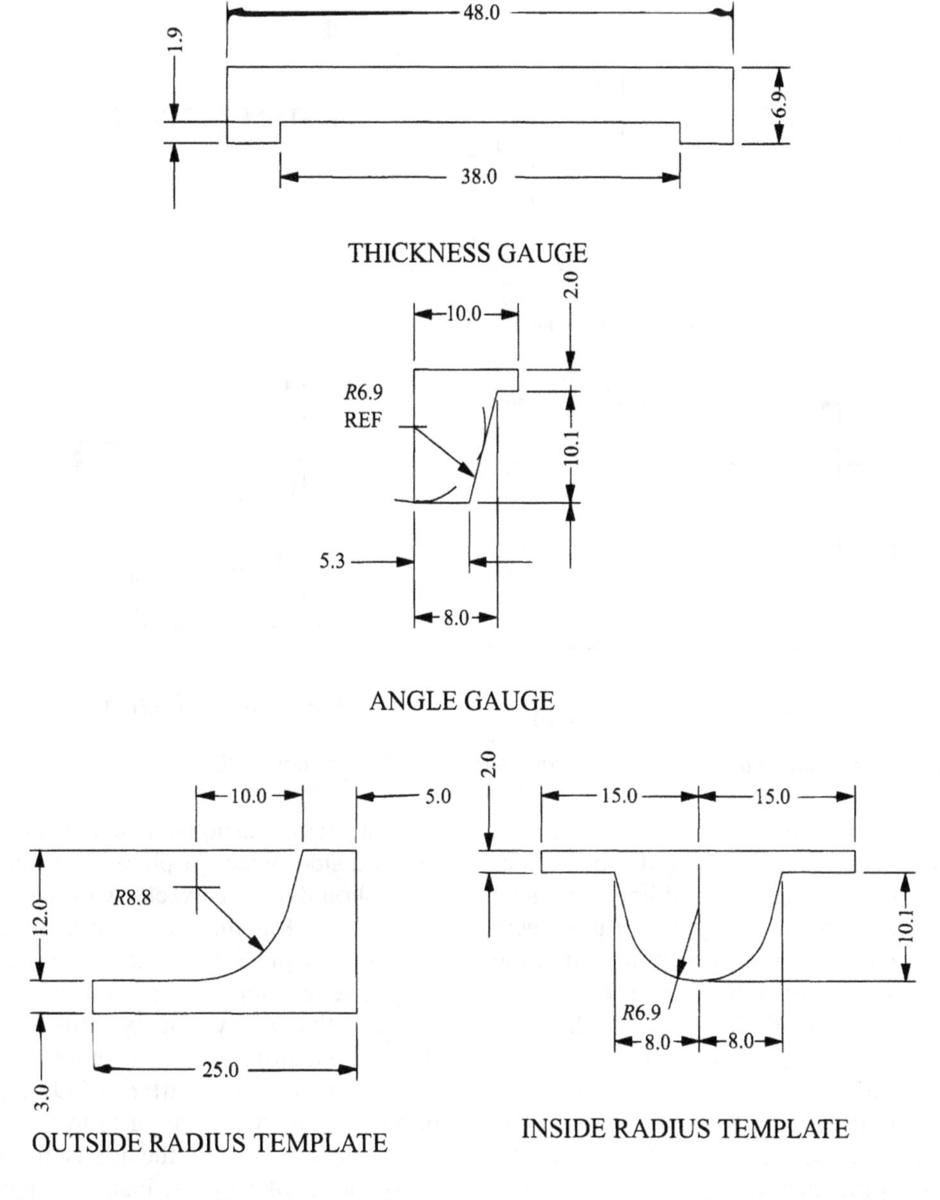

Fig. 5–2: Radius templates and thickness gauge for making the bucket pattern (Dimensions %PCD)

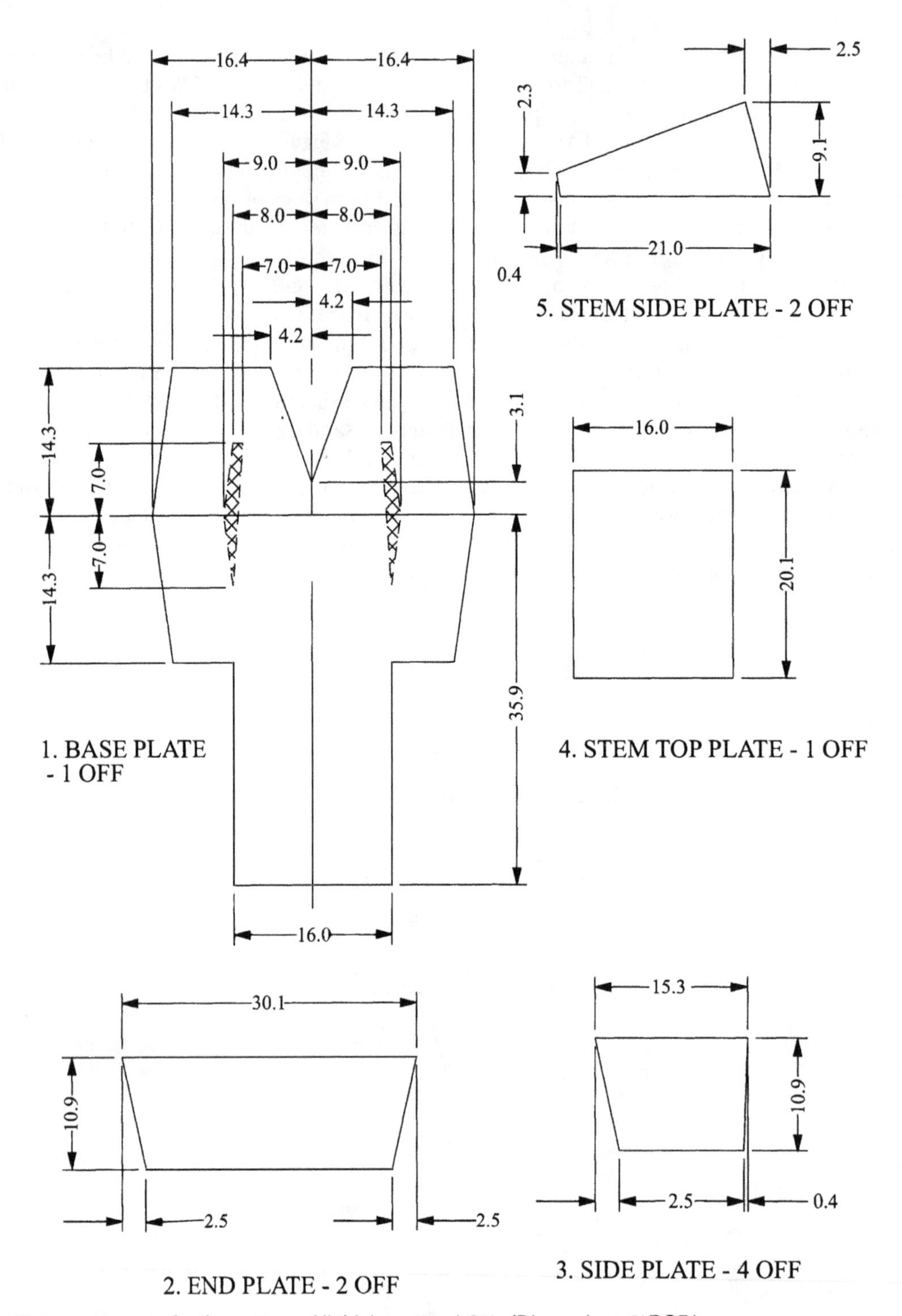

Fig. 5–3: Flat components for the pattern. All thicknesses 1.9%. (Dimensions %PCD)

development for this box is shown in Figure 5–5. Cut this out in plastic, score along the bend lines with a knife, crease along the bend lines, and glue together to make an open-topped box. Fill this with epoxy, so that the paste is level with the top of the box, and leave for 18 hours or more to set.

The parts are now ready for assembly. Start by locating the splitter ridge on the base plate. The sharp edge should be exactly central, and should cross the level of the top of the bucket 7% of PCD from the centreline – see Figure 5–6. Check that the end plate fits in the correct position and angle at the base of the ridge. When all is correct, glue the ridge in position using more epoxy mix. Now fix the end and side pieces in place. The outside bottom edges should line up exactly with the outside edges of the base. Put the angle gauge perpendicularly against each piece to set the angles correctly. Finally, glue the stem pieces in place.

Check that everything is sitting in the correct place. The major dimensions should be as shown in Figure 4–1 (using the pattern PCD). Apply a small amount of epoxy/talcum paste to the insides of all the corners to fix all the joints firmly together. Leave the epoxy to set. Place the pattern upside down on a piece of sandpaper or emery cloth on a

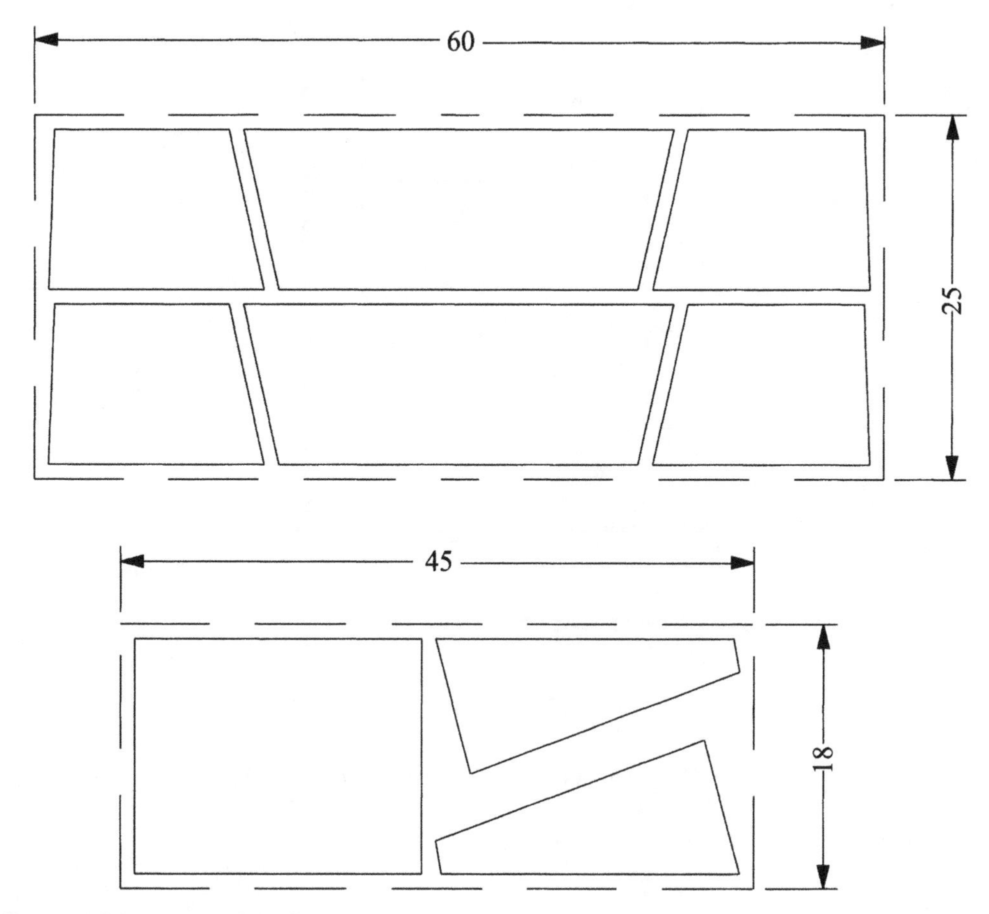

Fig. 5–4: One possible nesting of the flat components

flat surface, and rub it around to make all the sides and edges flat and at the same level, with a total height of 12% PCD.

The next stage is to produce the internal curves. This is done by adding epoxy paste and shaping it with the inside radius template. Hold the bucket at an angle, with the stem at the bottom. Place some paste into the bottom corners of the two sides, and smooth the paste out around the sides with the template until the desired shape is obtained. The template should be held upright and moved in a straight line along the edges of the shaded 'lands' on the base, and rotated about its inside edge at the corners of the lands. It is better to put in too little paste rather than too much; it is easier to add a thin layer later than to remove material at the end. Reposition the bucket and fill in the other corners.

The outside shape is made by filing off material. Place the pattern on a flat surface and run the outside radius template around it to check the shape. Keep the template perpendicular to the part of the surface being measured. When the outside shape is correct, add paste to the sharp corners between the bucket and the stem to give a radius of around 2% PCD. This is to prevent a stress concentration in these corners.

In Figure 4–1 the notch is shown as being formed by driving a radiused wedge through the top of the bucket. In practice it can be made in three steps: cutting a V-section groove at the correct angle, widening this out to a radiused channel, and then cutting away the sides at 50°. Make the first rough 90° V-groove at 52.7° to the back with a saw. Round out the base of this groove to a radius of 5.5% PCD with a file, and then cut the sides away at an angle of 50° – see the section B-B in Figure 4–1. This produces the basic notch, but some further shaping is required to produce the 14.0% notch width as seen looking on the face of the bucket (the upper left view in Figure 4–1). The notch should be symmetrical.

The next shaping operation is to make the 90° V-groove shown in the partial view on C in Figure 4–1 from the centre of the back of the bucket to the end of the splitter ridge, using a saw and a rectangular file. Do not make this groove too sharp, but leave a small radius at the bottom (about 0.5% PCD). Add paste if necessary to achieve this radius.

Cutting out the notch produces some sharp, thin edges at the end of the bucket (in the areas marked * * * in Figure 4–1), as shown in Figure 5–7(a). These edges need to be thickened up, otherwise stress cracks will occur. The material should be built up to make the edges slightly rounded, as illustrated in Figure 5–7(b).

The basic shape of the pattern should now be correct. Finish the pattern by sanding it all over with fine grade paper. Correct or fill in any flaws that are discovered. Any marks, holes or distortions in the

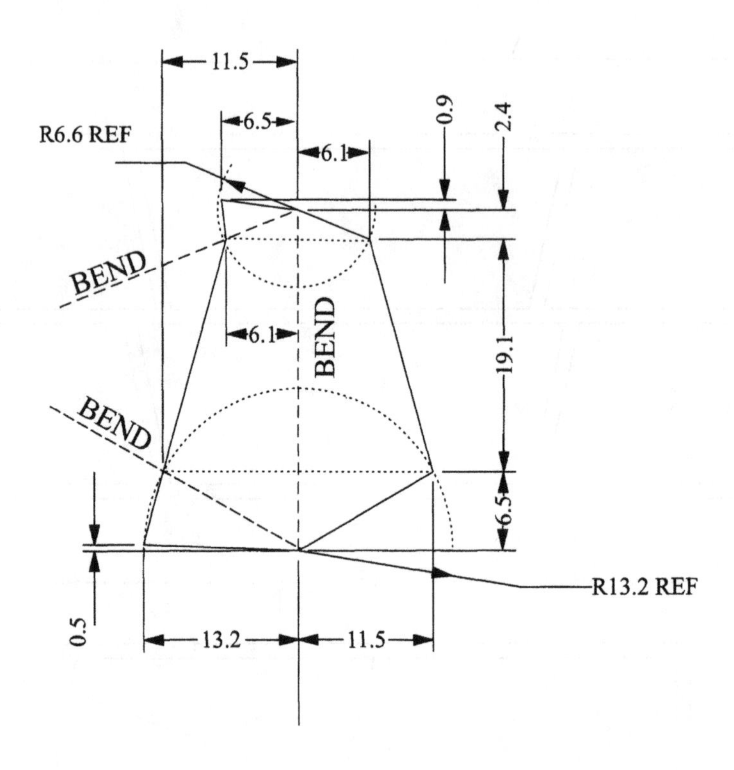

SPLITTER CASTING BOX DEVELOPMENT

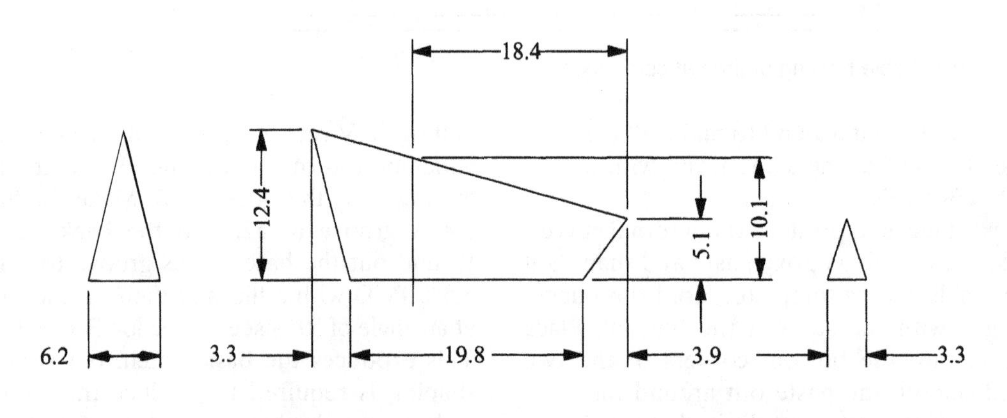

SPLITTER CAST SHAPE

Fig. 5–5: Development for making a box to cast the splitter ridge pattern, and the final epoxy splitter ridge (Dimension %PCD)

will be reproduced in the cast buckets, so the pattern must be as near possible perfect.

If aluminium working patterns are to be used, these should now be cast from the epoxy pattern. The casting process is the same as that for production buckets described below. The dimensions and shape of the resulting aluminium patterns should be checked thoroughly. Clean up any excess metal and sand all over with fine emery paper to obtain a good, clean, smooth surface. Small holes can be filled in with more epoxy filler if necessary.

5.1.2 Casting single buckets

This section deals in some detail with the process required to make single buckets. If the local technology is limited, this is the easiest, most reliable way of making a runner. Small foundries are found in most countries, but their capabilities and quality vary considerably. By describing the factors that go into making a successful casting, it is hoped that the reader will be able to work with a local foundry to obtain buckets of the required quality.

The following sections concentrate on low-technology casting. More advanced foundries will have varying degrees of mechanization, combined with process control and testing facilities. Such foundries will find making Pelton buckets relatively simple, and will need little help to produce good results.

Casting is a process that requires proper

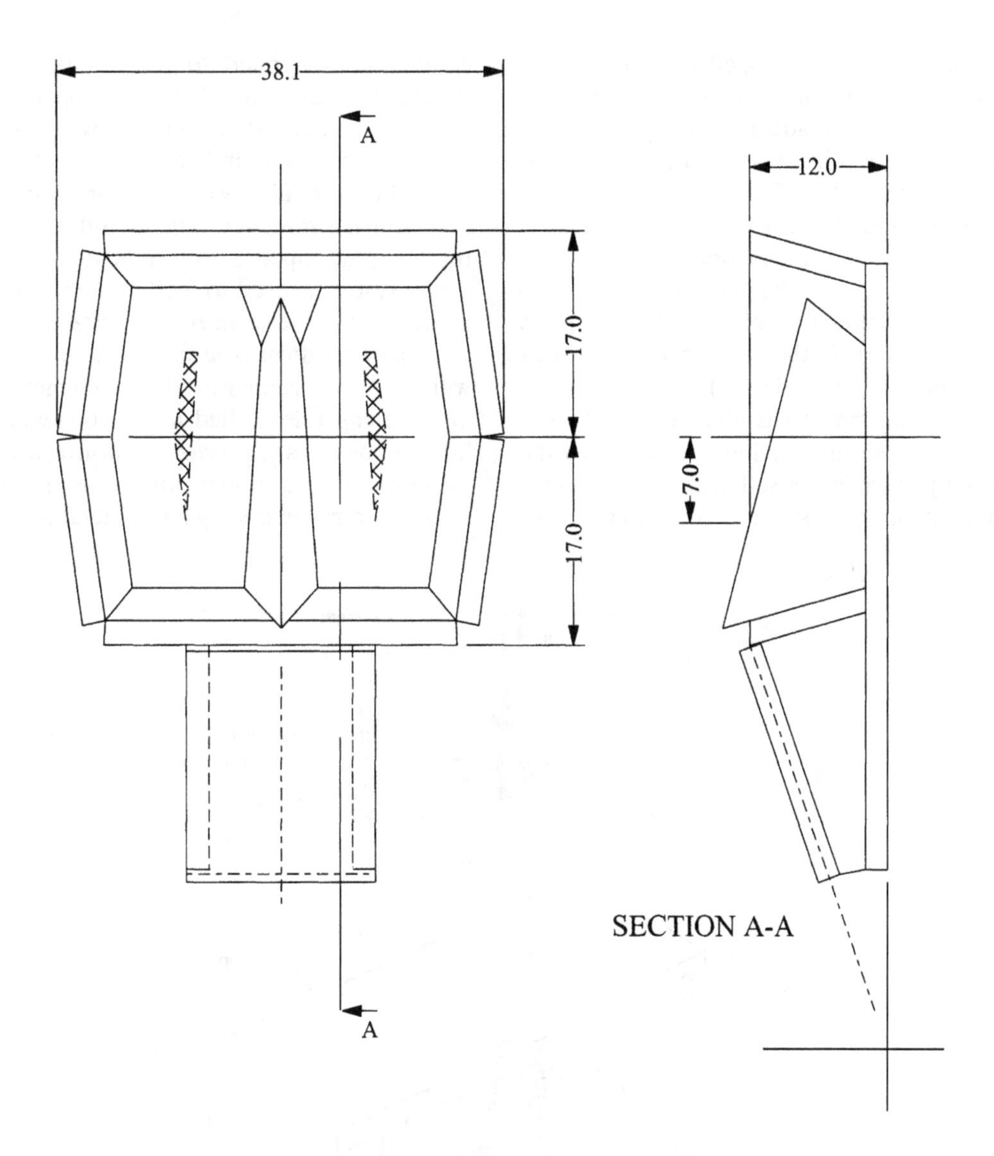

Fig. 5–6: Assembly of components to make the pattern (Dimensions %PCD)

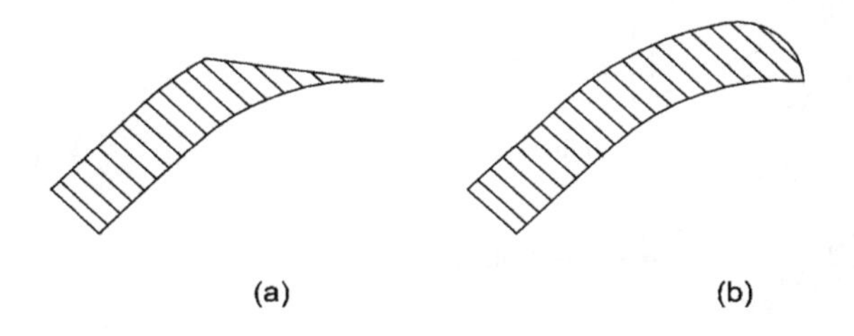

Fig. 5–7: Avoiding thin edges in the notch area

equipment and a lot of experience, and it is assumed here that a turbine manufacturer will in the first instance try to use an existing foundry. If this is not an option, the following sections provide a starting point for a manufacturer to produce his or her own castings, but they are not a comprehensive do-it-yourself foundry manual. For further instructions on general small-scale foundry practice, the reader is advised to consult a specialist book, such as Hurst (1996).

The basic casting process

Casting involves pressing the pattern into a special sand to make a cavity of the same shape. Metal is then melted and poured into the hollow or mould so that it takes the shape of the cavity.

Casting is an ancient process that has been transformed by developments in recent years. Casting was, for thousands of years, a craft, relying on experience and folklore handed down across generations. In recent years it has become a scientific industry. Through history, foundries have used sand with various natural binding agents: water, clay, and coal dust were common, but every foundry had its own special ingredients to add to the mix. Modern foundries still use sand, but many have switched to using synthetic resins as binding agents. These make the sand easier to work with, give better surface finish, and have controlled composition that can be adjusted for different types of casting. However, small foundries in less industrialized countries often still use old-fashioned materials. A turbine manufacturer may be dealing with a casting method unchanged for generations, or with a thoroughly modern process.

If the foundry has made the change to using resin-based sands, it will be able to produce castings of reasonable quality. The problems arise with more primitive foundries. Experimentation with different materials, and supervision of the process, may be required to get the desired result. Fortunately, a single Pelton bucket is a simple shape to cast, and the pattern is the same whether old or modern techniques are used.

A simple casting process is illustrated in Figure 5–8. Sand is packed into the lower box, called the *drag* and the pattern is pressed into it. The sand is then smoothed along the split surface, and a *parting* compound is applied to prevent the two halves sticking together later. The upper, open-ended box, the *cope*, is then fitted, together with a pattern for the feeder channels and venting system, and sand is packed into it. The two halves are separated, and the pattern is removed. The mould is completed by rejoining the top and bottom boxes. The *runner* or *feeder* system is used to feed molten metal into the mould. While the term *runner system* is more common, *feeder system* is used here to avoid confusion with the Pelton runner itself. Molten metal is poured into the *cup* (also called a *bush* or *basin*), down a hole, known as a *sprue* or *downgate*, into the opening into the mould, known as the *ingate*. On large or complex castings, the metal may also flow

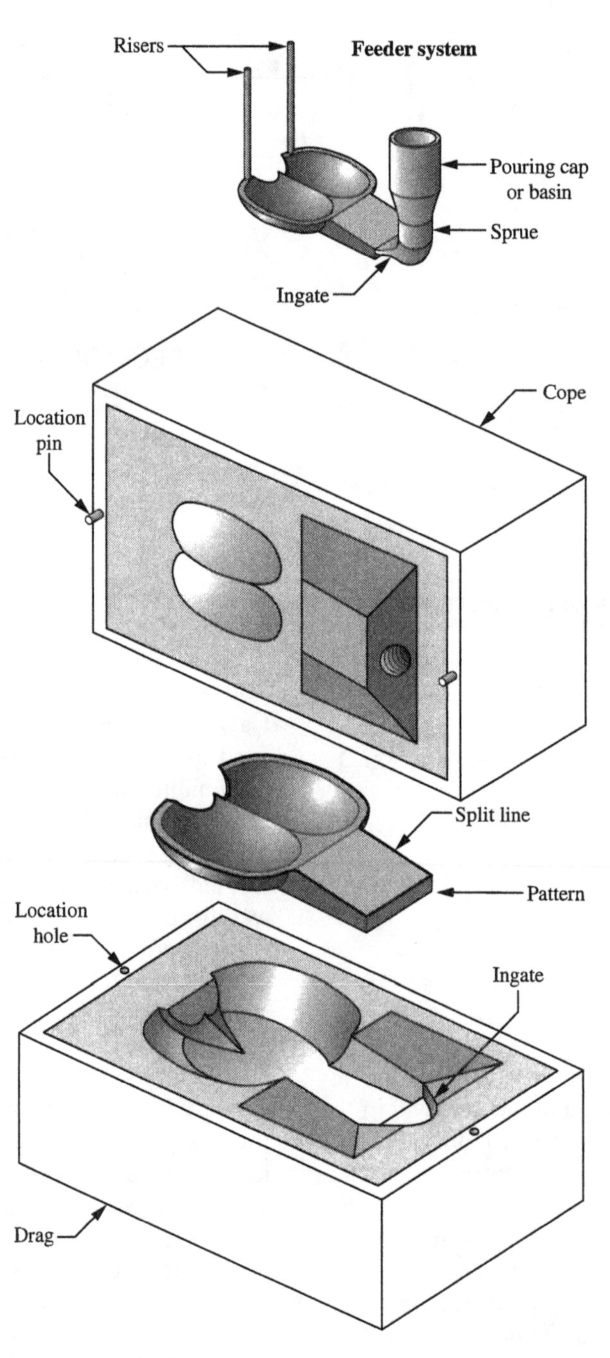

Fig. 5–8: Basic components for casting a bucket

from the mould into *feeder heads*. These are cavities attached to the mould that act as reservoirs and feed liquid metal back to the mould as the casting solidifies. They may be vented to the outside, or totally enclosed within the sand. There may also be small vents, called *risers*, coming out of the mould, which let the air and any gases generated escape, and which indicate that the mould is full. After pouring, the metal is left to cool and set. The sand and casting are shaken out of the box and separated. Finally, the casting is separated from the feeder and cleaned, and the sand is prepared for use again.

Casting a Pelton bucket

To cast a Pelton bucket, a metal pattern with the correct shrinkage allowance is required. Aluminium is the most common pattern material for this stage. A resin pattern should not be used, as it is likely to break as the sand is compacted. (Resin patterns are normally used to make the aluminium patterns.)

The most common moulding material is green sand, which is moist, graded, silica sand with binders and additives in it. The sand is placed in the drag (the lower mould box) and roughly pressed down. The pattern is then put into the sand and hammered down to compact the sand underneath it. The top face of the bucket should end up more or less on a level with the top of the drag box. More sand is placed on and around the pattern, and this is compacted down, first with rods pressed into it and then with a flat tool on top. When it is well compacted, but with the bucket still in place, the top layer of sand is scraped off to form the split surface between the two mould halves. The split line runs around the *outside* edge of the bucket and stem, and the *inside* edge of the notch (see Figure 5–9). The pattern is shown still in the sand. The extraction tool screws into the threaded hole to remove the pattern. Most of the split surface will be level with the top edge of the box, but it will sink down to meet the split line around the stem of the bucket. Note that the shape of the sand surface in the notch area is not important.

Spread a layer of *parting* compound or powder on the surface. This can simply be fine, well-dried sand, with no binders. Be careful not to let the parting accumulate in the corners of the mould. There should only be a thin sprinkling all over. Place the cope box on top of the drag. Location pins and holes, or a similar feature, are used to locate the cope relative to the drag. Fix, or hold, a pattern for the feeder sprue and cup just beside the base of the stem. The sprue should be short and have the same section as the end of the bucket stem. The cup should have a section as large as the section of the bucket stem where it joins the bucket. Rods can be fitted to make the risers at this stage (see Figure 5–8), but since these just make small diameter holes, it is possible to make the holes later. Fill the cope with sand, compress down with rods into it first, and then tap down on the surface with a flat tool. The porosity of the sand can be improved by pushing a small rod in several placed over the pattern, so that air from the mould can escape more easily through the sand.

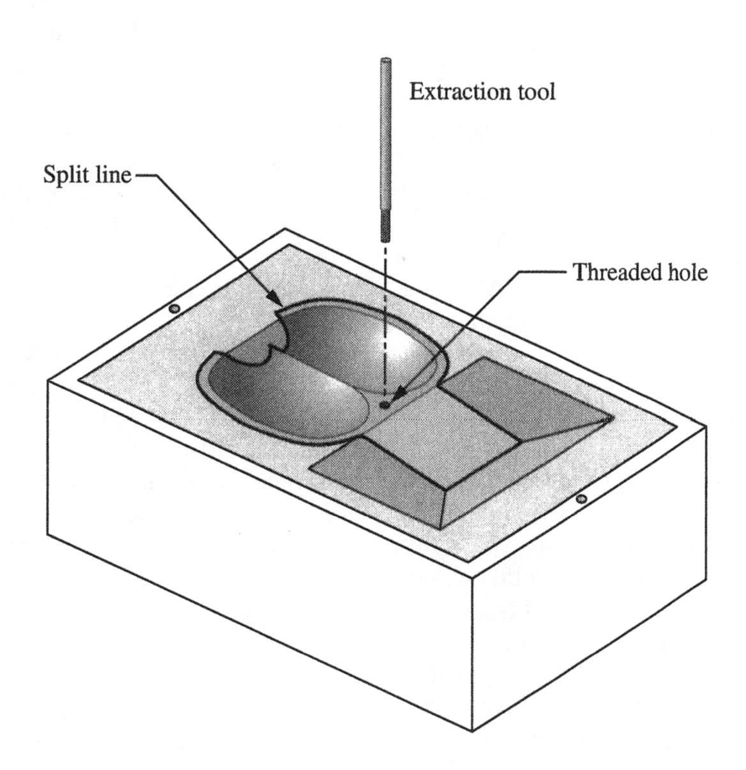

Fig. 5–9: The split-line and split-surface in the drag for casting a bucket. The pattern is still in the sand

Remove the feeder pattern, tap on the cope to loosen it, and carefully lift it off the drag. It should come away cleanly, with an imprint of the inside of the bucket on its underside. Check that the mould is clean and unbroken, especially the splitter ridge. Remove the pattern from the drag. It is convenient to put a threaded hole into the pattern, as shown in Figure 5–9, so that a rod with a thread on the end can be used to pull it out. A small quantity of water may be dripped on to the sand around the pattern to stabilize it before the pattern comes out. Tap on the rod first to loosen the sand, then gently withdraw the pattern. The inside of the mould can be tidied up with a flat tool, which should first be dipped in water. Cut an ingate into the sand of the lower mould from the sprue into the bucket stem. The ingate needs to have a minimum sectional area of at least half the base of the stem. For this moulding, the function of the ingate is to take the force out of the molten metal falling down the sprue, and to direct it gently into the mould. The sprue cannot discharge directly into the cavity as the metal hitting the bottom of the mould would damage the surface. Clean the upper mould and make holes for the risers if required.

Sprinkle a layer of fine graphite powder on to the surface of both halves of the mould. (The graphite should be nominally sieved through a 150 micron sieve; shaking it on from a cloth bag also works.) This heat-resisting powder helps to prevent the metal burning into the surface of the sand. The pouring cup should also be lined with powder. Refit the cope on the drag.

Do not leave the mould too long before casting, especially in cold, damp, conditions. Do not leave the mould unused overnight. The sand in the mould will absorb moisture, and the surface finish will be poor. Making a set of moulds is quite time consuming, and it may be advisable to make a set of buckets in two or more batches rather than leave the moulds too long. Just before pouring, open the moulds and heat the surfaces with a blow torch to completely dry the surface of the sand. Also heat the pouring cup, sprue and ingate. Refit the cope on the drag, and place a heavy weight on top of it. The molten metal is heavy, and the cope will float off the drag if it is not held down. The mould is now ready for pouring.

Melt the metal in a furnace, putting all the ingredients required in at the start. Bring the mix to the correct temperature, measuring it if possible. If not, it has to be judged from the way the metal behaves and the colour of the flame. If the metal is too cold, it will stick to a rod of metal dipped into it. The correct temperature is reached when it drips off.

Keep the moulds near the furnace. The crucible, and the amount of metal melted down, should not be too big, or it will be cumbersome and difficult to pour in a controlled manner. The pouring should be done immediately the crucible is removed from the furnace. Pour steadily, not from a great height, not too fast, but evenly and as quickly as possible. Pour until the cup stays full or, if risers are used, until metal is seen in them. The cup should be left full. After pouring, leave the metal to cool. This can take a number of hours. When cool, break out the sand, remove the castings, cut off the sprue and risers, and grind off (*fettle*) any extra bits of metal.

It is a good idea to cast more buckets than are needed. If the foundry is likely to produce a proportion of bad buckets (such as with blow-holes or poor surface finish) make sure that the batch size is big enough to give you plenty of good buckets. It is worth considering keeping cast buckets in stock for future turbines.

Casting quality

A variety of factors affect the quality of the finished casting:

- The mix and purity of the metal used.
- The melting and pouring temperature.
- The moulding sand and additives.
- The design of the pattern and the feeder system that takes the metal into the cavity.

a. Material

It might seem obvious to say that the metal used should be the one specified, but very basic foundries often melt down any old scrap metal, having no means of testing its make-up. It may be possible to distinguish between 'good quality scrap' and 'poor quality scrap', but the exact mix will remain a mystery. Unless the foundry is prepared to use pure ingredients to create the correct mix, or certified material is available, the buckets need to be designed for low-strength material. See the discussion on choice of material in Section 4.1.3.

Ideally, pure pig iron should be used, with the correct elements for the required composition. If scrap metal is being used, some extra ingredients will be needed to control the composition, such as silicon, and a fluxing agent such as limestone or ferro-manganese. If scrap cast iron is repeatedly remelted, its silicon content will be reduced, and the resulting castings will be very hard and difficult to machine. A certain amount of new scrap, 'pig iron', or plain silicon should be added each time. Conversely, if the silicon or carbon content is allowed to get too high, this results in the iron crystallising into coarse grains as it solidifies, and the casting will be weak and brittle.

As a general rule, the lower the melting point of a material, the easier it is to cast. It is much easier to work with brass or bronze than with ferrous metals, and iron is much easier to work with than steel. Any foundry will be able to handle brass, but small foundries will often not be able to get their furnaces hot enough to melt steel.

Metals have different behaviour in the mould. Cast iron tends to form a solid skin as soon as it touches the mould. It tends to bridge imperfections in the sand, and readily gives a smooth surface. Molten steel does not form a skin, and is a thicker liquid, so it does not flow so easily into corners. It is more difficult to mould a sharp splitter ridge on a bucket from cast steel. If a sharp lip cannot be obtained, then the lip on the pattern needs to be made thicker, so that material can be ground off it later to give a sharp edge.

b. Temperature

Correct melting and pouring are important. Cast iron is typically poured at 1400–1500°C, whereas stainless steels need 1580–1650°C. Some alloy steels may need to be held at a high temperature for a while, but then poured slightly cooler. When the composition is not known, it is difficult to predict the correct temperature, but the foundry should be able to tell by experience or by trial and error. If the pouring temperature is too low, the metal will be too viscous and will not flow properly into the mould.

Temperatures in the melting crucibles are best measured with optical or dipping pyrometers. If these are not available, the colour of the metal gives an indication of the temperature – see Table 5–2. Try not to add metal and additives to the crucible during casting as it both changes the mix, and brings down the temperature of the metal. If metal has to be added, use long, thin pieces, as these have less effect on the temperature. A wide piece of iron with a large surface area can significantly lower the temperature. When casting a particular composition of steel, the best practice is to charge the crucible with all the components before beginning to melt the mix.

Table 5–2: Colours of metals at various temperatures

Colour	*Temperature* (°C)
Barely visible	630
Visible	675
Dull red	775
Dark red	850
Bright red	990
Cherry red	1050
Orange	1150
White	1200
Bright white	1500

Pouring can be done straight from the crucible, or using clean ladles. Every attempt must be made to use clean metal without slag. Slag in molten iron can be made to coagulate by using common table salt, or dry sand. Glass acts as a coagulating flux for bronze and brass. Scrape off the slag from the crucible before using the metal. Ladles with spouts may be used to pour from below the surface of the metal. Pouring should be steady and fast, but not rushed. It is important that the mould is not damaged by heavy splashes of metal, or by pouring from too great a height. Non-ferrous castings especially are best when done quickly. Keep the crucible near the moulds.

c. Moulding sand

The basic mould material is sand. Moulding sand should be pliable and compressible. It needs porosity to allow the air in the mould to escape when the metal is poured in. The term *green sand* used by founders is not a description of a particular type of sand, but refers to sand that has not 'gone off', meaning that the binders have not been destroyed by heat or chemical reaction, and the sand retains its properties.

The basic mineral ingredient of silica sand is silicon dioxide (SiO_2) grains. These need to be sifted, or *graded* to give the correct size of grain. Sand sifted through a 300 micron mesh is appropriate for ferrous or bronze castings, but the grains must not be much finer than the mesh. Too coarse a sand gives a poor surface finish, and too fine sand lacks porosity and can trap gases in the mould.

Sand is made 'green' by the addition of water and binders. Natural deposits of sand all have some clay and minerals in them. While some deposits of sand naturally contain the correct binders, most have to have clay and other compounds added to them. The sintering temperature of silica sand (the temperature at which the particles start to fuse together) depends upon its clay content. The higher the clay content, the lower the sintering temperature. Most sands will withstand the temperatures required for casting non-ferrous metals and cast iron, but for steel, a sand with high silica content is required. Alternatively, a special type of sand with a different mineral base (for example, zircon sand) can be used. The sand has to be mixed well with the additives, and a mechanical mill is often used for this.

A simple foundry will typically use sand with bentonite clay, and water, and may add coal dust to the mix. The moisture content will typically be around five per cent. Used casting sand can be revived by pulverising it, sieving it clean, adding new sand, and replenishing the binding additives. A modern foundry will determine the correct additives for a sand by testing the composition of the sand, and its properties, and adding ingredients until the correct results are achieved. Without testing equipment, the process is a matter of trial and error, and relies on the experience of the founder.

Impure sand, or incorrect binding agents, can lead to poor surface finish or gas-related defects. If the molten metal reacts with the sand it can 'burn in', and the surface of the casting will have sand included in it, giving a terrible finish. The molten metal can also react with the sand to produce gases which get trapped in the casting, or prevent the metal from filling the mould. It may be possible to improve the finish by adding coal dust to the sand. A dusting of graphite put into the mould after the pattern is removed can improve surface finish too. If, after trying these improvements, problems persist, it is probably due to poor quality sand, which should be changed or refreshed. Excessive moisture also gives poor surface finish. Try re-mixing the sand, and do not leave the moulds for too long after making them. The surfaces of the mould can be dried with a blow torch before pouring.

It is quite possible to use different mixes of sand for different areas of the casting. For example, sand with a binder can be used to give a good surface finish for the inner surface of the bucket, cheaper binders used for the back and feeder system, and used sand for filling the remainder of the cope and drag. Some foundries may use special coatings painted on to the sand of the mould to improve the surface finish, particularly on the inside surfaces of the buckets. Pure silica sand, without binders or other additives, can be used as a parting compound. Other suitable parting agents are graphite and soapstone (the latter particularly for steel).

d. The pattern and feeder system

The final casting cannot be more accurate than the pattern, so it is important to check that the pattern is the correct size (allowing for shrinkage), undamaged, and with good surface finish.

After the molten metal is poured into the mould, it contracts in three stages. Firstly, the molten metal contracts as it cools. Secondly, there is contraction as the liquid sets to solid. Finally, the solid metal contracts as it cools. While the first two stages are occurring, metal has to remain liquid in the feeding arrangement so that the cavity remains full. This is easier to achieve with iron, as the total liquid plus solidification contraction is usually less than 1%. The equivalent figure for steel is about 5%, so it is much more important that the feeder arrangement is carefully designed.

If the feeding system is poor, the casting will emerge with cavities within it, or with sinkage *draws* on its surface. The feed arrangement needs to have a reservoir of molten metal that sets after the casting, so that it can keep feeding metal in as the casting cools and contracts. It should be possible to cast a shape as simple as a bucket without resorting to a *feeder head* built into the mould, and it is usually sufficient to make the cup and sprue reasonably large. The cup should be cylindrical (to keep the surface area to a minimum) with a diameter greater than the thickest section of the casting. It should feed into the thickest part of the casting, which is the stem. One possible arrangement for the feeder system is shown in Figure 5–8. Metal will flow through the stem, into the thick sections under the splitter ridge, and out into the end and sides of the mould. The thinnest sections, around the notch and the outside edges of the sides, will solidify first, and will be fed with molten metal from the material in the stem and in the middle. These in turn will be fed from the cup.

The bucket design used in this book does have a large section at the top of the stem. If shrinkage problems persist, it may be cured by putting the ingate at the side of the stem, near the thickest section, with a large cup that comes close to it. Covering the cup with sand after pouring can also help keep the metal in the cup molten. A '*riser*' is technically a small vent tube that leads from the mould to the air, that lets air and gases leave the mould. If the mould does not fill properly, risers should be put in to help vent the cavity.

Inspection and testing

Quite a lot can be learnt about the quality of a casting simply by looking at it. It should be the correct shape, not distorted or broken in any way. Hold it against the pattern to check this. It should be fractionally smaller than the pattern (by the shrinkage allowance). The surface should be even and reasonably smooth. There should be no sand embedded in the surface. Check that there are no holes, porosity, or cracks. There should be no signs of sinking or shrinkage on any surface, particularly near the thicker sections.

The next stage is to hacksaw through a sample number of buckets to check there are no internal cavities. Look particularly in the thickest sections of the stem. The metal of the casting should be even, without obvious patches inside or hard spots. Small surface defects (less than about 1% of PCD long, or around 2.5% of the bucket width) on the back of the bucket do not matter too much. Defect inside should be smaller (less than 0.5% PCD long, or around 1% of the bucket width). It may be possible to remove surface defects on the inside by grinding. More dangerous defects are shrinkage, cracks or holes. These can be invisible, but greatly weaken the bucket. If samples are found with these defects, it is best to have the batch re-cast. The quality that is acceptable varies with the application. A high-speed, high-power turbine needs very good buckets. Slow, low-power runners are more tolerant.

The only way of testing the composition of a casting is to send a sample to a laboratory, and such a facility may not be available. (One simple test for steel is to strike the bucket: steel buckets ring, cast iron ones give a dull note.) Tensile testing facilities are more common, and it is good practice to make a tensile strength test piece while the castings are being made. This is particularly necessary if the material composition is uncertain, if the quality of the foundry work is unknown, or if the application is critical. The cast size of test pieces for iron and steel are shown in Figure 5–11, together with the final machined test piece dimensions. The casting must be done from the same crucible and at the same time as the buckets are poured if it is to be meaningful. Poor quality buckets should be returned to the foundry, who ought not to charge for them.

Shrestha Industries process for casting a Pelton bucket

Figure 5–10 shows photographs of cast iron Pelton buckets being made in a simple foundry. There are a few minor differences in casting method which are discussed below.

- The bottom half of the mould is a bed of sand on the floor of the foundry rather than a separate drag box. Just before the cope is removed, curved metal rods are hammered into the bed of sand around the cope to locate it when it is put back on. These are visible in (f).
- As the cope is filled, long nails are put into the sand in the bucket cups, to strengthen the sand and make sure the cup moulds come out whole when the cope is removed. (Photograph (d))
- A single riser hole is made in the cope after it has been moulded by pushing a rod into the sand. The riser here is located at the top of the stem, not at the top edge of the bucket. Nevertheless, it seems to work.
- Note that an extra plate has been fixed on top of the stem. This is because this foundry always finds sinkage on the top surface of the stem. Rather than have the 'draw' go into the stem, they add material. When the bucket is machined, the depth affected by the shrinkage is machined off. It should be possible to overcome the shrinkage by improving the feeder system. Here the cup is quite small, the sprue long, and the gate fairly small. The thick section at the top of the stem is solidifying after the cup and sprue, forming the draw on the bucket. Increasing the size of the cup and putting it closer to the mould should stop this.
- The moulding sand is from a local deposit, mixed with bentonite and water only. The cast iron is made from scrap vehicle engine blocks with about 1.5% by weight of pure silicon and 0.25% by weight of ferro-manganese added to the mix. The material is poured from the same crucible it is melted in, after the slag has been scraped off. A charge of about 60 kg is used. This foundry has no facility for testing the sand, the metal composition, or the furnace temperature.

(a)

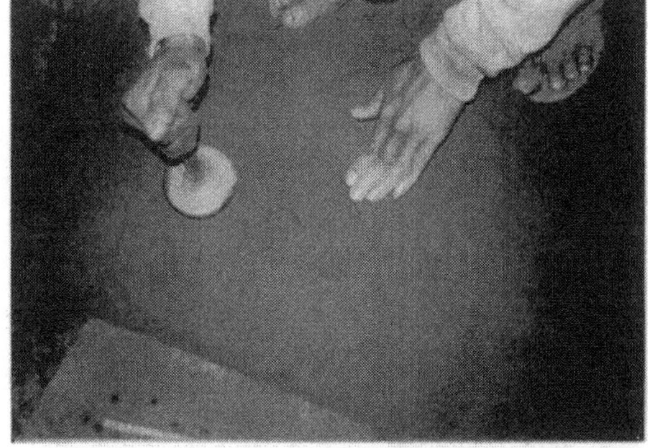

(b)

(c)

(d)

Fig. 5–10: Casting a grey cast-iron Pelton bucket

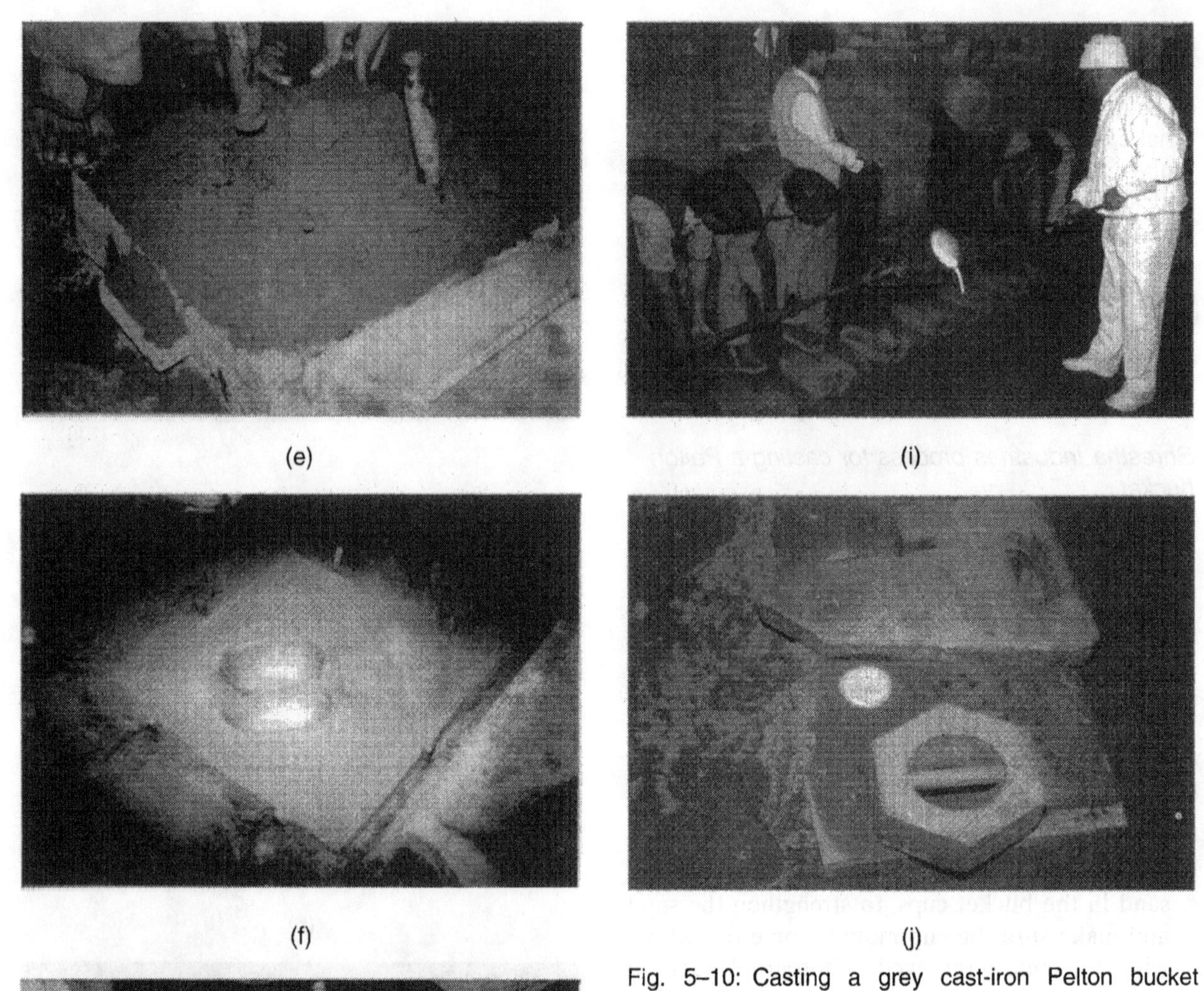

(e) (i)

(f) (j)

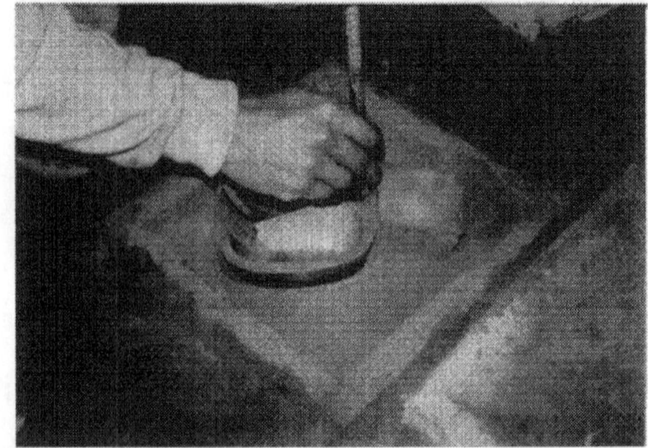

(g)

(h)

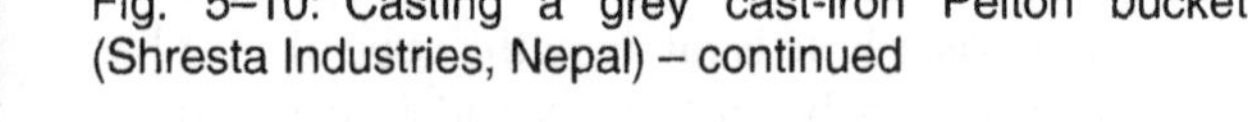

Fig. 5–10: Casting a grey cast-iron Pelton bucket (Shresta Industries, Nepal) – continued

Finishing the buckets

After casting, the buckets need to be finished. The best tool to use is a small grinding bit that can be fitted in an electric drill, or a special high-speed electric or air-powered hand grinding tool. The whole of the inside surface of the buckets should be ground to a smooth, polished finish. Also polish the notch and the area behind the notch. Remove any imperfections. Minor surface holes can be filled with weld and ground down to shape – provided the bucket material is weldable. Commercial runners are finished to Ra 0.8–1.6μm, and this is a good target to aim at. Final finishing should be done with emery paper. This whole process is time consuming, but it needs to be done. It is at this stage in the production that you discover the benefits of having good surface finish on the original castings.

Other casting processes

Die-casting is another way to make buckets. Hardened steel or graphite moulds are used, and these produce very accurate castings with excellent surface finish. The moulds themselves are very expensive, so this process is only worthwhile if a very large numbers of buckets is being made.

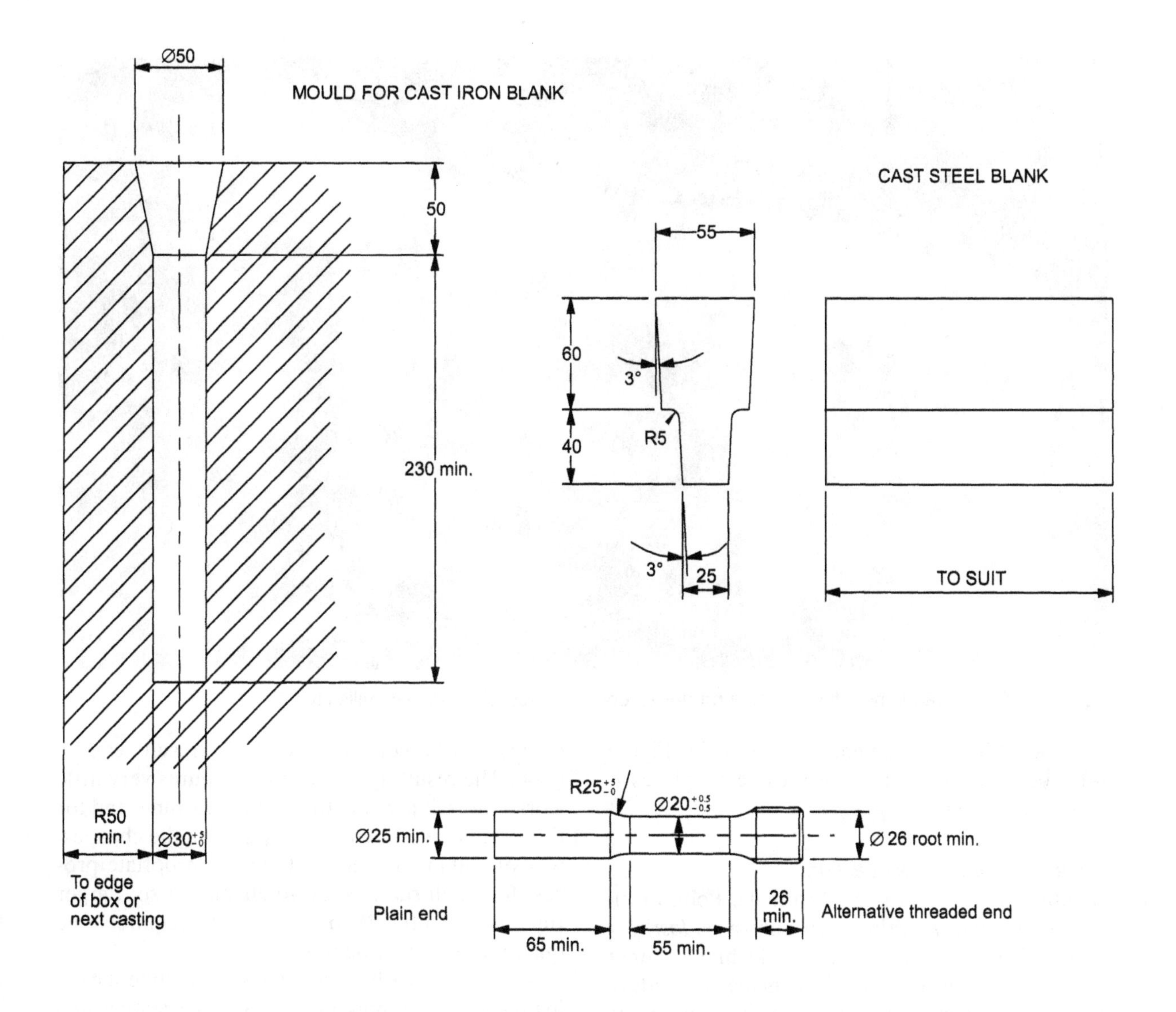

Fig. 5–11: Tensile test pieces for cast iron and cast steel. Based on IS:210-1978 and EN 1563

Investment casting, as described in Section 5.1.3, can be used to make individual buckets. It is a rather complex process to produce a simple shape, but if a good quality refractory coating is used the cast buckets can have a very good surface finish, so very little grinding is required.

5.1.3 Casting runners in one piece

The above section dealt with making single buckets, which can be machined or welded to fix them to the hub. Nearly all large, commercial turbine manufacturers have their runners cast in one piece, with all the buckets in place. Very large runners may be cast in sections, with several buckets on each piece, but the casting technique is similar. This section outlines the two processes used: multi-core patterns, and investment casting. Multi-core patterns can be used for almost any size of runner, though the cores become quite difficult to handle for very small runners. Investment casting is suitable for small runners, say up to about 350mm PCD, depending on the capabilities of the foundry.

Multi-core patterns

This method, which is the standard production technique for really large runners (several metres in diameter), is applicable to smaller runners too. It is just a variation of the standard sand-casting process. It is not possible to use a simple single-piece pattern the shape of a runner, as it cannot be withdrawn from the sand mould. Instead, the mould is made up from a complicated arrangement of patterns and cores, carefully assembled. Once the mould is made, casting is the same as making a single bucket. The method depends very much on the skill of the pattern maker and the foundry.

The basic idea is to make core pieces that are the shape of the spaces between the buckets on the runner. Each piece will have the shape of the inside of the bucket on one side of it, and the shape of the back of the bucket on the other side. These core pieces are put together to form a ring. The drag and cope have the runner hub moulded into them, and the core-piece ring fits between them.

Fig. 5–12: Core arrangement for casting a runner as one piece (courtesy of Newmills Hydro)

One possible arrangement is shown in Figure 5–11; the pattern for making these core pieces is shown on the right.

Investment or lost-wax casting

Another process that is used to make Pelton runners is *investment* casting, also known as *lost-wax* casting. For this, a pattern that looks like a runner is made out of wax. A mould is made to produce the individual wax buckets, and then these are joined to a wax hub. The wax is coated or *invested*, with a ceramic (refractory) slurry that sets at room temperature. The wax is melted out, leaving a ceramic mould that can be used to cast the runner. Some processes apply a secondary coating to the original layer to give better stability and strength, and may involve firing the mould in an oven, like pottery.

In industrialized countries, investment casting tends to be a highly-mechanized process, using sophisticated machinery. It can produce high-quality castings with very good surface finish. The weight of the casting is often limited by the size of the machinery, and so is useful only for smaller runners. A finished casting weight of, say, 20kg would be a normal maximum, limiting the size of runner that can be made to 300mm or so.

This section describes a simple investment casting process that can be used in less well-equipped workshops. However, as the following sections will show, investment casting is quite complicated and time-consuming, and it takes many steps to produce the final casting. A lot of equipment and different materials are required, and much of the work is best done by skilled craftsmen. If the quality can be controlled well, the results can be good. The resulting cast runner requires very little work before it is ready to go in the turbine, and for this reason investment casting is often the cheapest way of making a runner. It is an appropriate process for small runners, up to 200mm or so. Larger runners are difficult to handle, and require large quantities of investment material.

Some countries have indigenous investment casting industries for making, for example, statues and similar objects. These processes may be adapted to make turbine runners. In Nepal, it was demonstrated that indigenous investment casting could be used successfully to produce Pelton runners. Despite this, the process was not adopted, and manufacturers still cast individual buckets and machine them. While good quality could be produced in controlled conditions, it was not possible to achieve good results on a regular basis without closely supervising the foundries.

It is not possible to give comprehensive instructions for every detail here, and the reader is referred to a reference book such as Feinburg (1983) for information on how to proceed with investment casting.

a. Making wax patterns

All investment casting processes require a wax pattern. The pattern is often made by the turbine manufacturer, even if a foundry is doing the casting, so detailed instructions are given here. A resin pattern of one bucket is required, made with the correct contraction allowances (i.e. slightly larger than the design bucket, see Section 5.1.1). This pattern is then used to make a plaster mould.

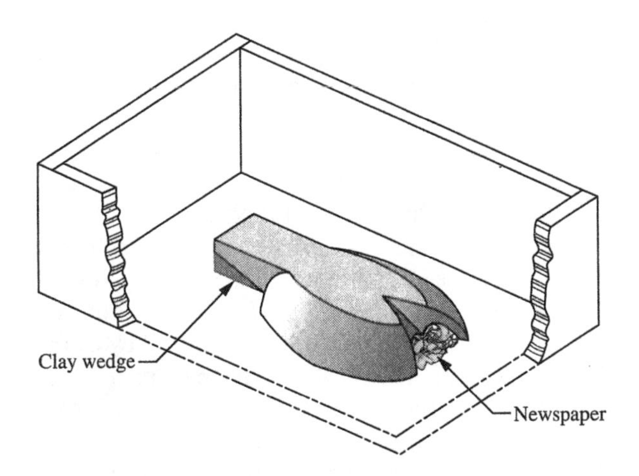

Fig. 5–13: Epoxy pattern on surface ready for making a Plaster-of-Paris mould

First put a wedge of clay or plasticine on the front of the pattern stem. Trim it so that its sides are parallel with the sides of the stem and its top is parallel with the front surface of the pattern. When it is finished, the pattern and wedge should lie flat on a surface, as shown in Figure 5–13. Use a paintbrush to coat the pattern with vegetable oil, to prevent plaster sticking to the pattern. Place the oiled pattern face-down on a flat piece of wood. Block off the notch in the bucket to prevent plaster from seeping inside the bucket (crumpled-up newspaper works well). Now place a frame around the bucket, leaving at least 30mm clearance all around.

Plaster-of-Paris (gypsum plaster) should be mixed 1:1 by volume to a creamy consistency. Add the plaster to the water, stirring continuously. When the plaster starts to form little islands on the mix that do not quickly dissolve, you are at the correct mix. Shake the container and leave it for a few minutes so that air bubbles can settle out. Then stir thoroughly with your hand until it just begins to set. Dribble the plaster on to the bucket pattern, building up from the pattern outwards. Be careful not to trap any air bubbles, particularly around the pattern itself. The plaster should be laid to a thickness of at least 30mm above the top of the pattern.

Leave for about half an hour to set, before carefully turning out the plaster mould. Leave the bucket in place, but remove the fill-in wedge. Trim the face of the mould to make a sharp split face. The split line should be around the *outside* edge of the bucket rim and stem, and around the *inside* edge of the notch (Figure 5–14). The surface needs to sink down at the bottom end of the pattern to meet the sloping section of the stem. Some location feature needs to be added to position the top half of the mould relative to the bottom. One suggestion is to scratch three radiused grooves (*guides*) around the top of the bucket, but not touching it (as in Figure 5–14). Another possibility is to put location pins in the frame, and leave the frame attached to the plaster mould. Either way, it is important to be able to split the mould quickly and easily. Make sure the split surface is smooth.

Paint a thin coat, or *slip* of clay on to the split surface, and follow this with a coat of oil. Clean out the inside of the bucket, and coat it with oil too. Mix up some more plaster as before, and build it up on top of the bucket pattern and out over the rest of the mould, till it is at least 50mm high. Leave for half an hour until the plaster has set.

The two halves of the mould can now be split, and the bucket pattern removed. Check that all the surfaces of the imprint of the pattern are well-formed and smooth. Repair if necessary.

Cut a feeder tube and feeder cup into the plaster, joining the stem of the bucket. The feeder should be at least 12mm diameter, and the cup at least 25mm diameter. Clamp the two halves of the mould together (rubber from a bicycle inner tube works well for this). Soak the mould in water for a

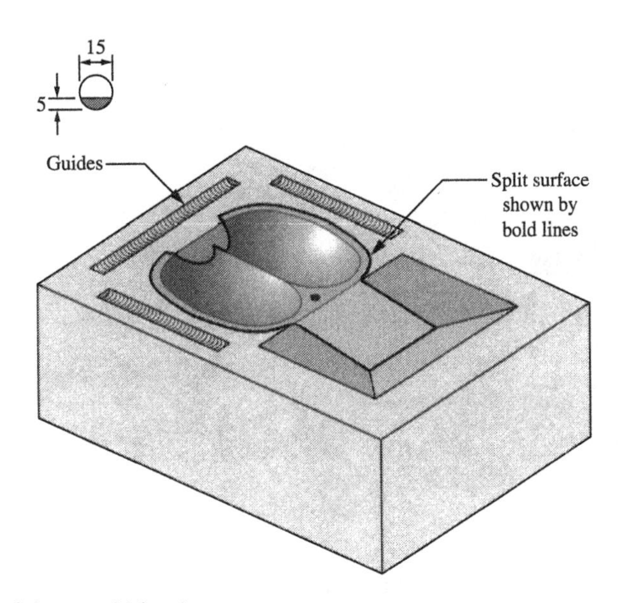

Fig. 5–14: The lower part of the mould for the wax pattern

few minutes, drain off the water, and the mould is ready to use.

Various mixes of wax can be used, and it will probably be necessary to experiment a bit with the ingredients available locally to get a suitable mix. A modelling wax that works is made from one part paraffin, one part beeswax, and one part resin. Various types of resin may be used. Some, such as rosin, mastic or damar, are soft resins derived from trees. Others, such as copal or amber, are hard resins derived from fossil plants and trees. The hardness of the wax is controlled by the quantity of resin. If the weather is hot, more resin is needed. Another recipe is six parts beeswax, six parts paraffin wax, three parts petroleum oil, three parts petroleum jelly, two parts clarified animal fat (or cocoa butter or coconut oil), and one part clear polythene (from polythene bags).

Melt the wax so that it flows easily, but is not steaming. Push the froth aside with a ladle and then scoop out clean wax, making sure it contains no bubbles. Pour gently but smoothly into the side of the feeder cup, allowing air to flow out of the mould from the middle. Wait for the wax to set and cool, and then pry open the mould. Pull on the sprue to release the wax from the mould, and then cut the sprue off with a knife. Clean up the wax pattern, filling in any holes and trimming off any excess wax. Remember that any imperfections in the wax pattern will appear in the final metal casting.

The mould can be used several times to make a set of buckets. It should be dipped in water every second time. Oil may be required on the inner surface of the mould to help release the wax mouldings.

A wax hub can be made in a similar way.

b. Assembling the wax runner pattern

When the hub and all the wax buckets are made, they need to be joined together to make the runner. It is essential to make a jig to hold the hub and buckets in the correct positions so that they can be joined together. The jig should have stops to locate the buckets at the correct outside diameter for the runner, and locations for the bucket slots to align them correctly. Stops on the jig can be used to fix the angular positions of each bucket. Be careful to design the jig so that the runner can be removed when it is complete!

When the buckets and hub are correctly positioned in the jig, they are welded together with strips of wax and a soldering iron. The wax should be rolled into strips 6mm diameter and about 150mm long. Virtually any sort of soldering iron can be used provided it has a reasonably large, flat end. Electrical soldering irons work, and the instant-heating type are particularly handy. A plain soldering rod can be made by flattening and bending over the end of an 8 mm bar, heating this from time to time in a flame. Lay the wax strips over the joints in the wax runner, and work over them with the soldering iron. Ensure that the heat works into the wax and does not just fuse the surface. Leave

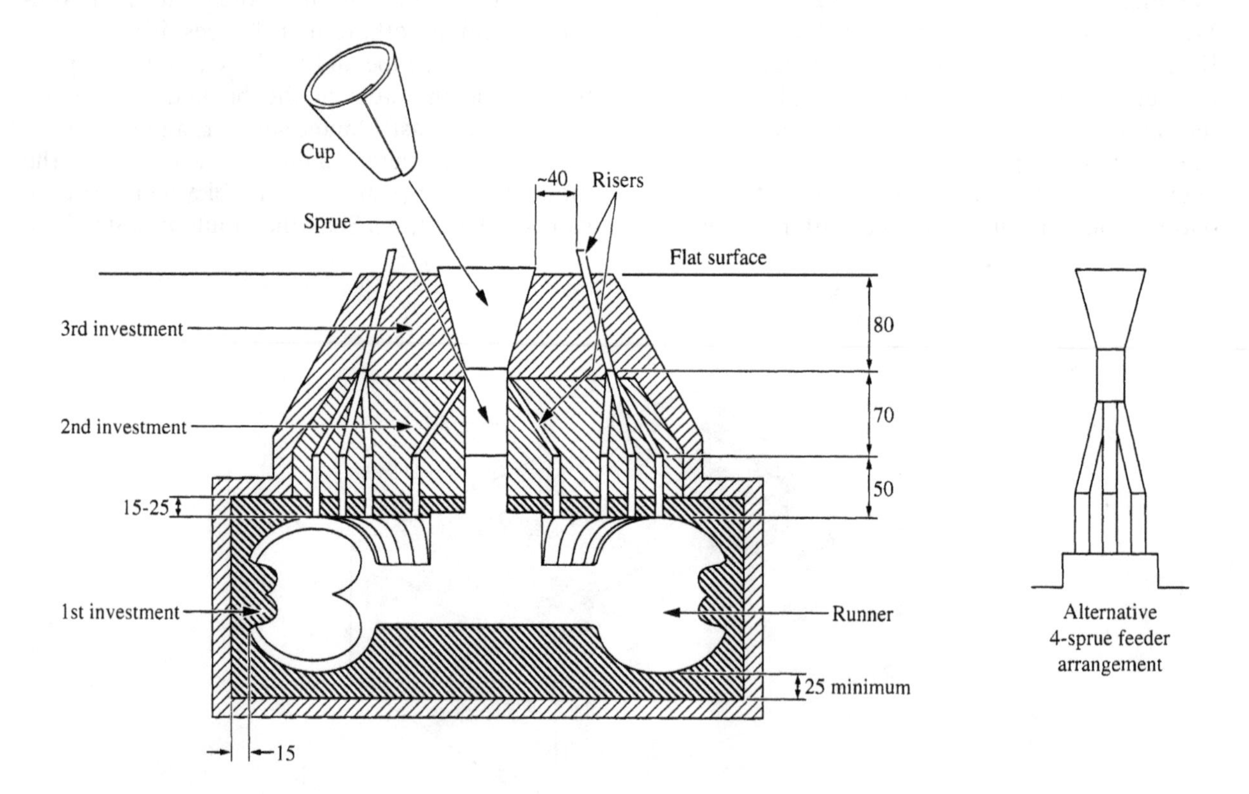

Fig. 5–15: Diagram of the wax runner with the completed investment, showing the arrangement of the sprue, cup and risers

the runner for at least an hour for the wax to set hard before turning it over to weld the other side.

If the jig makes it possible, take the runner out, turn it over and put it back in to weld the other side. If the jig cannot take a reversed runner, lay the runner on a flat surface but with a support under the hub so that this does not sag and distort during welding.

When the welding is finished and the runner has cooled, put it on a spindle and turn it gently by hand to check that the buckets are in line. Check particularly that ends of the splitter ridges are all on the same circle.

The foundry will generally decide the best arrangement for the feeder system.

c. Investing the pattern

This is the process of coating the wax pattern with a ceramic mix. This section describes a simple process for do-it-yourself investment casting. It is appropriate for brass or bronze runners, not really suitable for cast iron, and would not work for steel. Experienced foundries will have better techniques.

Begin by fixing the first stage of the feeder system to the wax pattern. It has been found that a single wax sprue with a diameter of about 25% of the runner PCD gives the best results, but it is possible to fit three or four smaller sprues around the hub. The sprues should be made about 50mm long. Risers can be made of strips of wax, with a 3–5mm section (they do not have to be round), 50mm long, welded to the tips of the buckets. Be careful to support the runner and hub so that it does not distort as the sprues and risers are welded to it. Casting is usually done with the runner laid flat (i.e. with its axis vertical), so that the feeder and risers are all on one side – most commonly the side from which the runner protrudes the most. It is best to feed the molten metal into the hub area through one or more sprues, and to have risers attached to the outside edges of each bucket to let the air out, remove impurities from the mould, and to improve the metal flow. One possible arrangement is shown in Figure 5–15.

The runner now needs to be de-greased to ensure a good bond between the investment and the wax. A mixture of half a teaspoonful of liquid detergent in half a litre of alcohol solution ('surgical spirit') can be used. The solution should be poured over every surface so that they are all wetted. Allow to dry for at least half an hour.

Take a piece of sheet metal with a width 50mm or so greater than the buckets, and bend it into a circle with a diameter 30mm larger than the runner. Hold it at the correct diameter with lengths of wire wrapped around the outside. This is used to mould the outside shape of the investment. It can be used on a flat wooden surface, but it is rather less messy if the steel band is placed inside a plastic bowl.

The investment is a mixture of refractory material and bonding agent. The refractory material withstands the heat of the molten metal, and the bonding agent holds the ceramic together. A suitable mixture can be made from one part gypsum plaster and two parts brick dust by volume, mixed with water. Do not make too much investment, as is sets quickly and any not used will be wasted. Three litres is a reasonable quantity with which to work. Mix the plaster and brick dust and sprinkle on to the water, stirring all the time. Allow a minute for the air to settle out, and then mix by hand. The mixture for the first coat should be quite thin.

Pour the investment mixture over the wax to build up a 2–3mm layer on the surface. If it is too thin it will peel off, but do not allow the liquid to accumulate in thick pockets. Pour about 25mm of the mixture into the steel hoop, and place the coated runner into this, with the hub and feeder system upwards. Pour more investment in so that all the buckets and the hub (but not the sprue and risers) are covered. Agitate the hub to help air bubbles escape, and be careful not to trap air under the runner.

The runner is now invested, and the remaining steps are to build up investment around the feeder system. Make some extra wax riser pieces 70mm long and join these to the wax risers already on the runner. The risers should be angled towards the centre of the hub, and joined together in threes. If multiple sprues were used, add pieces the same diameter to the tops of them, angled together so that they meet over the middle of the runner. Then fit a final sprue and pouring cup on top. If a single sprue is used, fix the sprue and pouring cup directly to the top of the runner. The cup can be made by rolling out a flat piece of wax in a circle 200mm or so in diameter, cutting along a radius and rolling it around into a cone. The joint can be soldered, and the bottom cut off to make a funnel shape.

The second investment mix is one part of plaster to one part brick dust and one part silica sand, mixed as before. If investment casting has been done previously, the remnants of the investment can be broken up and used instead of the sand. This mix should be fairly thick. Cover the top of the first investment, and then dribble the mix on so that it coats the sprues, risers and the pouring cup. As before, be careful not to trap air pockets. The whole of the cup and sprue system should be well covered with investment, and the risers should be just sticking out of the investment when finished. Leave the surface rough so that the third investment bonds well to it.

Add more pieces on to the risers so that they come just above the pouring cup, and 40mm away from it. Remove the metal band, and roughen up the sides where the metal has been. The final investment mix is now made of one part plaster, one part silica sand or previous investment. Put a good thick coating all over the hub and the feeder system. The risers should just protrude above the surface of the investment when finished. The final shape will be a big cylindrical block around the runner with a central column rising up around the risers and sprues. The top, around the cup and the risers, should be flat so that the investment can sit on it in the kiln. Leave the investment to harden overnight.

Before melting out the wax, wet the investment. It can be heated gently by steaming it over a pan of simmering water, with a container placed beneath it to catch the melted wax for re-use. When the bulk of the wax has come out, the investment should be baked in an oven or kiln.

There are many variations on the above process. Traditional lost-wax foundries in Nepal use a mixture of clay and cow dung for the first investment, adding coarse rice husks for the final two layers. They also coat the runner and sprues in a 25mm shell of investment, rather than making a solid block, and the investment is sun dried for several days before the wax is melted out.

d. Making the metal runner

Leave the moulds to cool. This takes quite a while. When they can be handled, remove them and take them to the furnace. The molten metal needs to be poured in steadily but quite fast to fill the mould before it begins to set (if poured too slowly it will not come up the risers). Pour till the cup stays full and metal is seen in the risers. Leave overnight to cool. Remove the burned investment with chisels, keeping the material for further use. Exercise great care around the runner to avoid damaging it. Cut off the risers and sprues, and grind the runner clean in the same way as for separate buckets. The investment casting should now be ready for use. It will need to be machined to fit to the turbine shaft, and should be checked for balance as described in Section 5.1.4.

Chill casting

A production method that seems to have great potential for micro-hydro, but as yet appears to be unproven, is chill casting. For this, individual buckets are cast with a basic stem as shown in Figure 4–3. These buckets are then held in the correct orientation in a mould, and a hub is cast on to them. The new metal fuses to the old, giving a runner that is effectively a single piece, though it may not have the strength of a casting made in one process.

The buckets should be cast as usual, but they can be cleaned up and polished before the hub is made; which is a significant advantage, as it is much easier to grind the buckets separately. The faces of the stem should be thoroughly cleaned back to bare metal. For casting the hub, a three-section box is ideal. The principle is to have the buckets fixed in the sand of the middle section, and the patterns for the two sides of the hub projecting into this from the top and bottom boxes. A jig is needed to ensure that the buckets are located accurately, and the hub patterns can be part of this. Sand is packed into all three sections, including around the buckets. By removing the upper and lower boxes, the hub patterns can be withdrawn, but the buckets are left in the sand. Flux is applied to the stems of the buckets to assist the molten metal to key to it. Molten metal is then poured into the cavity.

This process has been tried for prototype brass runners, but to the author's knowledge no such runners have actually been operated. In principle, it could be used for other metals too. Chill casting deserves further development.

5.1.4 Balancing

All runners (whether single-piece castings, bolted or welded assemblies) should be balanced before use.

There are two types of balancing: static and dynamic. Static balancing is a question of ensuring that the centre of gravity of the runner lies on the axis of the shaft. This can be done with the runner rotating slowly. Dynamic balancing involves spinning the runner at speed, and checking to see that there are no vibrations. For micro-hydro, static balancing it is generally sufficient, and this is the method described here.

Fix a shaft to the runner and set it on a pair of straight edges, as shown in Figure 5–16. Check that the edges and the shaft are horizontal, using a spirit level. Ideally the whole arrangement should be on a marking out table. It is possible to use a bearing arrangement to hold the shaft instead of the straight edges, but it must hold the shaft horizontal and have minimal friction.

Gently roll the runner several times, marking the location of the bottom point when it stops moving. If the marks are evenly distributed around the runner, then the unit is balanced. If the marks congregate on one side, then correction is required. The side around the marks is heavy, so either material needs to be removed from that side (by carefully grinding off material on the backs of some buckets – be careful not to make the buckets noticeably thinner) or by welding on small weights opposite. After adjusting, check the balance again, and repeat until no heavy spot can be detected.

Even the largest runners are normally only balanced statically. However, these runners are made from clean, accurate castings, carefully ground using templates. Poor castings may be much more uneven, and it is possible that vibration may be observed during high-speed operation. If so, it is possible to do dynamic balancing for a micro-hydro runner on a car wheel balancing machine.

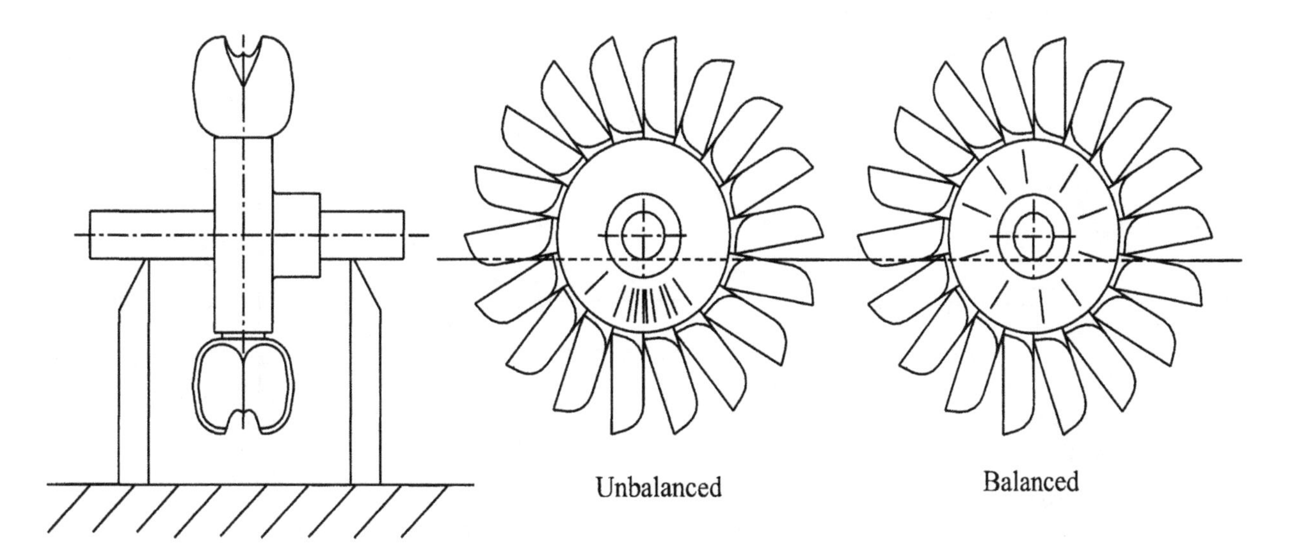

Fig. 5–16: Balancing a runner

5.2 Fabricated runners

5.2.1 Bolted assemblies

This section gives a possible machining procedure for the bolted runner assembly shown in Figure 4–8(a), which is shown in more detail in Figure 10–10(a), and uses this as an illustration of the steps required to machine any runner. The process can be adapted for other designs. How the job is tackled depends on the machinery available and the accuracy required. It is possible to do a number of the operations with hand tools, but at a minimum a lathe and a drill are required.

Machining the buckets

Before machining is started, the cast buckets should be cleaned up. Sand or slag should be removed from the surface, and the inside surfaces of the bucket should be polished to a smooth finish. Any buckets with holes, distortion, short or deformed splitter ridges, or with anything but minor flaws should be discarded. If blowholes are uncovered while a bucket is being machined, again, it should not be used. There are three stages:

- Cleaning-up, where the bucket surfaces are machined so that they can be used for location.
- Drilling of holes for fixing the buckets in the hub and the buckets.
- Machining the buckets and hub to fit each other.

For many designs (including the one here) a dummy hub needs to be made for an intermediate machining stage.

The various steps for producing the dimensions given in Figure 10–10(a) are illustrated in Figure 5–17.

Step (a)

The back of the bucket needs to be made flat. This can be done with a file, or, better, by rubbing the bucket on emery cloth on a flat surface. Alternatively, a shaper or milling machine can be used. The surface only needs to be skimmed flat – do not remove too much material. When complete, the bucket should sit firmly on a flat surface (such as a marking out table) without rocking. This step provides a reference or datum surface that can be used to locate the bucket for other machining operations. One of the advantages of this bucket design is that it has a large, flat back, and it is easy to fix in place.

Step (b)

Now clean up the front surface, around the edges of the bucket. Again, this can be done either by hand or in a milling machine, and the surface only needs to be skimmed. This provides a surface for clamping the bucket in step (d).

Step (c)

The front of the stem needs to be cleaned up. If a machining allowance was put on to the pattern, this will need to be taken off. It is best to make a block with a 20° slope on it to hold the bucket at the correct angle. The position of this surface is critical. If it is too high, the stems will be too wide and the

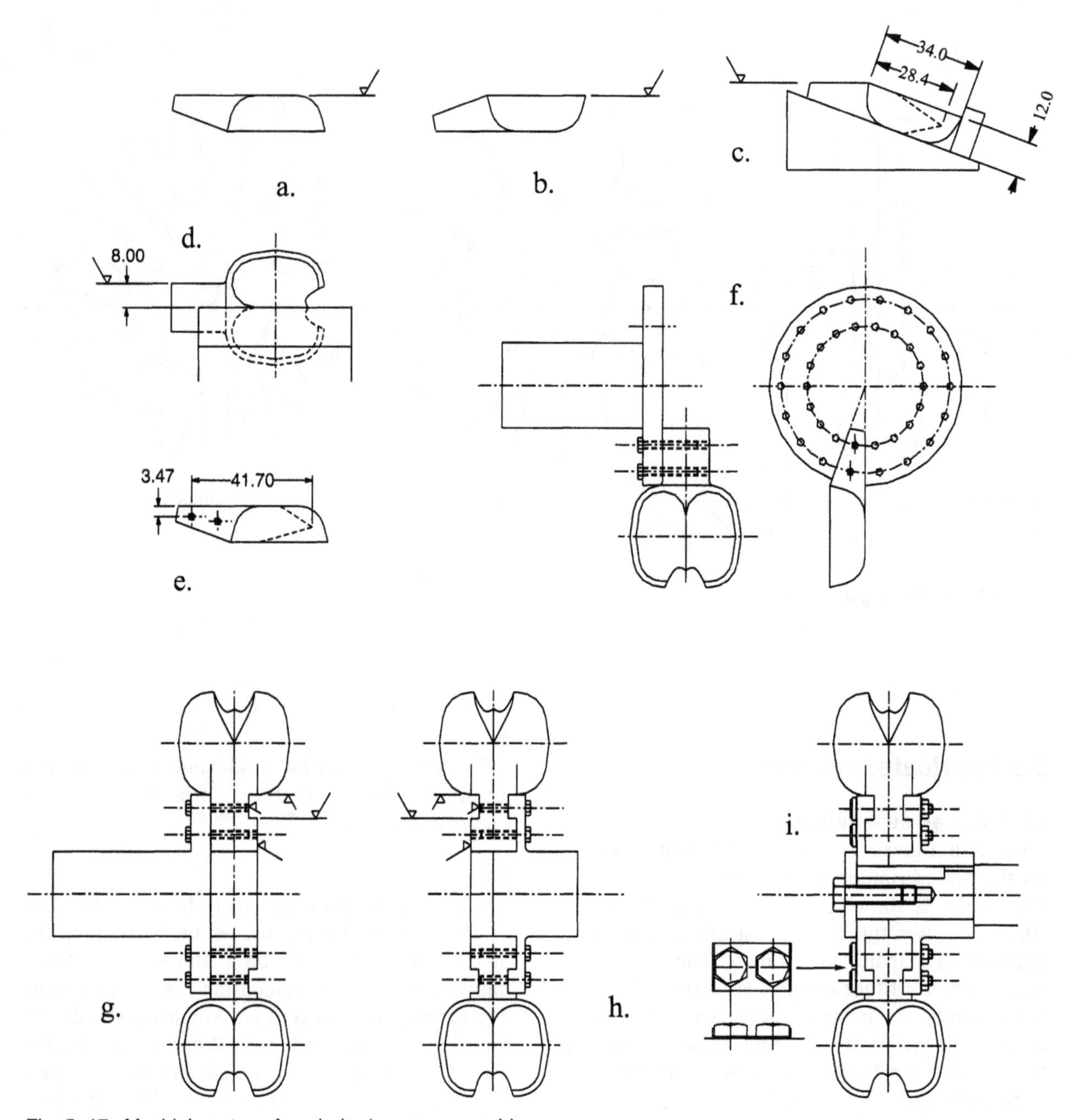

Fig. 5–17: Machining steps for a bolted runner assembly

buckets will not fit on to the hub. If it is too low, there will be gaps between the buckets. Two dimensions are shown in Figure 5–17(c) to the start of the surface, one from the tip of the splitter, and one from the end of the bucket. If the buckets are accurately made, and the splitter ridges are all the same shape and position, the best measurement is from the splitter end. However, if the splitter ridge ends vary between the buckets, it may be more practical to measure from the end of the bucket.

Step (d)

Clamp the bucket in a vice, and machine (or file) the sides of the stem. Align the splitter ridge with the tip of the vice. It is important to make the stem symmetrical about the splitter ridge, so that the ridges all end up in line on the runner. The maximum out of line of the splitter ends should be ± 0.5mm.

Step (e)

The holes in the stem need to be marked out and drilled. Because the sides of the stems need to be machined later, the buckets have to be held on one side only, and at this stage a smaller, tapped hole needs to be made. This is drilled out later to take the final assembly bolt. Either mark out the hole position carefully, or, better, make a drilling jig. The measurement should be from the back of the bucket and either the outside edge or the splitter ridge.

Step (f)

Make a dummy hub as shown. The hole centres should be accurately marked out on the correct

PCD and at 20° intervals on the disc before drilling. It is a good idea to drill the dummy hub and the final hub halves at the same time, so that the holes match exactly. The holes in the dummy hub should be a tight fit on the thread used for step (e).

Now fit the buckets to the hub. The line of the back of each bucket should pass through the centre of the hub – check this with a straight edge. It may be found necessary to remove a little more material from the front of some stems to make them fit. If there are gaps between the buckets, make shims that cover the whole of the stem fronts to fill them.

Step (g)
Tighten the bolts well, and machine the recess on one side of the bucket stems. This should be toleranced carefully to make it a press fit when the hub is finally assembled.

It is a good idea to make some spare buckets at this point. Remove one or two buckets, fit the spares in their places, and machine the recess in the spares.

Step (h)
Now fit the final hub to the machined side of the buckets, but still with the undersized bolts. Machine the recess on the other side. Do the same for the spare buckets if required.

Step (i)
Now assemble the other half of the hub to the runner. The holes should be drilled out to take the final size of bolts. Torque the bolts to the appropriate tightness. Use tag plates between pairs of bolts – as in the detail in Figure 5–17(i) (or some other locking system) – to make sure the nuts and bolts cannot come undone.

When the runner is assembled, it should again be put in a lathe to check it. The splitter ends should all be in line, to within a tolerance of about 0.75% PCD, and at the same diameter. It may be possible to skim the ends of the splitter ridges in the lathe to even them up a little, or to grind a few that are out of position. Poor alignment can lead to uneven running. Check the spaces between the buckets too: these should be equal.

The runner should then be balanced as in Section 5.1.4.

5.2.2 Welded assemblies
This section covers the welding of buckets to the hub. The buckets must be made of a suitable material for welding, as discussed in Section 4.1.3. The castings should have been cleaned, and the insides polished. This section is based on the bucket shown in Figure 4–11 and the assembly in Figure 10–10(b).

Welding the buckets
One of the advantages of welding is that the bucket stems do not need to be machined carefully. The back and front of the buckets should be cleaned up as in steps (a) and (b) above, but this is only to ensure that the bucket sits flat on these faces. There is no need to take material off the whole surface. Similarly, the front of the stem must be skimmed to make sure it is not too wide, but does not need to be fully machined.

Figure 5–18 shows a welding fixture for aligning the buckets as they are welded to the hub. The shaft is seated in two V-blocks, and another block is placed at the correct height for the front of the buckets. On this extra block is a lug which fits into the notch and has a location feature for the end of the splitter ridge. The hub has lines marked out on it at 20° intervals. The angle of the hub is first fixed in position by measuring from the table up to these lines, and then a bucket is placed in position on the block. After checking carefully that the splitter ridge is sitting in the right place, and that it is parallel to the sides of the hub, it is tack welded in place. Note that it is not important if there are small gaps between the buckets. The load is taken by the weld metal, not by the contact between the buckets, and small gaps make no difference. Large holes that will be difficult to fill with weld may need to be filled with shims. The process is repeated till all the buckets are in place. (Note that the fixture may need moving a little to fix in the last bucket.) The runner is now tacked together, but it should be checked carefully before it is fully welded. Check that the splitter ridges are in line, and at the same radius, by rotating the runner. Check that the bucket spacing is even.

Fig. 5–18: Welding fixture (NHE, Nepal)

The welding needs to be done carefully. If rods are being used they should be thoroughly dried before use, and it is best to preheat the runner to 70–80°C before welding. Weld between the buckets first to make the assembly rigid. Be careful of weld distortion, welding opposite buckets rather than adjacent buckets. Be careful to lay down good weld at the base of the splitter ridge, as this is a potential weak spot.

The welds around the hub should be done last. This will take several runs to fill the recess. It is *critical* that a root run is done first in the corner between the bucket and the hub, as shown in Figure 5–19. If this is missed, fatigue cracks can be initiated from this corner. Lay one run on each side in turn to avoid weld distortion. Tack welds and the edges of other welds must be re-melted during welding. When complete, allow to cool, and chip off any flux. The welds should then be ground, ideally with a high speed pencil grinder. This is to clean up any surface defects and cracks, and thus improve the fatigue life.

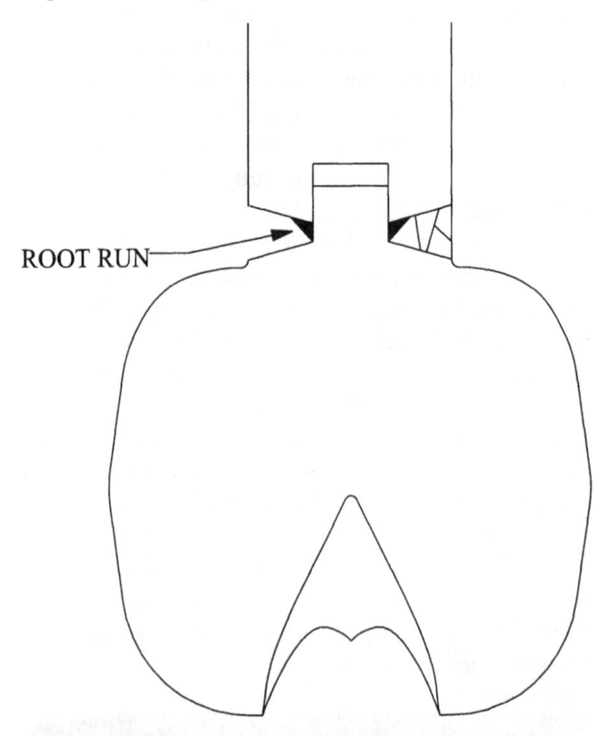

Fig. 5–19: Weld detail between the hub and the bucket

Fig. 5–20: Welding a 365mm PCD runner (NHE, Nepal)

Weld testing

For small runners at low to medium heads, weld testing is usually not necessary as there is plenty of safety factor in hand. If however, the weld stresses are anywhere near the design limit, the welds should be tested. The best method for this is using dye penetration. It is a little time consuming, but is far better than having a bucket fly off the runner in service. Dye penetration kits can be bought from welding materials suppliers. The kits will contain a solvent cleaner, the penetrant dye, and a developer.

First grind the weld surfaces, otherwise weld undercuts or overlaps show as cracks. Clean with a degreasing cleaner. Spray with red dye, leave for five minutes, and then wash off the dye with water and cotton wool. Then apply the special white paint developer. Now watch and see where the red dye emerges through the white. This will show where cracks and defects in the weld are located, and these should be marked with a pen. Cracks show as obvious red lines on the developer. Continue to watch for about 15 minutes. After that time, the dye will begin to spread, and its source will not be clear. After about an hour, the whole surface will have turned red.

Where cracks appear, grind out the weld till the crack disappears. The bottom of the ground-out groove should be tested again to see that the whole of the crack has been removed. Re-weld, and retest that area.

Other weld test methods may be used if they are available. Most of these require sophisticated and expensive equipment, and skilled operators, and are usually not available for micro-hydro. Suitable methods are ultrasonic testing and magnetic particle testing. Both of these allow inspection right through a section. The latter does not work on stainless steel. X-ray testing is useful for welded plates, and is not appropriate for inspecting runners.

Balancing

The finished runner should be balanced as in Section 5.1.4.

5.2.3 Testing

Casting is a process that can be reliable and repeatable, but unfortunately it can produce defects buried within the buckets, or use poor material quality that cannot be detected. Yet the runner is the critical part of a Pelton turbine, and it is often felt necessary to prove it before it is used. Unfortunately, it is difficult to ensure that fatigue failures are not going to occur.

Crack detection methods, such as those described above for weld inspection, can be used on runners. Surface crack detection using dye penetration gives some information, but it does not reveal sub-surface defects. Ultrasonic or magnetic particle methods are good, and both are used to inspect very large runners, but very few workshops will have access to such equipment.

Eisenring (1991) suggests a static test by hanging weights on each bucket in turn. The weight is attached so that it presses on the bucket at the same place as the centreline of the jet, and the runner is arranged so that the weight acts in the same direction as a jet would act when in the position that generates maximum bending moment. It is assumed that fatigue effectively reduces the tensile strength of steel or cast iron to approximately 0.3–0.45 of its original value, so the force applied should be 1/0.3 to 1/0.45 times the jet force, plus a safety factor of, say, × 2. The jet force can be calculated from Equation 10–23, Section 10.4.2. This test is certainly possible, and may give some confidence in the strength of the runner, but it is not a reliable guarantee that fatigue cracking will not occur.

One testing method that has been used for micro-hydro is an overspeed test. This is based on the same theory that fatigue effectively reduces the tensile strength. Since the centrifugal force on a bucket is proportional to the square of the speed, the test is run at √(1/0.3) to √(1/0.45), or 1.83 to 1.49 times the runaway speed, and often includes a further allowance for safety factor. The runner can be taken to five times its operating speed, which is highly dangerous, and needs a special rig to prevent the buckets flying out if they do break off. However, this only tests for centrifugal loads, whereas the fatigue loading on a runner comes from the jet force on the bucket. In certain places, the stresses may be in the same direction, but it is not true for all the welds. In view of the danger, and the fact that it may not predict failure, this author does not recommend the test.

In summary, highly technical, non-destructive inspection methods are very useful, but are unlikely to be available. Other tests may catch some defects, but will not be infallible. Whether you choose to do a test or not will depend on experience, your confidence in the manufacturing process, and how critical a failure would be.

5.2.4 Other methods of manufacturing runners

There have been some interesting developments in runner manufacture in recent years. They have been developed by large turbine manufacturers, and need sophisticated equipment. They are unlikely to be of use to the small micro-hydro manufacturer for some years to come, though it may be possible to buy runners made using these techniques in the future.

The Italian company, Fravit, is offering runners machined in one piece from a solid disk of steel. A forged disk of heat-treated X15 13/4 CrNi stainless steel is used, after a thorough programme of inspection and testing to check that it is free of defect. Numerically-controlled milling, which is integrated with the CAD system used for designing the runner, is then used to produce the shape of the buckets, followed by grinding and polishing. The process is claimed to give defect-free runners with superior mechanical properties, but a lead-time of 12 months is still quoted to produce a large runner (Fravit, 1999).

Fig. 5–21: The Sulzer Hydro *MicroCast* method of making Pelton runners (Schneebeli et al, 1996, courtesy of Sulzer Hydro Ltd., Zurich)

The *MicroCast* process developed by Sulzer Hydro (Schneebeli et al, 1996) starts with a forged disk. A numerically-controlled milling cutter is then used to produce the shape of the bucket roots around the edge of the disk. Robot arc-welding is used to deposit the remainder of the bucket shape on top of the roots, layer by layer (Figure 5–21). Further milling and grinding is used to obtain the final runner.

The machined roots of the buckets actually extend about one third up the bucket, with the welding adding the outer two-thirds of the bucket shape. About a third of the final weight of the runner is made up of weld, and for really large runners this can mean the welding takes around six weeks. It is claimed that careful control of the weld leads to material properties that are superior to cast buckets, with less chance of defects. The highly-stressed roots of the buckets fall in the forged/machined base area of the buckets, which has better fatigue properties than cast or welded material. The lead time for this type of bucket is said to be 30% shorter than for large cast runners. This process is already being used for production runners.

An alternative approach to runner fabrication (Chapuis & Fröschl, 1998) also starts with a forged and machined hub, with the shape of the inner portion of the buckets and the bucket roots cut into the solid hub. The remainder of the buckets are added as three separate pieces: a central section, containing the splitter ridge, the notch and the outer edge of the bucket, and two side pieces (Figure 5–22). These pieces are made as simple castings, and are welded to the hub and to each other, first the central section, and then the two sides. As with any runner, grinding is used to produce the final shape and finish. The welding needs to be carefully controlled, and ultrasonic testing is done on all the welds after they are laid down. This process has only been used for prototype runners to date, but may well become an accepted manufacturing technique.

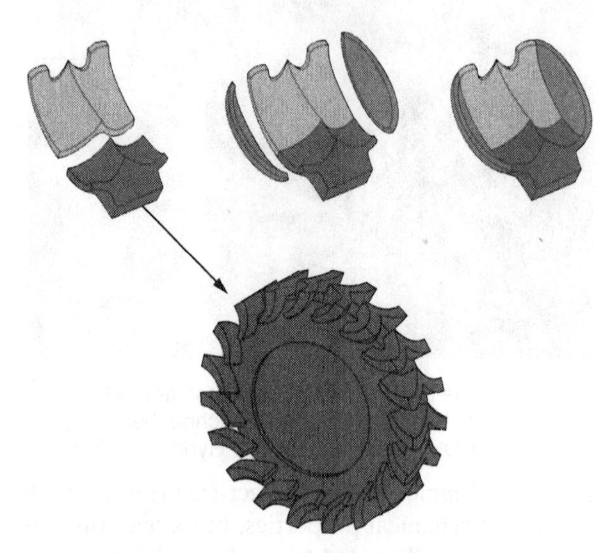

Fig. 5–22: Building a runner from a forged and machined hub and three cast pieces (after Chapuis & Fröschl, 1998)

5.3 Spear valves

The only point to watch for in fabricating a spear valve is that the front and rear bearings are in line with each other and the nozzle. This generally means that the assemblies have to be machined after welding the components together.

5.4 Manifolds

The manifold is a straightforward assembly, though some complex developments can be needed for bifurcations and bends. (Techniques for drawing the developed shapes required for such configurations can be found in most standard textbooks on engineering drawing or sheet-metal work.) The manifold needs to be made accurately, and ideally should be assembled to the turbine, valves and other components at the tack-weld stage, to ensure that it fits them.

It is good practice to pressure test the manifold in the workshop to make sure there are no weaknesses or leaks. There are a couple of reasons for this. Firstly, it is much easier to repair leaks in a workshop than at site. Secondly, pressure testing in a workshop is done with a pressure pump, and if a pipe leaks or bursts the pressure is released instantly, and little damage is done. At site, the manifold has the energy of a full penstock behind it, which can be dangerous and lead to catastrophic failure. The test pressure should be higher than the total design head (gross head plus maximum surge head) to allow for the corrosion allowance. In order to produce the same stress in a new manifold as would occur in the same manifold when the wall is thinner due to corrosion, the pressure needs to be:

$$H_{test} = H_{total} \cdot \frac{t}{(t - t_{corrosion})} \qquad \textit{Eq. 5–1}$$

H_{test} – test pressure (same units as H_{total})
H_{total} – total head, gross head plus surge
t – pipe thickness (same units as $t_{corrosion}$)
$t_{corrosion}$ – corrosion allowance

If spear valves are being used, these should be included in the test. However, unless a spear valve is made very carefully it will leak slightly, making it impossible to hold pressure. It can be made leakproof by inserting an annulus of thin rubber sheet or gasket material between the nozzle and the spear valve and closing the valve tightly. Any other open ends will need to be blocked off with thick flanges. The manifold then needs to be filled with water, bleeding off all the air. Pressure can be applied with a proprietary testing machine. Hand-pumped machines are in many ways better than electric pump versions, because even the slightest leak will make it impossible to reach the test pressure. It is not too difficult to fabricate a home-made a hand-pumped pressure tester.

5.5 Turbine Housing

Housings are usually made of welded sheet steel, and this is the simplest, most flexible method. It is possible to have cast housings, or to use existing boxes or containers. The author has seen very neat Peltric sets made from circular aluminium saucepans. This section assumes sheet materials are being used.

The most important part of making the housing is the marking out of the sheets before any cutting or drilling is done. The first thing to do is to mark an accurate rectangle the size of the plate required. The rectangle should be checked by measuring across the diagonals; these dimensions should be equal, as shown in Figure 5–23. If a set square is to be

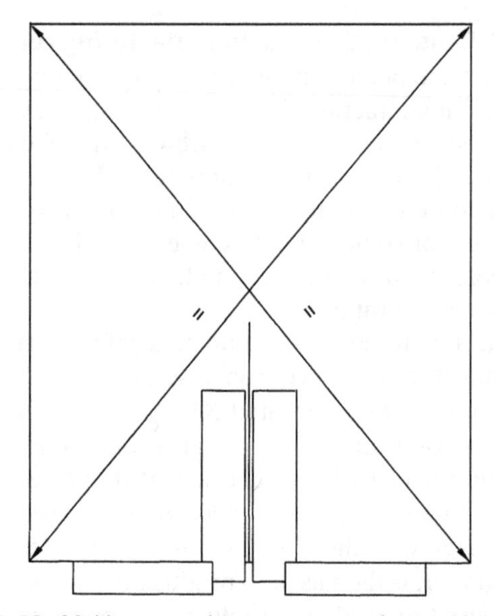

Fig. 5–23: Making sure plates are square for marking out

used, first check it by marking a line with it both ways round (again as shown in Figure 5–23); if the lines are not on top of each other, the square is inaccurate. Once the rectangular plate is made, the hole centres and other details can be marked out on it.

When welding the plates together, it is suggested that edge to edge fillet welds are used, as shown in Figure 5–24. This gives a strong, neat, leakproof joint, and allows a small amount of angular adjustment of the plates that hold the nozzles. Initially the housing should be tacked together with the nozzles and the shaft in place. The alignment should then be checked and adjusted. For spear valves, the easiest way of doing this is to make a long pointed rod to replace the spear. It should be a good fit in the spear bearings, and long enough to come out as far as the runner shaft. (On some designs, the spear itself can be slid out far enough, if the nozzle is removed.) The distance from the shaft to the end of this rod should be checked for all nozzles, as shown in Figure 5–25(a). This verifies the PCD. The axial alignment of the nozzles can be checked by fitting a magnetic base to the shaft with a pointer on it. This should be lined up with the end of one spear rod. If the shaft is now rotated, this should line up exactly with the ends of all the other spear rods (Figure 5–25(b). This pointer is then where the centreline of the runner should be.

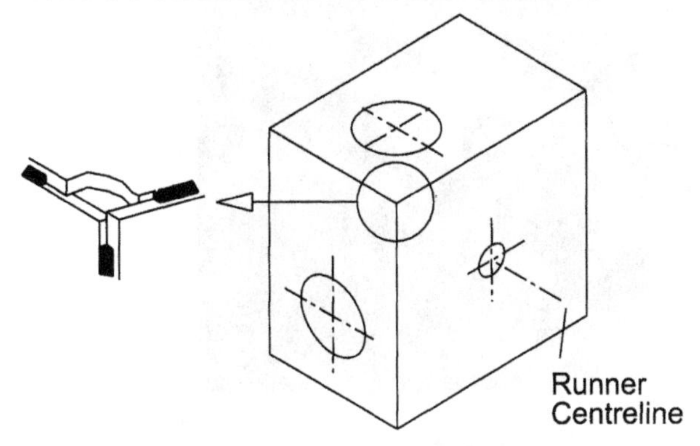

Fig. 5–24: Edge welding for turbine housings. Detail shows top plate cut away for clarity

If the design has nozzles instead of spear valves, alignment is not quite so easy. The best way is to put cross-hairs over both ends of the nozzle (Figure 5–26). Site between these on to a steel rule placed on the runner shaft to check the turbine PCD. To check the axial alignment, either fit a pointer to the shaft (as for spear valve alignment), and site on to this, or fit the runner in place and site on to the splitter ridge ends.

When it is absolutely certain that the alignment is correct, the housing can be fully welded. Be careful that the tacks are strong enough to hold the casing in shape, and weld alternately in small areas on opposite sides of the housing to prevent heat build up in one zone causing distortion.

For larger casings, it is a good idea to go around the casing hitting it with a hammer. Try especially in the middle of large, unsupported plates. If there is a resounding booming noise, then the casing is not stiff enough, and some extra stiffening pieces need to be added to the outside in the noisy areas. If this is not done, the turbine can be very noisy in operation. If self-aligning bearings are not being used (see the discussion in Section 4.7.1) the bearing housing faces will need to be machined after welding.

5.6 Assembly

After fabrication, the turbine should be assembled, together with the baseframe, generator and any drive system, in the workshop before transportation to site. Check that the runner can be fitted and removed smoothly – if the runner is a press fit on the shaft it may be necessary to use a puller or make a special tool for this. Check the nozzle alignment by eye and by simple measurements; if there is any doubt that it is correct, measure it again using the methods described in Section 5.5. For small runners it is particularly important that the centreline of the jet lines up with the splitter ridge, and the alignment should be ±1mm or better. Make sure the runner turns smoothly, and listen to the bearings to make sure there are no noises as they turn. Make sure that the deflectors operate correctly and that they do not touch the runner. Fit every component, every seal, nut, bolt and gasket, and then pack these away as the turbine is dismantled for shipping to site. This ensures that nothing is missing when the turbine is installed on site.

One area of the assembly that requires particular attention is the bearings, and for this reason the following section gives some detailed instructions for their assembly. If taper sleeves are being used to locate the bearings on the shafts, special care needs to be taken in their adjustment, and this is covered separately.

5.6.1 General bearing assembly

The following instructions apply to all rolling element bearings. Bearings are delicate, precision-made pieces of machinery, and they need to be handled carefully. Small amounts of dirt or incorrect fitting can drastically reduce their life. Try and work in a clean area if possible, or otherwise clean the working area and all the mating components before beginning. Any protective coating on new bearings can be removed with a little paraffin on a rag. If old bearings are being re-used they must first be thoroughly cleaned with paraffin and dried.

Unless the bearings are the sealed-for-life kind, grease should be packed into the spaces inside the bearing before it is fitted. The free space within the bearing should be completely filled, and the housing filled 30–50% with clean, fresh grease. Do not overfill, or the bearing will run hot.

(a) Checking the PCD

(b) Checking the axial alignment

Fig. 5–25: Checking the alignment of the spear valves with the runner (NHE, Nepal)

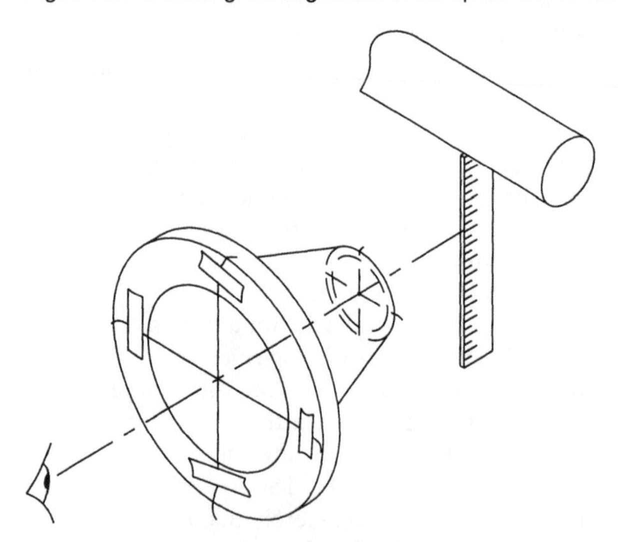
Fig. 5–26: Checking the alignment of the nozzles using cross hairs

Never hit or press a bearing in such a way that the force is transmitted though the ball or rollers. If the bearing is being fitted to the shaft, only press or hit the inner race. If the bearing is being fitted to the housing, only hit the outer race. If the rolling elements take the assembly force, they will become damaged and quickly fail. This instruction also applies to fitting a runner on to a generator shaft. Do not hammer it into place, because the forces are taken by the bearings inside the generator. Use the thread in the end of the shaft to pull the hub into place.

Check that any seals required are in place, and undamaged. Plummer block housings often use felt seals. It is recommended that felt seals are soaked in hot oil for a few minutes before they are fitted. If this is not possible at site, they should at least be soaked in cold oil or grease for an hour or so.

After assembly, rotate the shaft to check how it feels. It should rotate freely. Adjust it again if it feels stiff or if there are rough patches as the bearing is turned.

5.6.2 Fitting taper-sleeve bearings to the shaft

With taper sleeves, the internal clearances in the bearing are set by how hard the bearing is pushed on to the taper. If the bearing is driven too far on to the taper, the inner race expands and pushes the rollers on to the outer race. The bearing becomes difficult to turn, gets hot when it is run, and fails quickly.

Note that a bearing with a taper sleeve can be fitted either way round on the shaft. Fitting it with the nut facing inwards on the shaft can make it much easier to knock the bearing off the taper for removal. If the sleeve is fitted with the nut on the outside (which seems the most natural way round) then a bearing puller may be required to get it off.

There are two basic ways of setting the bearings. One way is to push the bearing on to the taper by a measured amount. This is straightforward and quick, but not very accurate. A better method is to measure the internal clearance (the gap between the rolling elements and the races) in the free bearing, and then to tighten it on to the taper until the clearance reduces by a measured amount.

Method 1: measuring the axial displacement

Loosely fit the taper sleeve on the shaft, and push the bearing on to the sleeve by hand until it just sits tightly on the tapers. Fit the shaft loosely to the housing and check that everything is correctly positioned. With everything in the correct place, fit the lock nut and locking washer. Measure the position of the bearing relative to the sleeve (x in Figure 5–27) and tighten the nut to push the bearing up the taper. It should move up the taper by the 'drive-up' distance given in Table 5–3, i.e. drive up = $x - x'$. When correct, bend the locking washer tags over to fix the nut.

Table 5–3: Axial drive-up and radial internal clearance setting dimensions for spherical roller bearings on 1:12 (diameter) taper sleeves

Shaft diameter d – (mm)		*Axial drive-up* $x - x'$ (mm)		*Reduction in radial internal clearance* (μm)		*Min. radial clear.* (μm)
over	incl.	min.	max.	min.	max.	min.
40	50	0.40	0.45	25	30	20
50	65	0.45	0.60	30	40	25
65	80	0.60	0.75	40	50	25
80	100	0.70	0.90	45	60	35
100	120	0.75	1.10	50	70	50
120	140	1.10	1.40	65	90	55
140	160	1.20	1.60	75	100	55

Method 2: Setting the internal clearance

It is more accurate to set bearings by measuring the internal clearances between the rollers and the races. This method should be used whenever possible and is illustrated in Figure 5–28. A set of feeler gauges with a minimum thickness of 0.03mm (30μm) is required.

Before fitting the bearing, measure the internal clearance of the bearing with a feeler gauge. For taper-roller bearings this is done by first rotating the bearing a few times to settle the rollers, and pushing the top roller back until it touches the guide ring in the middle of the bearing. Now put the feeler gauge between this roller and the outer race. Measure both sides of the bearing, and read off the 'Reduction in Internal Radial Clearance'

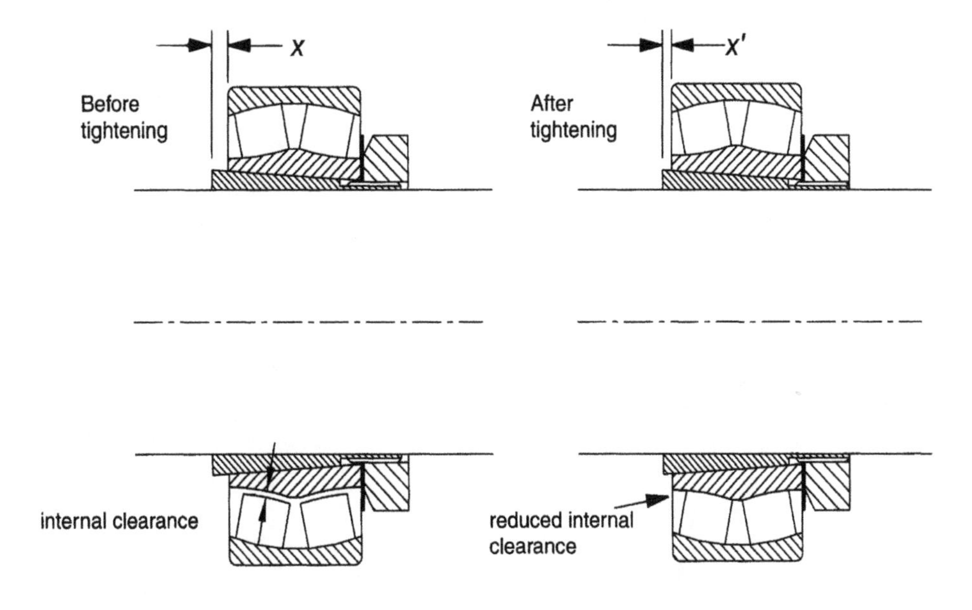

Fig. 5–27: Setting a taper-roller bearing using axial drive-up

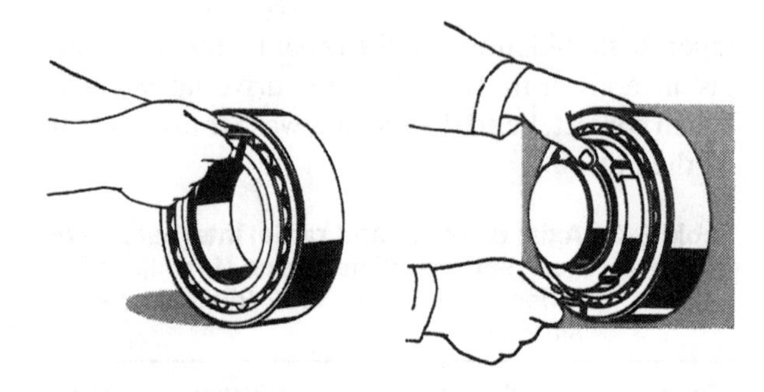

Fig. 5–28: Setting a taper-roller bearing by measuring internal clearance (© SKF)

from manufacturers' information or Table 5–3. Fit the bearing, and tighten the nut, measuring the internal clearance periodically. The clearance should reduce by the *Reduction in Internal Radial Clearance* given in Table 5–3. If it is outside the limits given, adjust again. As a final check, the installed internal clearance should not be less than the 'Minimum Radial Clearance' given in Table 5–3. For example, if the shaft diameter is 80mm nominal, the reduction in internal clearance is 0.040–0.050mm. Suppose the internal clearance in the bearing is measured at 0.080mm before it is installed. After tightening on the taper the clearance should be between (0.080 – 0.040) = 0.040mm maximum, and (0.080 – 0.050) = 0.03mm minimum.

5.7 Packing and transport

When the turbine is complete, it needs to be packaged ready for transport to site. Unless it is being delivered personally, it is a good idea to make sturdy wooden cases for each part. If the housing is too large for this, a wooden pallet or frame around the base stops the open end from being damaged. If the runner is being sent separately, it should be well wrapped and put in a wooden box to prevent bucket ends being chipped off.

As was mentioned in Section 4.7.1, turbine bearings can be damaged during transportation, particularly if a heavy turbine is taken on a rough road in a vehicle with poor suspension. If this is likely to happen, it is better to assemble the bearings to the turbine at site, after transportation.

6

INSTALLATION, COMMISSIONING AND TESTING

6.1 Installation

Installing the turbine is just one part of the construction of the power plant, and for large plants, the civil work is usually completed before the turbine is put in place. With micro-hydro plants, the procedure can be reversed, with the turbine being installed first and the rest of the scheme built up around it. The choice of order depends on the site conditions, and on personal preference. Starting with the turbine, the penstock can be erected from the bottom upward, knowing that everything fits together. However, for plants with a lot of civil work to do, work may be needed on the intake, headrace, penstock and powerhouse at the same time to complete the project on schedule.

A convenient way of fixing the turbine is to cast it in place. The tailrace and turbine pit need to be dug out first, and the area under the turbine should be dug down to solid, well compacted earth or rock. On top of this should be laid rubble or stones to a depth of 50–100mm, and these should be well compacted into the soil beneath. Any formwork required to shape the sides of the turbine foundation or the turbine pit is put into place. The baseframe is stood on piles of bricks or stones, in the correct position – which will be 100mm or so above the prepared level. The turbine housing and manifold are assembled to the baseframe, and joined to the penstock, if this is in place. Anchor bolts are fitted into their holes, and supported in the correct orientation. After checking that everything is correctly aligned, concrete can be poured in to make the foundation block. This flows around the baseframe and the anchor bolts, fixing these in the correct positions as it sets. A reasonable mix for this concrete is:

Cement	1 part
Coarse (sharp) sand	2–3 parts
Ballast: gravel or crushed stone, 6–25mm	4–6 parts

Leave the concrete for several days to achieve full strength before doing further work on the turbine. It is possible to make the foundation block first, complete with the turbine pit and tailrace in it, and then drill into this to fix the baseframe. This is a little more time consuming, and means that the concrete work must be made accurately.

At an appropriate point, the runner, spear valve, and the rest of the equipment can be assembled to the housing. Take great care when assembling the bearings (see Section 5.6.1), and make sure the runner is aligned correctly. The turbine housing will almost certainly need to be opened a few times during the commissioning procedure, so the final touching up of paintwork should be left till everything is working. When the runner has been proved satisfactory, the exposed parts of the shaft and hub should be painted to prevent corrosion and corrosion fatigue. A bitumen paint or two coats of red-lead paint should be adequate, unless some proprietary system is being used (see Section 4.8.4). Repeated opening of the joints may damage the gaskets and seals, and it is a good idea to replace the gaskets on access panels when the turbine is assembled finally.

6.2 Commissioning

The primary consideration for commissioning is safety. For turbines of a few kilowatts, the dangers are not great, but with high heads and larger powers, there is the risk of injury to the operators and serious damage to the plant if there is a major failure. The procedure should be thought through in advance, and the commissioning should be done in steps. The safety features are checked first.

The notes in Section 6.2.1 list possible basic steps when starting a new plant. They assume that deflectors are fitted, and the procedure will be somewhat different if there are no deflectors. Other changes may be necessary for differing designs, and extra tests should be added for equipment not covered in this book. Testing cannot be done until the intake, headrace, forebay and penstock are complete, for obvious reasons. All these works should be thoroughly inspected before commissioning the turbine.

6.2.1 Commissioning procedure

1. Check the turbine to make sure it is complete. Typical checks are:
   - Spear valve seals are in place and adjusted correctly.
   - Spear valves and shut-off valves are closed tight.

- ○ Shaft seals/flinger seals are fitted.
- ○ Deflectors are fitted, and sit in the correct position in relation to the nozzles.
- ○ All bolts are fitted and correctly tightened.
- ○ Visual inspection of all welds has been completed.

2. Check the driven machinery. For electrical systems, check the power circuit wiring and the safety circuits.
3. Fill the penstock, purge the air, and check for leaks. Make sure that water does not come out of the nozzles (look in the tailrace).
4. Check that the deflector mechanism works, putting the deflectors in front of the nozzles when the safety circuits trigger.
5. With deflectors shutting off the jets, slowly open the valve on one jet a small amount. (If a shut-off and spear valve are fitted, fully open the shut-off valve first, then open the spear valve.) The runner should not turn at all. If it does, the deflectors need to be adjusted.
6. Continue opening the valve until it is fully open. The runner still should not turn. Make sure there are no leaks from the housing, and especially from the shaft seals.
7. Shut the valve.
8. Repeat steps 5. to 7. for each jet in turn.
9. Prepare the generator or other driven machinery for operation.
10. With the deflectors covering the nozzles, open one jet a small amount. Move the deflector a little so that the turbine starts turning. Check that everything turns smoothly, and that there are no leaks.
11. Move the deflector further until the generator excites, or the driven machinery runs correctly. Check the speed, voltage, load, etc. Put the deflectors into their fully open position, and see that they are held there by the safety system. Listen to the bearings with a rod pressed between your ear and the bearing housing (be especially careful not to get you clothing, hair or fingers caught by the moving shaft). They should run with a deep 'purring' noise. Operate the deflector trip, and check the turbine comes to a halt.
12. Start the turbine again under reduced flow, and slowly increase the flow to the design opening. Check that the power is as expected.
13. Shut all the valves.
14. Repeat steps 10. to 13. for each jet in turn.
15. Start the turbine with one jet, and add the other jets in turn until full flow and power is achieved. Check the speed and power. Trip the deflectors at full power, and see that the turbine stops.
16. Shut all the valves.
17. Run the turbine at full power for a few hours. Monitor the bearings to make sure that they do not overheat. Normal grease can only function up to 60°C, which is about the highest temperature which can be touched without getting burnt.

A good test is to run the turbine slightly over and under the rated speed and measure the power. If an ELC is fitted this can be done by adjusting the frequency control. If the turbine is designed to run at optimum speed, the power at higher or lower speeds should be less than the power at optimum speed. If it is not, it means the turbine is operating on the up or down side of the efficiency curve. There is a mismatch between the turbine speed and the generator. If this is severe, it will lead to reduced power, and it is worth finding the cause.

The turbine should be watched carefully for the first few days of operation, and preferably should only be run when an operator is present.

7
MAINTENANCE AND PROBLEM SOLVING

7.1 Maintenance

One advantage of Pelton turbines is that they require very little maintenance. The only regular servicing is to re-grease the bearings. Apart from this, there should be a regular running inspection, plus a major periodic inspection to check for wear or damage.

7.1.1 Regular inspection

The operators should keep an eye on the equipment when working, but it is good practice to inspect it thorough every week or so.

- Look for leaks. Leaks on the manifold should be dealt with urgently as they can be a sign of incipient failure. Similarly, leaks that are letting water get on to or near the bearings or electrical equipment need to be sealed before this equipment gets damaged. It may be possible to tolerate other leaks until the turbine is next stripped down, though it is good practice to correct them straight away.
- Check that all bolts are secure and tight.
- Check the bearings. With the turbine running, listen by placing one end of a screwdriver or a piece of wood on the bearing casing and the other on your ear. The bearing should run with a deep 'purring' noise. If you can hear whistling or screeching noises, the bearing needs to be dismantled and checked. Check that the seals are working, and no grease is leaking out. Make sure no water is getting into the bearings.
- Check the spear valves and shut-off valves. These should open smoothly, without any unusual noises. When shut, there should be no flow in the tailrace, and the runner should stop even if the deflectors are in the running position.
- Check the operation of the deflectors and the safety circuitry associated with them. Make sure all the levers move freely, and that the deflectors stop the turbine when operated.

7.1.2 Bearings

Most micro-hydro turbines use grease-lubricated bearings. If sealed-for-life units are being used, no maintenance is possible. Most bearings require periodic re-greasing, and this is covered below. Some machinery uses oil lubrication, but the maintenance of this depends on the system; refer to the manufacturer's literature if possible, or to a specialist publication such as the *SKF General Catalogue* (1994, or the current equivalent).

Use a good quality grease of the same variety as the initial fill (Section 4.7.5). Cheap greases will degrade and need changing more frequently. Try not to mix different types of grease. It is usually possible to add grease through a grease nipple. If too much grease is put in, the bearings will run hot. As a rough guide, the bearings should be relubricated every 500 hours of operation with a mass of grease:

$$m_{grease} = 0.005 \times D.B. \qquad \textit{Eq. 7-1}$$

m_{grease} – amount of grease to be added for relubrication (g)
D – bearing outside diameter (mm)
B – bearing width (mm)

This is a conservative lubrication interval for micro-hydro-sized bearings with water present; small bearings may be able to go much longer between lubrications. More accurate intervals can be found in bearing manufacturers' literature, such as the *SKF General Catalogue* and the *SKF Bearing Maintenance Handbook*. However, it is not very convenient to tell an operator to put in so many grammes of grease. A better method is to weigh how much grease comes out of a particular grease gun each time it is squeezed, and calculate how many squeezes are needed. (If manufacturers' literature tells you the volume of grease to add, the same method can be used, based on an average density of grease of $0.87 g/cm^3$.)

If the bearing housings have to be opened up to add grease, this is a good chance to inspect the bearings. If there is any visible damage, a new bearing should be fitted. Check the grease too.

If the grease:

- is excessively dirty
- has metal particles in it
- has gone solid
- has water in it

then it should be cleaned out. Brush down the housing and the bearing with paraffin to clean them, and dry everything thoroughly. Push grease into the space inside the bearings as much as possible, and fill the housing about half full. (Manufacturers recommend the housing is filled 30–50% with grease, but because it is difficult to fill the inside of the bearing itself, this procedure leads to a true fill of around 40%.) Check the seals, and re-assemble.

Note that, when removing a bearing, never apply force through the rolling elements. That is, never hit the outer race to get the inner race off a shaft, or hit the inner race to remove the outer race from its housing. Similarly, a bearing puller should only be used directly on the race being removed. If force is applied through the rollers it will damage the bearing surfaces, and may lead to early failure. If (because of a poorly designed arrangement) the rolling elements have to be loaded to remove the bearing, replace it with a new one.

7.1.3 Major inspection

At periodic intervals, the turbine should be shut down, opened up and thoroughly inspected. This could be done once a year, or more frequently if it is felt necessary.

The runner, nozzles and spears should be inspected periodically for signs of erosion. This is wear caused by silt particles suspended in the water. The first signs are usually on the splitter ridges, which lose their sharp edges. A badly-rounded splitter can lead to a loss of efficiency of a few per cent. More serious abrasion leads to a wavy appearance on the inside of the bucket, that looks like ripples on the sand of a beach after the tide has gone out. In very serious cases, the erosion eats into the corners of the curves in the buckets, and reduces the outlet angle, so that water leaving one bucket starts to hit the back of the next bucket. This sort of erosion is a sign of relatively coarse, abrasive particles in the water, and the de-silting arrangement needs to be improved. Mild abrasion can be rectified by grinding out the buckets till they are smooth. Bad areas can be built up with weld (provided the bucket material is weldable) and ground back to shape.

Wear in the spear valve usually only occurs at very high heads. This is caused by fine abrasive particles, which give a pitted appearance to the surface of the spear, and eats away the tip. This is not a common problem at micro-hydro heads, but if it does occur it is difficult to overcome. The spear valve surface should be hardened, and the spear tip will need to be replaced periodically.

Check the housing to see that it is not rusting. If areas of paint have been worn away by water inside the housing, touch these up. Touch up the paint on the runner and shaft too (Section 4.8.4).

7.2 Problem solving

There is the tense moment when the turbine is first run, with the fear that something might break or go wrong. It is an intense relief when it starts, and runs smoothly. If the instructions in this manual have been followed, there is every reason to expect the turbine will work. There are times, however, when it does not. The following sections attempt to analyse various symptoms, to give tests that can be carried out, and to suggest remedies.

Finding the cause of a turbine problem usually requires some detective work. It is not possible to see clearly what is going on inside a Pelton, so one must look for clues. What are the symptoms? When are they present? Is anything else unusual happening? The discussion below lists most possible problems, and should help to give a diagnosis.

When a problem is discovered, do not rush to make major changes. It pays to analyse carefully what is happening, and to make as many measurements as possible. Look inside the turbine and see if anything is obviously wrong. Check the alignment. Look too at the paint on the inside of the housing to see if it has been worn away; this can say a lot about where the water is going. Having formulated a theory for what is wrong, see if there are any further tests to prove whether this is correct. Changing the runner, re-aligning the jets or increasing the manifold size are big jobs, and can be expensive. It is best to be sure about what is going on before acting. Table 7–1 summarizes the potential problems and their solutions.

7.2.1 Low output power

Power is usually lost in a Pelton because the water is not doing what is expected. There can be excessive losses in the penstock or or the manifold; the nozzle or spear valve can fail to produce a good jet; the runner can fail to utilize the jet; the exhaust water or tailrace can interfere with the runner. Note the relative magnitude of the various effects. Only a few problems give very large power losses; most give a power reduction of a few per cent. If the power is well below what is expected, concentrate first on the major possible causes: pipework losses, or the jet missing the runner.

Penstock or manifold problems

Penstock or manifold losses manifest themselves as normal power at small flows, but low power at higher flows. For example, in a two-jet Pelton, the first jet might produce the design power, but two jets together produce hardly any more power. If this is occurring, check that the effect is the same whichever order the jets are opened. If it is, the problem is almost certainly losses in the penstock or the manifold. What is happening is that at the lower flows the velocity in the pipework is small, so the losses are small, but at higher flows the losses are becoming excessive. This can give a major loss of power. The only real solution is to increase the size of the offending pipework. A small change in diameter can make a huge difference, since losses are roughly proportional to d^5 (where d = diameter). Check the

Table 7–1: Summary of problems and solutions

Problem	*Possible causes*	*Additional diagnostic tests*	*Remedy*
Low power	Blocked nozzle	○ Listen for hissing or screeching noise	Un-block; Section 7.2.1 – Jet problems
	High losses in pipework	○ Only occurs at higher flows ○ Check loss calculations ○ Measure pressure in pipework	Increase pipe size Altering pulley ratio may improve power; Section 7.2.1 – Penstock or manifold problems
	Speed mismatch	○ Plot power against speed ○ Check manifold losses	Adjust drive ratio; Section 7.2.1 – Runner problems
	Jet misaligned	○ Check alignment ○ Vibration on casing opposite	Adjust runner position; Section 7.2.1 – Runner problems
	Worn nozzle	○ Inspect	Renew; Section 7.2.1 – Jet problems
	Eccentric spear	○ Inspect	Replace bearings, replace spear or re-machine body; Section 7.2.1 – Jet problems
	Interference between jets	○ Put fixed cover over jets	Correct jet/runner alignment, or get better bucket design; Section 7.2.1 – Runner problems
	Tailrace water touching runner	○ Check tailrace level	Un-block tailrace or increase channel size; Section 7.2.1 – Exhaust water problems
	Casing flooding	○ Check inside of casing	Remove obstructions in casing, or increase size; Section 7.2.1 – Exhaust water problems
	Swirl in manifold	○ View jet ○ Add temporary vanes in nozzle pipe	Put straightening vanes before nozzle; Section 7.2.1 – Jet problems
	Turbulence in manifold	○ Put pressure gauge on manifold	Redesign; Section 7.2.1 – Penstock or manifold problems
	Poor bucket finish	○ Inspect	Grind and polish; Section 7.2.1 – Runner problems
	Air in penstock	○ See if running is intermittent ○ Check inlet for sucking air ○ Check joints/valves for sucking	Submerge inlet deeper, or realign penstock; Section 7.2.1 – Penstock or manifold problems
Noise and vibration	Bearing failure	○ Listen to bearings	Change bearings; Section 7.2.2
	Flinger seals touching	○ Turn runner by hand and listen	Adjust clearance; Section 7.2.2
	Out of balance runner or pulleys	○ Stop the turbine! ○ Check for broken buckets etc.	Balance; Section 7.2.2
	Damaged driven machinery	○ Disconnect drive belt	Repair; Section 7.2.2 & Section 7.2.1 – Runner problems
	Jets missing the runner	○ Feel opposite the jets ○ Check paint inside casing	Correct runner alignment; Section 7.2.2
	Large, thin housing plates	○ Feel for vibration	Reinforce unsupported plates; Section 7.2.2

calculations for the penstock and manifold losses to see if these are correct (see Section 4.10.2). The losses can be located by attaching a pressure gauge at various positions along the pipework, seeing where the pressure losses occur, and comparing the measured figures with the calculated design losses. (Remember that the pressure at any point in the system is equal to the static head minus the head loss down the pipework to that point.) If there are excessive losses along the penstock, it may be a big problem, because it could mean the whole of the penstock is undersized. However, the pressure drop may be traced to a small section of the manifold, which can be replaced. Another possibility is that a sharp bend or junction is causing losses, though this rarely gives a significant drop in power.

If losses are causing the problem, but it is too expensive to contemplate changing the pipework, another change can make some improvement. Losses reduce the head at the nozzle, making the jet speed lower. This shifts the turbine power curve to the left (and down), making the optimum speed lower, as shown in Figure 7–1. Trying to run the turbine at the design speed now means operating on the steeply-sloping right side of the power-speed curve. By changing the pulley ratio of the drive system to reduce the operating speed, the turbine can be matched to the actual optimum speed. This is not ideal, but it makes the situation better.

A different type of pipe problem occurs when air gets into the system. This can result in the turbine starting and stopping, often at quite regular intervals. Air can enter the system by being sucked into the penstock inlet. This occurs when the pipe is too close to the surface: a vortex can form that pulls air into the pipe. To avoid this, as a rule of thumb, the top of the pipe needs to be at a depth of at least 1.5 times the velocity head in the penstock ($V^2/2g$).

Air can also leak into the penstock if the slope of the pipe is not sufficient. This can create negative pressure at some points, and air may be sucked in through air vent valves or poor gaskets. This problem is not common with penstocks (it is in headrace pipes), but it might occur if the penstock goes up and down, or if there is a long, shallow-sloping section at the beginning of the penstock. The solution is to ensure that the pipe always lies below the hydraulic gradient line; the theory is beyond this brief discussion, but can be found in books on water supply systems, such as Jordan (1996).

Jet problems

A Pelton jet should be smooth and cylindrical, and if the spear valve or nozzle is carefully designed and made, it can be. Micro-hydro jets are often a little less perfect, and may be a bit ragged, but they should still stick together till they hit the runner. A dispersed jet leads to much reduced power.

The most common jet problem is blockage. Leaves, twigs, stones, pieces of concrete, all can sometimes get into the nozzles. Even when there is a good trashrack in place, debris that got into the penstock during construction can make its way down to the turbine. Blockages are very common in the first few days of operation. A large stone in a spear valve can reduce the power by a half. A partially blocked nozzle often makes a distinctive hissing or screeching noise, especially for high heads, and this helps to identify the problem. Small stones in a spear valve can sometimes be removed simply by fully opening the valve with the water flowing. If this does not work, and a blockage is suspected, open up the nozzle and see. If shut-off valves are

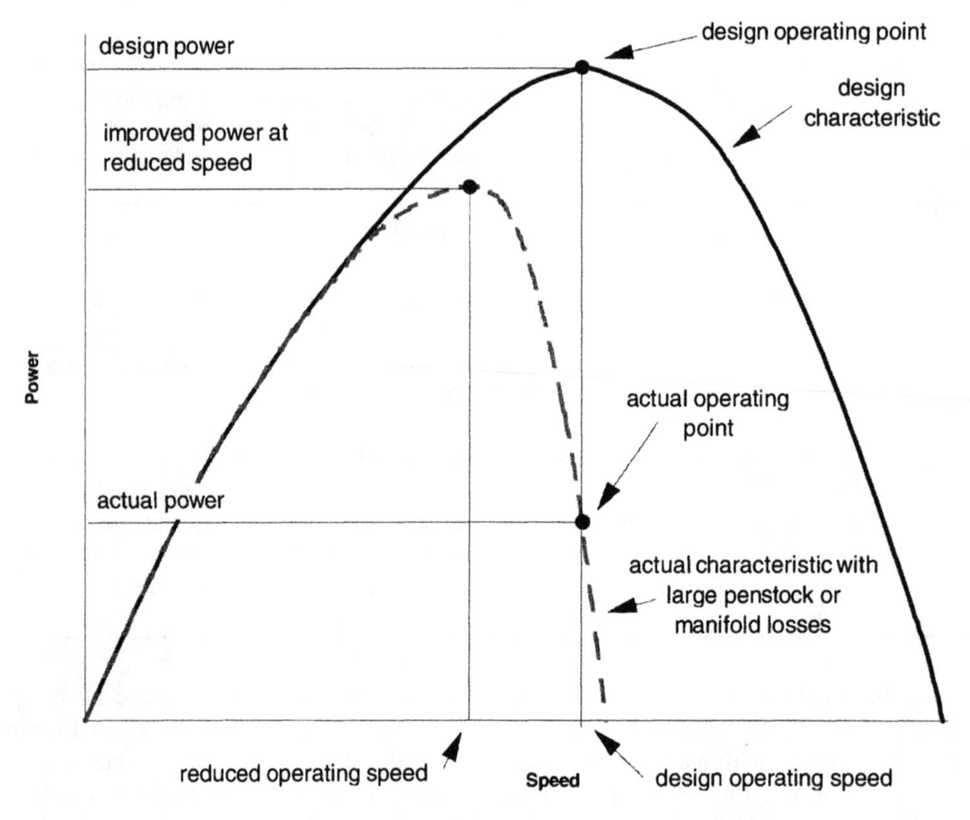

Fig. 7–1: Changing the operating speed to improve the power output of a turbine with excessive manifold or penstock losses

fitted, this can be done without having to drain the penstock, though it is always safer to do so.

A spear valve can give a poor jet if the spear is eccentric to the nozzle. This can happen if the shaft gets bent. It is normally noticeable because it would not seal very well when shut. It is also possible for the spear bearings to wear so that there is excessive free play, and the spear moves off centre as it is pulled back from the nozzle. Check for this by emptying the spear valve and inspecting.

A worn nozzle can give a poor jet. The edge of the nozzle should be sharp. It can be damaged during assembly, or it can be worn away by the water over time. Nozzles made as a simple, sharp-edged hole in a plate are especially prone to wear. Damage or wear can be checked by simple visual inspection. Mild wear might lead to a power loss of a few per cent, and severe wear might typically give a power reductions of 10–20%.

Another problem is turbulence or swirl. If there is excessive swirl at the inlet to the nozzle this can lead to conical spray rather than a coherent jet. This is discussed in Section 4.3.3 and Section 4.10.1. This is not a common problem, as the swirl has to be severe to be noticeable, but it is possible. One way of checking this is to look at the jet: if it is coming out as a smooth cylinder, it is good. In practice, it is very difficult to view the jet. If an access panel is removed in order to look inside a turbine, anyone using it will be confronted by a dense fog of spray that is likely to soak them. In some circumstances it may be possible to discharge the jet into the open to see it. If swirl is the problem suspected, the simplest test is to put some flow straightening vanes just upstream of the nozzle. These can be just tack-welded in for a test.

Badly designed manifolds can lead to large-scale eddies in the pipework, which again disturb the jet. This would lead to irregular running, and oscillating power. This is, thankfully, rather uncommon. The solution is to improve the bends or junctions, making the transitions more gradual and rounding off the internal corners as described in Section 4.10.

Runner problems

If the jet is coherent and at the correct speed, the problem could be with the runner. The most likely problem is that the jet is not hitting the runner in the correct place. If the jet is pointing too far out, water will miss the runner and hit the casing. The power falls off quickly as the jet is moved outwards. This will be noisy, and it may be possible to feel the vibration on the plates opposite the jet. Look inside the casing and see if the paint has been blasted off in the area opposite the jets. If the jet is too far in, the water will collect at the inner ends of the buckets and choke the flow. Small inward displacement changes the power very little, but if the effective jet PCD is much less than the runner PCD the losses become noticeable. If the jet is displaced sideways, it may cause the flow around that side of the buckets to disintegrate. If misalignment is suspected, check it as described in Section 5.5. Minor misalignment might reduce the power by 5%. Note that it is often much easier to adjust the alignment by moving the runner on its shaft or shifting the runner shaft bearings than it is to alter the angle or position of a nozzle. If there is a large distance between the nozzle and the runner, the jet may have diverged considerably before it hits the buckets, giving reduced power. There can be room for a deflector, but otherwise the nozzle should be as close to the runner as possible. However, unless the nozzle is ridiculously far from the runner, the power will only be improved a few per cent by moving it closer.

There can be a mismatch between actual runner optimum speed and the design speed. This has been mentioned in regard to runner losses (Figure 7–1), but it can occur from calculation errors. Instead of working at the peak efficiency, the turbine may be well to one or other side of it, and the power may easily be 10–20% less than designed. To check this, run the turbine a little over and under speed, and measure the power. If the power decreases either side of the working speed, then it is working at the best point. If the power increases sharply as the speed increases or decreases, it is running in the wrong place. The gear ratio of the drive system needs to be changed. If the runner is directly coupled to the generator, it is difficult to get over this problem.

A good runner should take the energy out of all the jet water. If there is a fault with the notch design, or if the alignment is wrong, some water may pass through the runner, being deflected slightly, but still maintaining its speed. If this water hits another jet, it can 'explode' it so that it gives little or no power. Consider, for example, a horizontal axis, two-jet turbine, as shown in Figure 1–12. Suppose that each nozzle works fine by itself, but when both jets are on together, the power is little more than that of one jet. This could be a manifold loss problem, as described above, but might be because water from the upper jet is hitting the lower jet. A way to test for this is to put a cover over the lower jet, projecting out as far as possible. (These covers are a good idea anyway, as they protect the jet from droplets thrown off the runner.) The cover will significantly improve the power, but longer term one should find out what is wrong with the bucket design.

A badly finished runner will have low efficiency. A rough surface finish on the inside of the bucket, blunt splitter ridges, mis-shapen notch areas, poor alignment of the splitter ridges; each of these will reduce the efficiency by a few per cent. Quality control during manufacture is important. Once a bad runner reaches a site, some rectification can be done by grinding, but it is really rather late. A word of caution, though: do not expect too much from a micro-hydro runner. While large power-stations may get over 90% efficiency from Pelton runners, 85% is an excellent figure for micro-hydro, and 80% is good. For the smallest Pelton runners, with PCDs of 100–150mm, then 70–75% is quite reasonable (Baines & Williams, 2000). Poorly made runners can have efficiencies as low as 55–60%.

Exhaust water problems

The runner exhaust water very rarely causes major problems, but when it does it can virtually stop a turbine. The worst case is when the tailrace capacity is insufficient, and the water rises up to touch the runner. The 'windage' losses become enormous, as the runner has to paddle through the water. Check that the tailrace is not blocked. Look up the tailrace when the turbine is working, if possible, and see how full it is.

Poor internal casing design can lead to a large amount of water dropping back on to the runner or the jets. This is hard to diagnose, as it is very difficult to see into a casing when the turbine is working, even if a see-through panel is incorporated in it. Make sure that there are not obvious obstructions on the inside, and that the water has paths to flow out down the walls. This sort of problem tends to reduce the efficiency by a few per cent.

Driven machinery

Low power is almost never due to the driven equipment, because if mechanical power is being lost somewhere, something would soon get very hot. Suppose 8kW of electrical power is expected from a generating set, but only 6kW is realised. It is tempting to suspect that the drivebelt is slipping, that there is friction in the bearings, or that the generator is inefficient. But this would mean that at least 2kW of power is getting lost. A 2kW input into the bearing, into a V-belt drive, or into the generator, would quickly make the item overheat: the bearing would start smoking, the V-belt would burn and snap, or the generator would get too hot to touch. It is not possible to run a small system for long with 2kW of heat being lost – something will soon break.

7.2.2 Noise and vibration

All turbines make some noise. There are two main sources: water movement and mechanical vibration. A well-designed turbine should run smoothly, but will make a noise varying from a hum to a quiet roar. Spear valves and shut-off valves will make a screeching noise when just cracked open. Any unusual loud noise is an indication that something is wrong, and the turbine should be shut down immediately. Check that nothing is broken, or that nothing is hitting the runner. Turn the runner by hand and see if it is free.

Severe vibration can be caused by imbalance. This will not be obvious at low speeds, but can shake turbines apart at high speed. (It is more likely to be a problem for directly coupled turbines, which run at generator speeds of around 1500rpm.) The only solution is to balance the runner as in Section 5.1.4. If a bucket breaks off or a bolt falls out of the runner, this has the same effect, and can be very damaging if the turbine is not stopped.

Damaged bearings can give loud clanking or grinding noises. Listen to the bearing with a rod to the ear to find where the problem is occurring. If it is uncertain whether the failure is in the turbine or the generator, disconnect the drive belt and run the turbine briefly up to speed. Bearings do have a finite life, and so may need replacing occasionally. Check, though, that they have not failed because of lack of grease, or because water or dirt has got inside. Make sure the bearing seals are in good condition. If bearings are failing early or regularly for no apparent reason, check the life calculations using the method described in Section 4.7.3. The bearing may be undersized. Failure can also be induced by side loads, caused by the runner being offset from the centre of the jet.

If part of a jet is missing the runner, and hitting the casing, this can give a loud whining or roaring noise. Vibration may be felt on the panel opposite the jet. The solution is to adjust the runner so that it catches all the water – see Section 7.2.1, Runner problems. There is another normal water noise, from the exhaust water leaving the buckets and being thrown onto the side plates, and from droplets being flung out from the runner on to the top and end plates. If these plates are too thin, or have large, unsupported sections, the turbine housing amplifies the sound. This can be unpleasant. It may be possible to deaden the sound by reinforcing the vibrating plates on the outside.

A less serious noise comes from the flinger seals. If these are rubbing badly, they can make a fearsome sound. Check by turning the runner by hand, and adjust the flinger clearance if necessary.

7.2.3 Leaks

Micro-hydro Peltons often leak a little. Drips can come from the spear valve, from the joints in the casing, from the baseframe, or from the seals. It is good practice to mend leaks as soon as they occur. Leakage from the shaft seals may run on to the bearings or into the generator, and these must be stopped before damage occurs. Similarly, leakage from high pressure pipework should be attended to, in case it is a sign of impending failure.

Leaks can normally be cured by dismantling the offending parts, cleaning up the seal surfaces, and renewing the gasket or seals. Silicon sealant, which is available in tubes, can be used for added protection in awkward or irregular areas.

8
CONCLUSION

Since their invention over a hundred years ago, Pelton turbines have proved so successful that many competing turbine designs have disappeared completely. For very high heads, there is no other choice. Peltons may not be the most efficient turbines, but they are only fractionally behind the best, and they hold this efficiency over a wide range of flow. They are used extensively around the world, and have been running in many sites for generations. Single Pelton wheels are producing hundreds of megawatts, and yet small Peltons can be used for just a few watts. Few machines are so versatile.

The major competitors of the Pelton for large schemes, the Francis and Kaplan turbines, are rather difficult both to design and manufacture. This makes them expensive for small machines. Peltons do not suffer in this way. As has been seen in this manual, Peltons can be made with very basic workshop technology. The selection procedure is straightforward, and once a design has been completed for a given size, it can be used at a variety of different heads and flows. The complex shape is restricted to the runner, which is easily and repeatably made by casting, once a pattern has been produced. All the other components are straightforward to make. The end result is an inexpensive, robust, reliable turbine. By following the designs and instructions in this book, it should be possible to produce a working turbine. With experience, the reader will almost certainly want to modify the design or try different manufacturing techniques.

8.1 Further developments

The efficiency of modern large Pelton runners is such that very little futher improvement can be expected, and efficiency can only be increased by a fraction of a per cent at most. Development is focusing on increasing the head that can be used, on reducing erosion and cavitation, and reducing the interference between jets in multi-jet turbines. New materials are beings looked at, such as ceramic coatings for spear valves, and new production methods, such as weld deposition of buckets. Companies are using CAD systems to speed up the design process, and linking this to computer modelling to predict performance and stresses. Work is continuing on computer flow analysis, and it is hoped that there will soon be agreed methods of predicting full size turbine behaviour from scale model studies (Brekke, 1994; Grein & Keck, 1988; Grein, 1988).

All of these are interesting developments, but have limited application for micro-hydro. If computing power is available, the repetitive nature of Pelton designs lends itself to tailoring turbines to sites on CAD. Most improvements to micro-hydro are likely to come from introducing the standard practices of large turbines into small machines, as the technologies become more widely available.

Most beginners start by casting individual buckets, and machining these to fit them to a hub. As a foundry gains experience, it may be possible to change to single-piece runners, and this should lead to stronger, more reliable, and cheaper runners. With experience, again, the foundry may wish to try other designs of bucket, to improve the efficiency. Note that, while very large turbines achieve over 90% runner efficiency, scale effects mean that anything in excess of 85% is good for micro-hydro. Once a good shape, a consistent and accurate casting process, and a good surface finish have been achieved, there will be few returns from extra work spent on the runner.

Once the foundry can produce reliable Peltons, a next step is to make the design standard. It is possible to produce turbines for most sites by choosing a few standard runners, with sizes increasing in steps of, say 25mm, and only using a few standard spear valves (only the orifice size needs adjusting). This will reduce cost and lead times, and makes it possible to hold stock which can be assembled into a variety of turbines.

Hydro-power is a cheap, clean, renewable and versatile energy source, and Pelton turbines are an excellent way of exploiting this. I wish you well as you work in micro-hydro.

9

APPENDIX: STANDARD UNITS, FORMULAE AND NOTATION

9.1 Standard units

Equations in this book are presented in the metric or SI (Système International d'Unités) system of units. To use them, quantities should be converted to the base or derived units shown in the table below, and then the output will also be in these units. Some books specify different units for the quantities given in an equation, even when the book uses metric units. In such cases the units given have to be used, but the author will have included the necessary conversion factors within the equation.

In theory, basic equations may be used for any self-consistent system of units, provided that both the input and output quantities are in the base units for that system. So it is possible to use imperial units in the equations, provided one is consistent and only use base foot-pound-second units. However, in the author's opinion, the metric system is by far the easiest, and it is recommended that the reader uses it exclusively.

Table 9–1: Metric system of units

Quantity	*Unit*	*Symbol*	*Equivalent*
Base units			
Length	metre	m	–
Mass	kilogramme	kg	–
Time	second	s	–
Electric current	Ampere	A	–
Temperature	Kelvin	K	–
Luminous intensity	candela	cd	–
Supplementary units			
Plane angle	radian	rad	–
Solid angle	steradian	sr	–
Derived units			
Frequency	Hertz	Hz	1/s
Force	Newton	N	$kg.m/s^2$
Work/energy	Joule	J	N.m
Power	Watt	W	J/s
Electric potential	Volt	V	W/A
Electric capacitance	Farad	F	C/V
Electric resistance	Ohm	Ω	V/A
Magnetic flux	Weber	Wb	V.s
Magnetic flux density	Tesla	T	Wb/m^2
Inductance	Henry	H	Wb/A
Luminous flux	lumen	lm	cd.sr
Illumination	lux	lx	lm/m^2
Note also:			
Temperature	Celcius	°C	K

The interval of 1°C is the same as 1°K, but the zero point of the Kelvin system (Absolute Zero) is at –273.2°C. Despite K being the standard unit, Celcius is much more common.

9.2: Multiplication factors

The above units can be multiplied by the following factors to make the numbers more easy write. The factors should be removed before putting quantities into equations. So, 1 MW ≡ (is equivalent to) 1 000 000W, but only W will be used in equations.

Table 9–2: Multiplication factors

Prefix	*Symbol*	*Multiplication factor*	
Multiples			
Tera	T	1 000 000 000 000	$\equiv 10^{12}$
Giga	G	1 000 000 000	$\equiv 10^{9}$
Mega	M	1 000 000	$\equiv 10^{6}$
kilo	k	1000	$\equiv 10^{3}$
milli	m	1/1000	$\equiv 10^{-3}$
micro	μ	1/1 000 000	$\equiv 10^{-6}$
nano	n	1/1 000 000 000	$\equiv 10^{-9}$
pico	p	1/1 000 000 000 000	$\equiv 10^{-12}$
femto	f	1/1 000 000 000 000 000	$\equiv 10^{-15}$
atto	a	1/1 000 000 000 000 000 000	$\equiv 10^{-18}$
Sub-multiples			
hecto	h	100	$\equiv 10^{2}$
deca	da	10	$\equiv 10^{1}$
deci	d	1/10	$\equiv 10^{-1}$
centi	c	1/100	$\equiv 10^{-2}$

9.3 Other SI units

Table 9–3: Other SI units are also used by convention or for convenience

Quantity	*Unit*	*Symbol*	*Equivalent*
Volume	litre	l	10^{-3} m^3
Note that 1 litre ≡ 1 decimetre^3, dm^3			
Mass	tonne	t	1000kg
Pressure	Pascal	Pa	$\equiv N/m^2$
	bar	bar	10^5 N/m^2
Dynamic viscosity	centipoise	cP	10^{-3} $N.s/m^2$
Kinetic viscosity	centistoke	cSt	10^{-6} m^2/s
Electrical conductivity	Siemens	S	1/Ω

9.4 Basic formulae

The following two formulae are given as they are fundamental to all hydro-power derivations, and also to illustrate the use of correct units.

9.4.1 Turbine output power

The power output from a hydraulic system is:

$$Power = Pressure \times Flow \times Efficiency \qquad \text{Eq. 9–1}$$

Or:

$$P = p.Q.\eta_{total}$$
$$= (H_{gross} \times \rho_{water} \times g).Q.\eta_{total} \qquad Eq.\ 9\text{–}2$$

P	– power (W)
p	– pressure (N/m^2)
Q	– flow (m^3/s)
η_{total}	– total system efficiency (unit-less, fraction)
H_{gross}	– gross hydraulic head (m)
ρ_{water}	– density of water (kg/m^3)
g	– acceleration due to gravity (m/s^2)

This is true for any turbine. If we convert this to take power in kW and flow in litres/s, and substitute values for ρ_{water} and g it becomes:

$$1000.P = H_{gross}.\rho_{water}.g.\left(\frac{Q}{1000}\right).\eta_{total}$$
$$\therefore P = \frac{H_{gross}.\rho_{water}.g.Q.\eta_{total}}{10^6}$$
$$= \frac{H_{gross}.1000 \times 9 \cdot 8.Q.\eta_{total}}{10^6}$$
$$= \frac{9 \cdot 8.H_{gross}.Q.\eta_{total}}{1000} \qquad Eq.\ 9\text{–}3$$

where: P – (kW)
Q – (litres/s)
H_{gross} – (m)

This form is often found in books. It is quite valid, but it is not the root form of the formula, and only works with the specific units given for the quantities.

Below is given a form of the formula which is very easy to remember. It assumes a η_{total} of 51%, which is quite a good initial estimate for a micro-hydro scheme generating electricity.

$$P_{electrical} = 5H_{gross}.Q \qquad Eq.\ 9\text{–}4$$

P – (W)
Q – (litres/s)
H_{gross} – (m)

9.4.2 Power/torque relationship

For any shaft:

$$Power = Torque \times Angular\ Speed \qquad Eq.\ 9\text{–}5$$

or:

$$P = T.\omega \qquad Eq.\ 9\text{–}6$$

P – mechanical power (W)
T – torque (N.m)
ω – angular speed (rad/s)

This equation uses the standard SI units. Note that ω is in rad/s, not rpm. It is possible to convert this equation to use rpm:

$$P = T.\omega$$
$$= T.\,2\pi.\left(\frac{N}{60}\right)$$
$$= \frac{\pi.T.N}{30} \qquad Eq.\ 9\text{–}7$$

N – rotational speed (rpm)

Note the final equation has $\pi/30$ in it. The 2π converts revolutions to radians, and the 60 converts rpm to revs/s. So the unit conversion is included in the equation, and non-standard units can be used.

9.5 Unit conversion

Below are listed some conversion factors commonly used in hydro-power work. Some are between metric units, the rest are from the old British, Imperial, or inch, units to the SI system.

Table 9–3: Conversion factors

To convert *From*	*To*	*Multiply by*	*Notes*
Length			
in	mm	25.4	
in	m	0.0254	
ft	m	0.3048	foot
yard	m	0.9144	
Area			
mm^2	m^2	10^{-6}	square millimetre
cm^2	m^2	10^{-4}	square centimetre
in^2	m^2	645.2×10^{-6}	square inch
ft^2	m^2	0.0929	square foot
yard^2	m^2	0.8361	square yard
Volume			
litre	m^3	10^{-3}	
cm^3	m^3	10^{-6}	cubic centimetre, cc
mm^3	m^3	10^{-9}	cubic millimetre
in^3	m^3	16.39×10^{-6}	cubic inch, cu.in
ft^3	m^3	28.32×10^{-3}	cubic ft, cu.ft
yard^3	m^3	0.7646	cubic yard
pint	m^3	0.5683×10^{-3}	
gallon	m^3	4.5461×10^{-3}	Imperial gallon
US gall.	m^3	3.7851×10^{-3}	US gallon
Angle/angular speed			
deg	rad	$\pi/180$ or 0.01745	
rpm	rad/s	$\pi/30$ or 0.1047	
Mass			
tonne	kg	1000	metric tonne
lb	kg	0.4536	pound
cwt	kg	50.8	hundredweight (1 cwt ≡ 112 lb)
ton	kg	1016	Imperial ton
ton	t	1.016	So a metric tonne is nearly the same as an Imperial ton!
(1 ton = 2240 lb = 20 cwt)			
Force			
lbf	N	4.448	pound force (the force exerted by 1 lb mass under standard gravity).
tonf	N	9964	Imperial ton force

(continued over)

To convert From	To	Multiply by	Notes
Torque			
daN.m	N.m	10	
kN.m	N.m	1000	
N/mm	N.m	1000	
lbf.in	N.m	0.1130	
lbf.ft	N.m	1.3558	foot-pound (often incorrectly written simply as lb.ft)
Pressure/stress			
Pa	N/m^2	1	i.e. Pascal = N/m^2
bar	N/m^2	10^5	
kN/m^2	N/m^2	1000	
kgf/cm^2	N/m^2	98 066	standard gravity
m_{water}	N/m^2	9800	metres of water, i.e. the pressure exerted by a head of this many metres of water.
psi	N/m^2	6890	psi = lbf/in^2
psi	bar	14.5	
ft_{water}	N/m^2	2987	
atm	N/m^2	1.013×10^5	atmosphere
So a 100m column of water generates a pressure of: $100\ m_{water} \equiv 10$ kgf/cm^2 $\equiv 9.8$ bar $\equiv 9.8 \times 10^5$ N/m^2 $\equiv 9.8 \times 10^5$ Pa $\equiv 0.98$ MPa $\equiv 142$ psi			
MPa	N/m^2	10^6	
MPa	N/mm^2	1	usual unit for material stress.
N/m^2	N/mm^2	10^6	
kN/mm^2	N/mm^2	10^9	
GN/m^2	kN/mm^2	1	usual units for specifying material modulus properties, such as Young's Modulus.
ton/in^2	N/mm^2	15.44	
ton/in^2	N/m^2	15.44×10^6	
psi	N/mm^2	6.895×10^{-3}	
Power/energy			
kW	W	1000	kilowatt
MW	W	10^6	Megawatt
hp	W	745.7	Imperial horsepower
mhp	W	735.5	metric horsepower
(1 mhp = 75 kgf.m/s)			

9.6 Greek alphabet

Many of the equations in this book use Greek letters for variables.

Table 9–4: Greek alphabet

Lower case	Upper case	Name	Lower case	Upper case	Name
α	Α	alpha	ν	Ν	nu
β	Β	beta	ξ	Ξ	xi
γ	Γ	gamma	ο	Ο	omicron
δ	Δ	delta	π	Π	pi
ε	Ε	epsilon	ρ	Ρ	rho
ζ	Ζ	zeta	σ	Σ	sigma
η	Η	eta	τ	Τ	tau
θ	Θ	theta	υ	Υ	upsilon
ι	Ι	iota	φ	Φ	phi
κ	Κ	kappa	χ	Χ	chi
λ	Λ	lambda	ψ	Ψ	psi
μ	Μ	mu	ω	Ω	omega

9.7 Accuracy and the acceleration due to gravity

Many of the quantities used in the calculations for turbines are not known accurately. Surveying techniques may be approximate, and if an Abney level is used to measure the head then an accuracy of ±0.2m would be quite good. Unless values of such quantities as the velocity coefficient for the nozzle and the strength of the steel being used can be tested, average values have to be assumed. Given this inaccuracy in the inputs, it must be realised that the calculated figures will only have limited accuracy. As a rule, the results of calculations should only be given to the number of decimal places of the least accurate input. It is useless to quote the output power to five decimal places if one only knows the head to within a metre or so.

One quantity that is usually assumed to be known accurately is the acceleration due to gravity. Most book consider this to be a fixed and immutable 9.81 m/s^2, or 9.80655 m/s^2 (the International Standard Gravity) if greater accuracy is required. However, while these figures are true for most of Europe and North America, for many other parts of the world they are not so accurate. The tables below give the variation in gravity from the Equator to the Poles, and the approximate variation with height above sea level (Harris, 1983).

Table 9–5: Variation of gravity with latitude and height above sea level

Latitude	Gravity (m/s^2)	Height above sea level (m)	Reduction in gravity (m/s^2)
0°	9.780	1000	–0.0031
10°	9.782	2000	–0.0062
20°	9.786	3000	–0.0093
30°	9.793	4000	–0.0124
40°	9.802		
50°	9.811		
60°	9.819		
70°	9.826		
80°	9.831		
90°	9.832		

These figures are not given so that super-accurate values of g can be used in calculation – quite the opposite – but to illustrate that even g is not fixed. If a micro-hydro site is anywhere south of 40° latitude, using a value of 9.81m/s^2 is actually worse than using 9.8m/s^2. A figure of 9.8 is quite adequate for most hydropower calculations anywhere. Indeed using g = 10m/s^2 only gives an error of 2%.

10 APPENDIX: DERIVATION OF FORMULAE

10.1 Nozzles

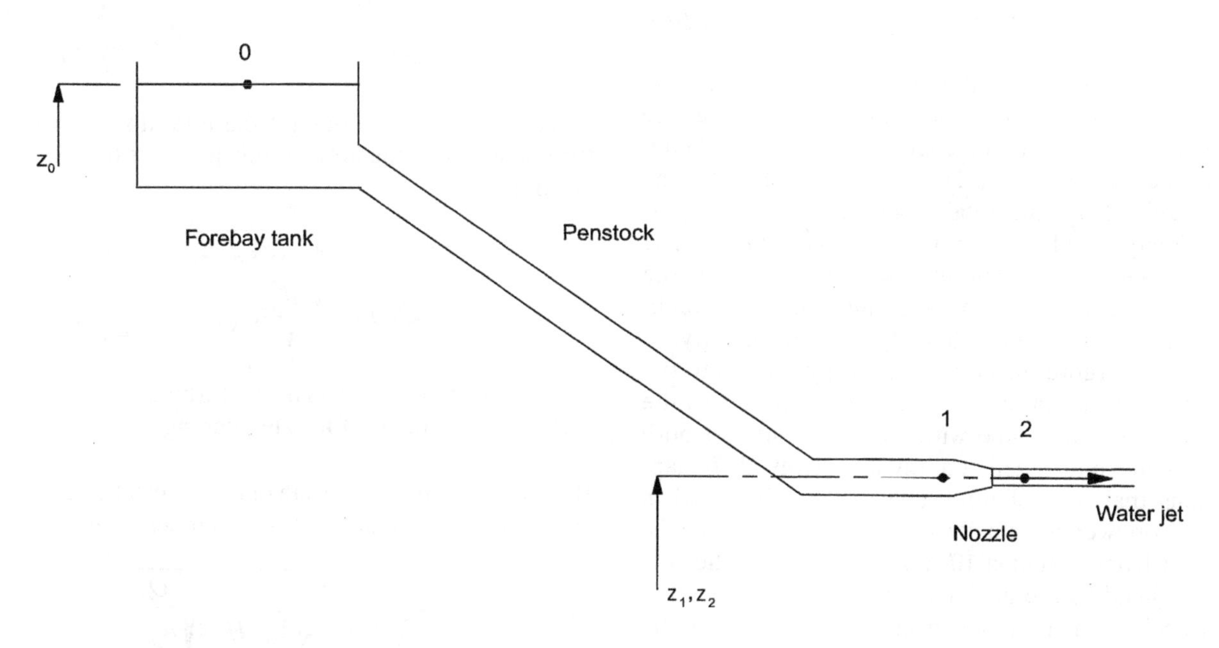

Fig. 10–1: Simplified forebay tank, penstock and nozzle

Consider the simplified system shown in Figure 10–1. A forebay tank feeds a penstock which takes water down to a nozzle, where it comes out as a jet. Bernoulli's equation for an incompressible fluid (such as water), modified to take account of frictional losses, between the surface '0' of the water in the forebay tank and the inlet to the nozzle '1' is:

$$h_0 + \frac{v_0^2}{2g} + z_0 = h_1 + \frac{v_1^2}{2g} + z_1 + \Delta h_{pen}$$

Eq. 10–1

h_0	– pressure head at the surface of the forebay tank (m)
v_0	– water velocity in the forebay tank (m/s)
z_0	– height to water surface in forebay tank (m)
h_1	– pressure head before the nozzle (m)
v_1	– water velocity before nozzle (m)
z_1	– height of nozzle centreline above datum (m)
Δh_{pen}	– frictional head loss in penstock and manifold to nozzle (m)
g	– acceleration due to gravity (m/s^2)

Now $h_0 = 0$, atmospheric pressure, $v_0 \approx 0$, since the water is effectively still in the forebay, and $z_0 - z_1 = H_{gross}$, the gross head. So:

$$h_1 + \frac{v_1^2}{2g} = z_0 - z_1 - \Delta h_{pen} = H_{gross} - \Delta h_{pen} = H_n$$

Eq. 10–2

where H_n is the net head at the nozzle. We can write a similar equation between the nozzle inlet and the jet; assume for the moment that the nozzle is perfect, with no losses.

$$h_1 + \frac{v_1^2}{2g} + z_1 = h_2 + \frac{v_2^2}{2g} + z_2 \qquad \textit{Eq. 10–3}$$

h_2	– pressure head in the jet (at the vena contracta)
v_2	– water velocity in the jet (at the vena contracta)
z_2	– height of jet centreline

Now $z_1 = z_2$ and $h_2 = 0$ (atmospheric pressure) so, using the result in Equation 10–2 this becomes:

$$h_1 + \frac{v_1^2}{2g} = H_n = \frac{v_2^2}{2g} \qquad \textit{Eq. 10–4}$$

Rearranging this:

$$v_2 = \sqrt{2g.H_n} \qquad \textit{Eq. 10–5}$$

This result is known as Torricelli's theorem, from Evangelista Torricelli (1608–47) who established by experiment that the flow from an orifice was proportional to the square root of the head. In practice the nozzle is not perfect, but by convention, the losses are accounted for by a velocity coefficient, C_v, so that the actual velocity in the jet is:

$$v_2 = C_v.\sqrt{2g.H_n} \qquad \textit{Eq. 10–6}$$

There are two points to note about the definition of C_v. Firstly, it gives the velocity of the jet at the narrowest section of the jet (the vena contracta described in Section 2.2.1). Secondly, it calculates the velocity from the total head, pressure head plus velocity head, at the inlet to the nozzle. The idealized experiment for measuring C_v has the nozzle in the side of a large tank. The inlet conditions are the pressure head but at zero velocity, so the head represents the total energy going into the nozzle. In a real nozzle, the water approaches the nozzle down a pipe with some velocity, so both pressure and kinetic energy are involved. It happens that the definition of net head, H_n, used in hydropower is actually the total head, as can be seen from Equation 10–2. H_n is therefore the correct head to use in Equation 10–7. In practice the velocities in the pipes in micro-hydro are usually kept below 3m/s, and the velocity head is less than:

$$h_{vel} = \frac{v^2}{2g} = \frac{3^2}{2\times 9{\cdot}8} = 0{\cdot}46 \qquad \textit{Eq. 10–7}$$

So even if measured head is used rather than total head, the difference is not that significant.

The flow through an ideal, perfectly efficient nozzle would be:

$$Q = A_{noz}.\sqrt{2g.H_n} \qquad \textit{Eq. 10–8}$$

In practice, it is less than this, and the discharge coefficient, C_D, is defined as a fraction of the ideal flow:

$$Q = C_D.A_{noz}.\sqrt{2g.H_n} \qquad \textit{Eq. 10–9}$$

Using the definition of the contraction coefficient C_c in Equation 2–2:

$$\begin{aligned} Q &= A_2.v_2 \\ &= C_c.A_{noz}.C_v.\sqrt{2g.H_n} \\ &= C_c.C_v.\sqrt{2g.H_n} \end{aligned} \qquad \textit{Eq. 10–10}$$

Comparing Equation 10–9 with Equation 10–10 it can be seen that:

$$C_D = C_c \times C_v \qquad \textit{Eq. 10–11}$$

Separating the sides of Equation 10–4 and using Equation 10–6, the efficiency of the nozzle is:

$$\begin{aligned} \eta &= \frac{\textit{energy input}}{\textit{energy output}} \\ &= \left(\frac{v_2^2}{2g}\right) \Big/ \left(h_1 + \frac{v_1^2}{2g}\right) \\ &= \frac{C_v^2.H_n}{H_n} \\ &= C_v^2 \end{aligned} \qquad \textit{Eq. 10–12}$$

If there is more than one jet the total flow for the turbine is simply the flow for one jet times the number of jets:

$$\begin{aligned} Q &= C_D.n_{jet}.A_{noz}.\sqrt{2g.H_n} \\ &= C_D.n_{jet}.\frac{\pi.d_{noz}^2}{4}.\sqrt{2g.H_n} \end{aligned} \qquad \textit{Eq. 10–13}$$

n_{jet} – number of jets in the turbine
d_{noz} – diameter of nozzle opening

Rearranging this, the nozzle diameter can be calculated for a given head, flow and number of nozzles:

$$d_{noz} = \sqrt{\frac{4}{\pi.C_D.\sqrt{2g}}.\frac{1}{H^{\frac{1}{4}}}.\sqrt{\frac{Q}{n_{jet}}}} \qquad \textit{Eq. 10–14}$$

10.2 Spear valves

The diameter of a jet emerging from a spear valve can be calculated as follows. Consider the spear valve shown in Figure 10–2. The flow above the spear passes through the line BC, which is drawn perpendicular to the spear across to the corner of the orifice. The spear valve is shown opened by a distance 's'. The direction of the flow is very nearly perpendicular to BC.

Now the line BC is actually part of a cone, with its apex at A and a base diameter of d_{noz}. The area of conical surface of a cone of side length l and base diameter d (not including the base) is:

$$A = \frac{\pi.d.l}{2} \qquad \textit{Eq. 10–15}$$

The area of a section of cone between l_2 and l_1 is:

$$A = \frac{\pi.d_1.l_1}{2} - \frac{\pi.d_2.l_2}{2} \qquad \textit{Eq. 10–16}$$

Also:

$$\frac{d_2}{d_1} = \frac{l_2}{l_1} \qquad \textit{Eq. 10–17}$$

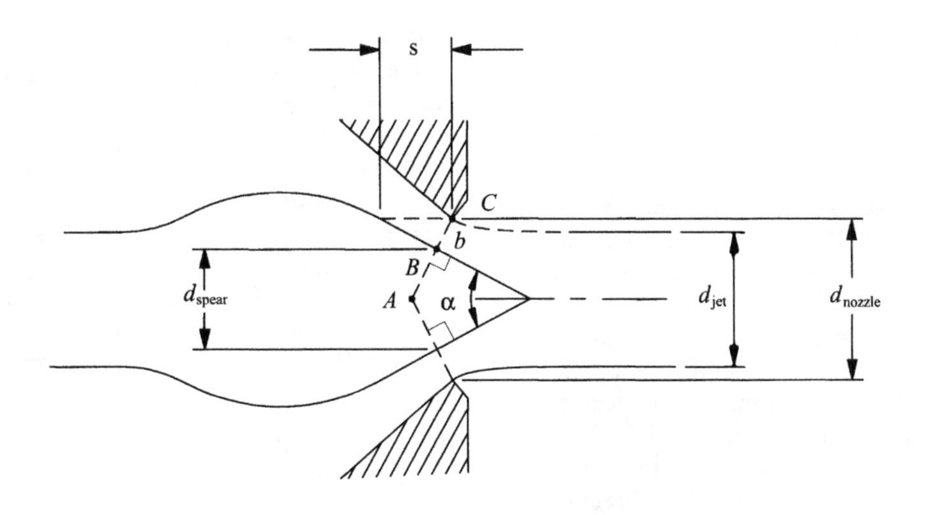

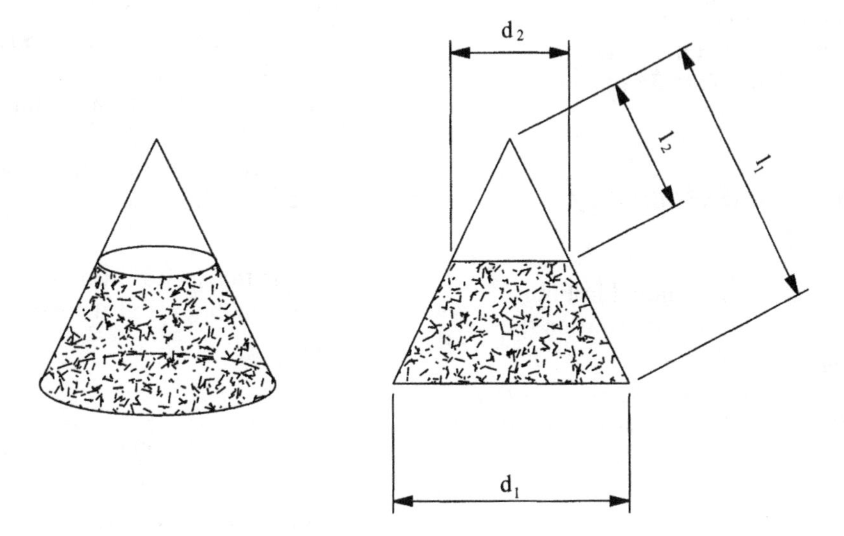

Fig. 10–2: Spear valve geometry

Using this result in Equation 10–16 the area of the cone section can be written:

$$A = \frac{\pi . d_1 . l_1}{2} - \frac{\pi . d_2 . l_2}{2}$$

$$= \frac{\pi}{2} . (d_1 . l_1 - d_1 . \frac{l_2^2}{l_1})$$

$$= \frac{\pi . d_1}{2} . \frac{(l_1^2 - l_2^2)}{l_1}$$

$$= \frac{\pi . d_1}{2} . \frac{(l_1 - l_2).(l_1 + l_2)}{l_1}$$

$$= \frac{\pi}{2} . (l_1 - l_2).(d_1 + d_2) \qquad \textit{Eq. 10–18}$$

Putting the length of BC as b. the following formulae can also be written from the geometry in Figure 10–2:

$$b = s . \sin\frac{\alpha}{2}$$

$$d_{spear} = d_{noz} - 2b . \cos\frac{\alpha}{2} \qquad \textit{Eq. 10–19}$$

It is now possible to write the equation for the area through which the water is leaving the spear valve, using Equation 10–18 and Equation 10–19:

$$A_{spear} = \frac{\pi}{2} . b . (d_{noz} + d_{spear})$$

$$= \frac{\pi}{2} . s . \sin\frac{\alpha}{2} . (d_{noz} + d_{noz} - 2b . \cos\frac{\alpha}{2})$$

$$= \frac{\pi}{2} . s . \sin\frac{\alpha}{2} . (2 d_{noz} - 2s . \sin\frac{\alpha}{2} . \cos\frac{\alpha}{2})$$

$$= \frac{\pi}{2} . s . \sin\frac{\alpha}{2} . (2 d_{noz} - s . \sin\alpha)$$

$$= \pi . \sin\frac{\alpha}{2} . (s . d_{noz} - \tfrac{1}{2} s^2 . \sin\alpha) \qquad \textit{Eq. 10–20}$$

If the speed at the conical section is assumed to be the same as the jet speed, then by continuity the effective area of the spear valve must be the same as the jet area, and:

$$A_{spear} = \frac{\pi . d_{jet}^2}{4} = \pi . \sin\frac{\alpha}{2} . (s . d_{noz} - \tfrac{1}{2} . s^2 . \sin\alpha) \qquad \textit{Eq. 10–21}$$

This leads to the result:

$$d^2_{jet} = 2.\sin\frac{\alpha}{2}.(2s.d_{noz} - s^2.\sin\alpha)$$
Eq. 10–22

Some books have a different formula for d_{jet}, for instance, Nechleba (1957). They argue that the flow is not parallel to the spear, but varies between the line of the spear and the line of the nozzle. They therefore calculate A_{spear} using a cone that bisects the angle between the spear and the nozzle. The resulting equation is much more complicated, but gives practically the same value for d_{jet}. The justification for the method used here is that the spear has the greatest influence on the flow direction because the water continues to travel along the spear even after it has left the nozzle.

For a given head, flow is directly proportional to the jet area, so Equation 10–21 can be used to calculate the flow. The results are plotted on the graph in Figure 2–3.

10.3 Turbine power and efficiency

10.3.1 Simple Theory

Consider the single Pelton bucket shown in Figure 10–3. The jet velocity is v_j, and it chases after the bucket, which is moving at the slower speed v_b. The jet enters the bucket at a relative velocity $(v_j - v_b)$, as shown in the vector diagram. The jet is split in two by the splitter ridge and each half travels around one side the bucket, emerging at edge. The sides are angled outwards γ, and the jet leaves the bucket at this angle. The exit velocity is slightly less than the entry velocity due to friction and turbulence losses as the water flows around the bucket: $v_2 = \zeta.(v_j - v_b)$ where ζ is an efficiency factor for flow in the bucket.

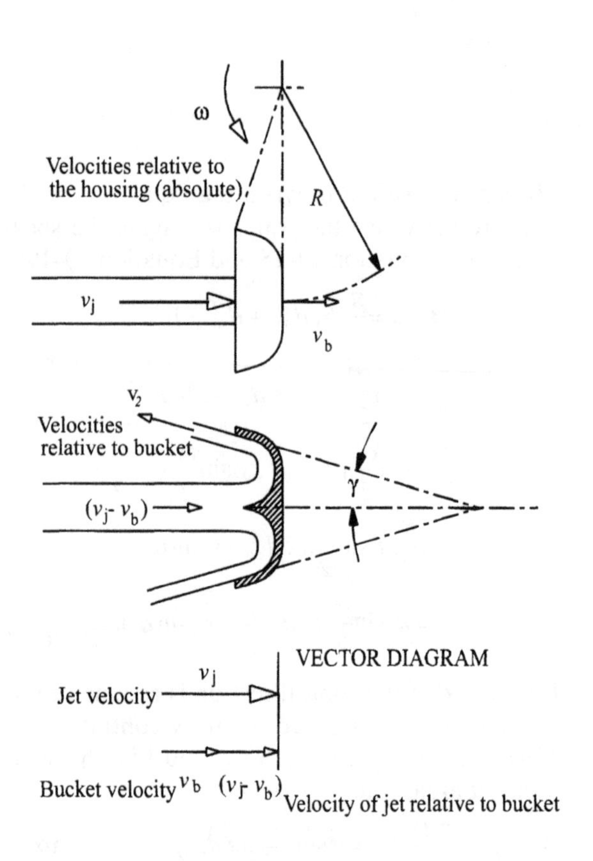

Fig. 10–3: Simplified diagram of flow in a Pelton

The force on the bucket in the direction of travel is found from the rate of change of momentum between the water entering and leaving the bucket.

$$\begin{aligned} F &= \dot{m}_b.[(v_j - v_b) - \{-\varsigma.(v_j - v_b).\cos\gamma\}] \\ &= \dot{m}_b.(v_j - v_b).(1 + \varsigma.\cos\gamma) \end{aligned}$$
Eq. 10–23

F – water force on the bucket (N)
$\dot{m}_b$ – mass flow rate into bucket (kg/s)
v_j – jet velocity (m/s)
v_b – bucket velocity, runner tangential velocity at PCD (m/s)
ζ – efficiency factor for flow in bucket
γ – angle of bucket sides

If we define x as the speed ratio between the tangential speed of the bucket at the runner PCD and the jet speed:

$$x = \frac{v_b}{v_j}$$
Eq. 10–24

Then Equation 10–23 becomes:

$$F = \dot{m}_b.v_j.(1 - x)(1 + \varsigma.\cos\gamma)$$
Eq. 10.25

The power input P to the runner is:

$$\begin{aligned} P &= F.v_b \\ &= \dot{m}_b.v_j.(1-x).(1+\varsigma.\cos\gamma).v_b \\ &= \dot{m}_b.v_j^2.x.(1-x).(1+\varsigma.\cos\gamma) \end{aligned}$$
Eq. 10–26

Now the total kinetic power in the water hitting the bucket is:

$$\tfrac{1}{2}\dot{m}_b.v_j^2$$
Eq. 10–27

So the efficiency of the turbine η (as a fraction) is:

$$\begin{aligned} \eta &= \frac{\dot{m}_b.v_j^2.x.(1-x).(1+\varsigma.\cos\gamma)}{\frac{1}{2}\dot{m}_b.v_1^2} \\ &= 2x.(1-x).(1+\varsigma.\cos\gamma) \end{aligned}$$
Eq. 10–28

Differentiating this to find the maximum:

$$\frac{d\eta}{dx} = 2(1-2x).(1+\varsigma.\cos\gamma) = 0$$

$$\therefore \quad x = \tfrac{1}{2} \qquad \textit{Eq. 10–29}$$

So the efficiency is maximum when x = 0.5. Interestingly, this is independent of the friction in the bucket, ζ, or the angle, γ.

Force on a bucket

To calculate the force on the bucket we need to know the mass flow entering the bucket:

$$\begin{aligned}\dot{m}_b &= A_{jet}.(v_j - v_b).\rho_{water} \\ &= \rho_{water}.A_{jet}.v_j.(1-x) \\ &= \dot{m}\,(1-x) \qquad \textit{Eq. 10–30}\end{aligned}$$

A_{jet} – cross-sectional area of the jet (m^2)
ρ_{water} – density of water (kg/m^3)
$\dot{m}$ – total mass flow in jet (kg/s)

Substituting this into Equation 10–25:

$$F = \dot{m}.v_j.(1-x)^2.(1+\varsigma.\cos\gamma) \qquad \textit{Eq. 10–31}$$

This is maximum when the runner is stationary, x = 0.

10.3.2 Water missing the runner

Figure 10–4 shows a diagram of buckets in a jet. Though the whole jet is shown, we shall only consider what happens to the jet on its centreline. For the bucket in position A, the end of the splitter ridge P is just on the centreline. This leaves a section of the jet l_1 cut off from the main jet but continuing on to the next bucket, bucket B. B' shows the position at which the rear end of the length l_1 finally hits the bucket. (To simplify the calculations, the splitter ridge is drawn as a radial line, which is not usually the case.)

In moving from B to B' the bucket turns through an angle of ($\theta_B + \theta_A - \delta$). This takes a time:

$$t_1 = \frac{\theta_B + \theta_A - \delta}{\omega} \qquad \textit{Eq. 10–32}$$

ω – angular velocity of runner (rad/s)
θ, δ – angles (rad)

Now from the definition of the velocity ratio x above:

$$\omega \,.\, R = x.v_j \qquad \textit{Eq. 10–33}$$

R – radius of pitch circle = PCD/2 (m)

So:

$$t_1 = \frac{(\theta_B + \theta_A - \delta).R}{x.v_j} \qquad \textit{Eq. 10–34}$$

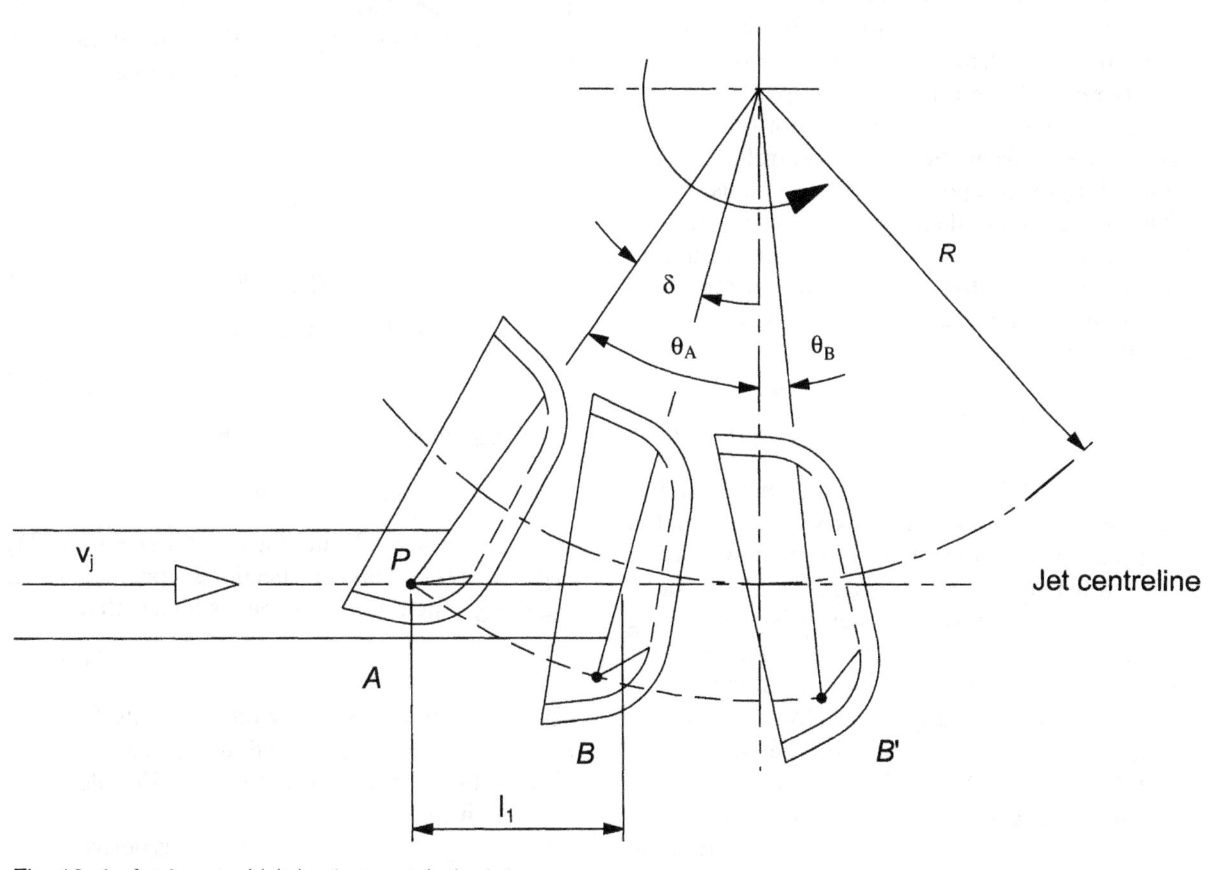

Fig. 10–4: Angles at which buckets catch the jet

In this time the jet has to cover a distance in a straight line of ($R.\tan\theta_B$ + $R.\tan\theta_A$), which takes time:

$$t_1 = \frac{R.(\tan\theta_B + \tan\theta_A)}{v_j} \qquad \textit{Eq. 10–35}$$

Since the jet and the bucket arrive at B' at the same time, these two times t_1 are equal and:

$$\theta_B + \theta_A - \delta = x.\ (\tan\theta_B + \tan\theta_A) \qquad \textit{Eq. 10.36}$$

As x increases, the runner turns further before the last bit of cut-off jet enters bucket B. There is a limiting value of x when $\theta_B = \theta_A$, when:

$$x_{lim} = \frac{(2\theta_A - \delta)}{2\tan\theta_A} \qquad \textit{Eq. 10–37}$$

For the bucket design used in this book, $\theta_A = 36°$, $\delta = 20°$, and $x_{lim} = 0.62$. What this equation says is that at high runner speeds, when x is greater than x_{lim}, a portion of the jet never catches up with the bucket, and this water passes straight through the runner.

Some qualification needs to be made. The analysis has been done for the jet centreline only. The lower portion of the jet is cut off by the splitter ridge at a lower θ, and starts to miss the runner at a lower value of x. The top of the jet does not pass through the runner until a higher value of x. The x_{lim} derived here represents an average value for the jet.

Note that a modified version of Equation 10–37 can be used to calculate the minimum number of buckets required on a runner. The part of the jet that misses the runner first is the bottom, the element furthest from the runner centreline, and a modified θ_A for this position needs to be used.

The reduction in efficiency due to water missing the runner can now be calculated. The maximum amount of the jet that each bucket can receive is the amount leaving the nozzle in a time δ/ω, which has a length:

$$l_b = \frac{\delta}{\omega}.v_j = \frac{\delta.R}{x.v_j}.v_j = \frac{\delta.R}{x} \qquad \textit{Eq. 10–38}$$

Up to x_{lim}, all the water enters the bucket, and so Equation 10–38 holds. For $x > x_{lim}$ some of the water misses the bucket. Figure 10–5 shows bucket A just touching the jet centreline, and the same bucket later at position A' when it is just moving off the centreline. The particle of water at P' for position A' was a distance l_2 back when the bucket was in position A, and l_2 can be found as the jet velocity multiplied by the time it takes the runner to rotate from A to A'. The length of water that has gone into the bucket is l_2 minus the distance PP'. The fraction of the total possible water hitting the bucket is therefore:

$$\frac{(l_1 - PP')}{l_b} = \frac{(v_j.t_{A\to A'} - 2R.\tan\theta_A)}{\left(\frac{\delta.R}{x}\right)}$$

$$= \frac{x.\left(v_j.\left[\frac{2\theta_A}{\omega}\right] - 2R.\tan\theta_A\right)}{\delta.R}$$

$$= \frac{x.\left(v_j.\left[\frac{2\theta_A.R}{x.v_j}\right] - 2R.\tan\theta_A\right)}{\delta.R}$$

$$= \frac{2(\theta_A - x.\tan\theta_A)}{\delta} \qquad \textit{Eq. 10–39}$$

So, multiplying Equation 10–28 by this additional loss factor, the efficiency for $x > x_{lim}$ becomes:

$$\eta = 4\,x.(1-x).(1+\varsigma.\cos\gamma).\frac{(\theta_A - x.\tan\theta_A)}{\delta} \qquad \textit{Eq. 10–40}$$

This is the major cause of the asymmetry of a real Pelton efficiency curve. The effect is also discussed in Section 2.3.2.

There is another factor that also causes the jet to 'miss' the runner. Above a certain value of x, the bucket actually travels faster than the jet at high values of θ. In Figure 10–6 point D is on the splitter ridge. The length CD is R.tanθ. The apparent velocity of the bucket along the path of the jet centreline (the line of CD) can be found by differentiating.

$$v_b = \frac{d(CD)}{dt} = \frac{d(R.\tan\theta)}{dt} = R.\sec^2\theta.\omega \qquad \textit{Eq. 10–41}$$

Using Equation 10–33 this becomes:

$$v_b = x.v_j\sec^2\theta \qquad \textit{Eq. 10–42}$$

From this equation $v_j = v_b$ when:

$$x = \cos^2\theta_{lim} \qquad \textit{Eq. 10–43}$$

The jet cannot catch the bucket at a given x if θ is larger than this. This only becomes a problem when $\theta_{lim} < \theta_A$, which occurs at speeds higher than:

$$x_{\theta lim} = \cos^2\theta_A \qquad \textit{Eq. 10–44}$$

Above this limit, the water only does useful work on the bucket when θ is less than θ_{lim}, either side of vertical. For $x > x_{lim}$, Equation 10–40 holds with θ_{lim} used instead of θ_A.

For $x \le x_{lim}$ the efficiency equation is derived from a multiplier similar to Equation 10–39, with θ_{lim}

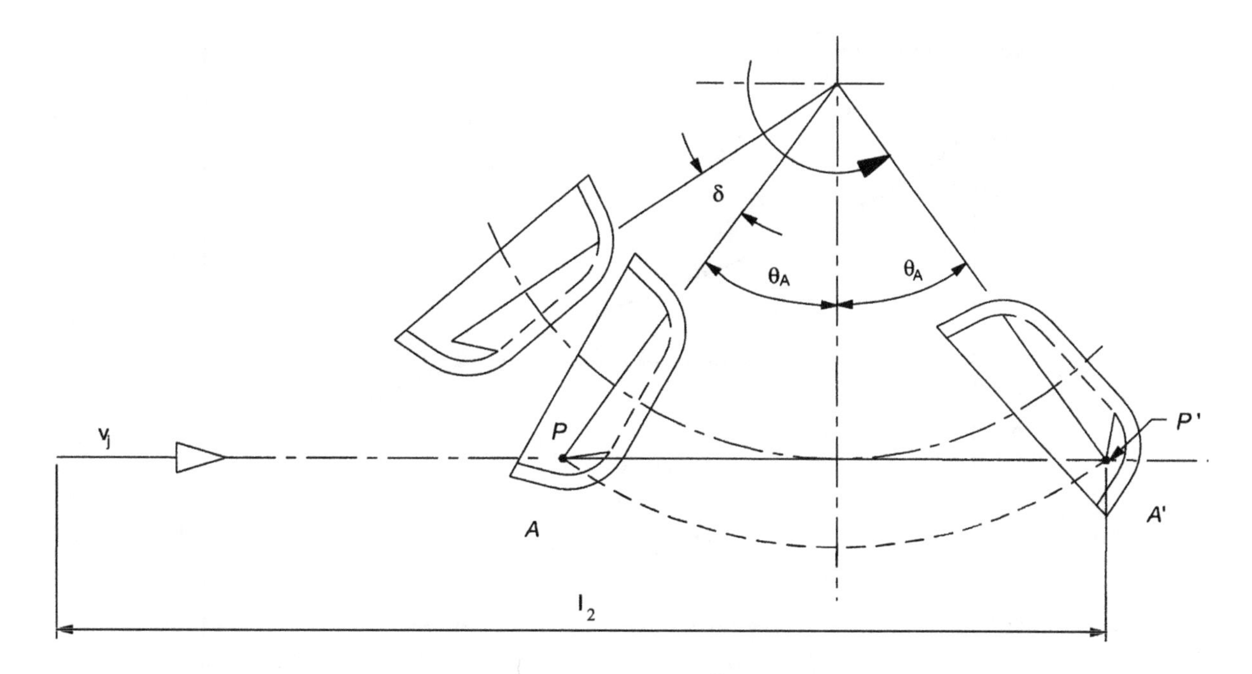

Fig. 10–5: Water missing a bucket for $x > x_{lim}$

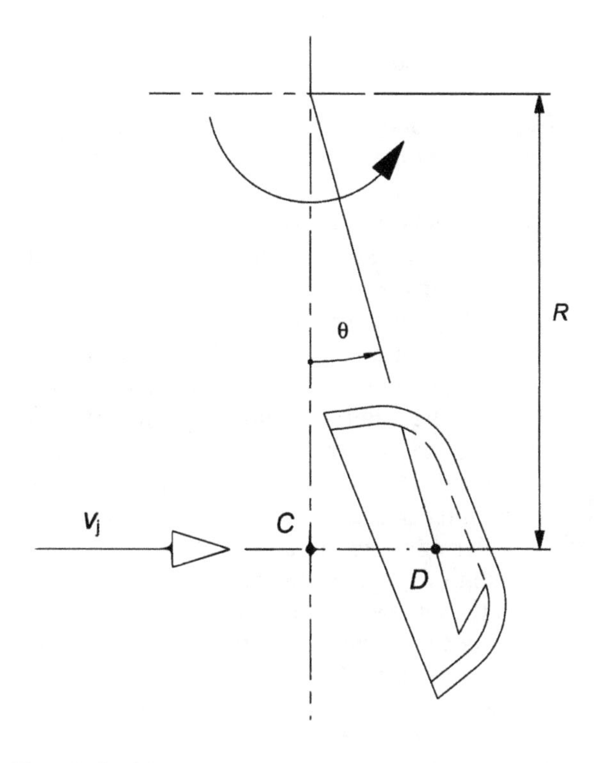

Fig. 10–6: A bucket moving in the line of the jet

substituted for θ_A or θ_B if θ_{lim} is less than either of these values. Note, however, that for the bucket geometry used here $x_{\theta lim}$ is (just) greater than x_{lim}. It should also be noted that when $x_{\theta lim}$ is exceeded, the jet will be acting on the rear of the buckets, slowing them down, though this effect is ignored here.

Figure 10–7 shows how θ_B and θ_{lim} vary with x. This is drawn for the splitter-end radius of the bucket design given in this manual. The angle θ_A, at which the splitter first enters the jet, is fixed by the geometry of the runner until it reaches the θ_{lim} line. Above this speed the splitter goes faster than the jet, and only slows down enough to be caught at an angle less than θ_A. The angle θ_B starts at a negative value (before vertical); at very low speeds it tends to a value of –18°, and is only cut off when the next bucket occludes the jet. As the speed increases, the position at which the last water hits the bucket moves around, past the vertical, until it reaches the value θ_A. This is at x_{lim}, and beyond here θ_B has the effective value θ_A; water starts to pass through the runner without being caught. At a slightly higher speed, θ_B also meets the θ_{lim} line and follows it down, as the splitter end again exceeds the speed of the tail end of the cut-off length of jet.

The conclusions thus far can be summarized in the following formulae:

For $x \leq x_{lim}$:

$$\eta_{x \leq x_{lim}} = 2x.(1-x).(1+\varsigma.\cos\gamma).\frac{\left[(\theta_{B'}+\theta_{A'})-x.(\tan\theta_{B'}+\tan\theta_{A'})\right]}{\delta}$$

Eq. 10–45

For $x > x_{lim}$:

$$\eta_{x \geq x_{lim}} = 4x.(1-x).(1+\varsigma.\cos\gamma).\frac{(\theta_{A'}-x.\tan\theta_{A'})}{\delta}$$

Eq. 10–46

Where:

$$x_{lim} = \frac{(2\theta_A - \delta)}{2\tan\theta_A}$$

Eq. 10–47

Using $\theta_{A'} = \theta_A$ for $\theta_A \leq \theta_{lim}$
$\theta_{A'} = \theta_{lim}$ for $\theta_A > \theta_{lim}$

$\theta_{B'} = \theta_B$ for $\theta_B \leq \mathit{lim}$
$\theta_{B'} = \theta_{lim}$ for $\theta_B > \theta_{lim}$

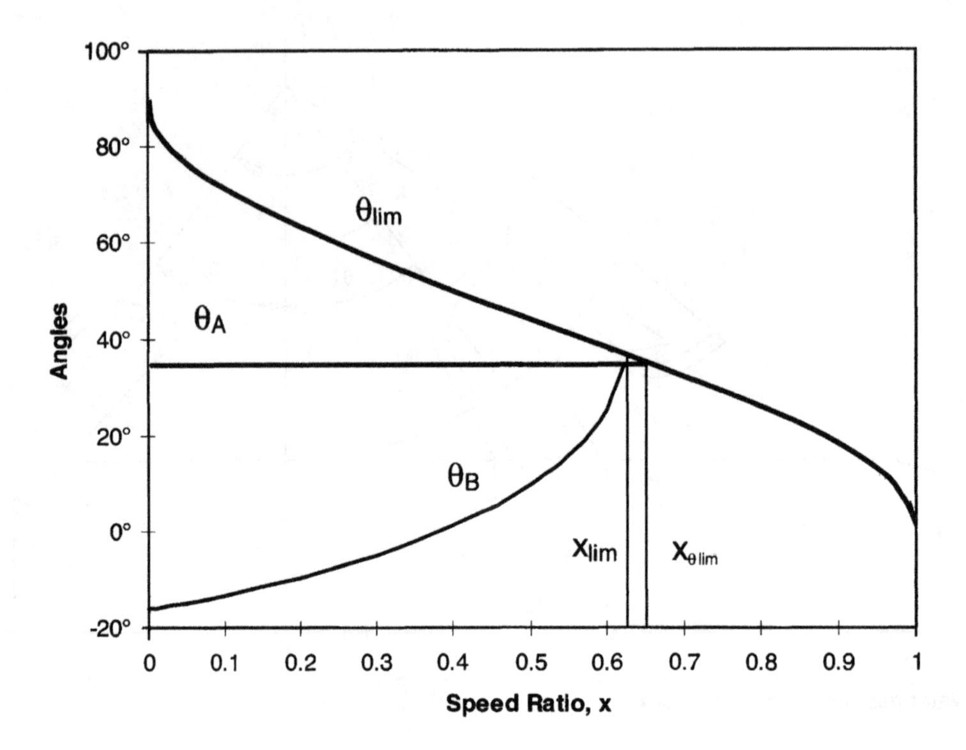

Fig. 10–7: Graph of the variation of θ_B and θ_{lim} with x; drawn for a radius to the splitter ridge end of 62% PCD

With θ_{lim} derived from the equation:

$$x = \cos^2\theta_{lim} \qquad \textit{Eq. 10–48}$$

The implications of this are discussed in Section 2.3.2.

10.3.3 Three-dimensional theory

The analysis in Section 10.3.1 is a gross simplification of what is really going on in a Pelton. The flow into and out of the bucket was analysed only with the bucket at 90° to the jet, and all the water in the jet was assumed to follow the path of the water on the jet centreline. The real flow pattern is much more complex.

When the bucket is at an angle θ before the bottom the jet strikes it at an angle, the water flows around the curves of the bucket, and it emerges at a range of completely different angles. This is illustrated in Figure 2–8.

A relatively simple analysis can take some account of the varying inlet and outlet angles as the runner turns through the jet. This is presented below. It still assumes that all the water follows the path of the element on the jet centreline, but it gives some interesting results that are quite close to what is found in real turbines. It can be shown that, even if each part of the jet is analysed separately, the resulting average corresponds quite well to the values calculated for the centre element (Kisioka & Osawa, 1972). The case analysed is shown in Figure 10–8. The main assumptions are given below:

- Viewed in side elevation, the jet leaves from the same point it enters. This is not true, as the water sweeps around the curves of the bucket and comes out some way along from where it enters, but it makes the calculations much more complicated to include this effect. However, since as a fraction of the radius from the runner centre the displacement is quite small, it does not have a large effect on the results.
- The jet emerges at the same instant as it enters. This again is not true, but it is extremely difficult to allow for the difference. Not only is there a time delay, but the bucket orientation changes during this period. The runner actually turns about 7–8° (at optimum speed) while the water goes round it.
- Considering a section across the bucket, the jet leaves at the angle of the side wall γ. This is true when the jet leaves near the centre of the bucket but is not so accurate for the water that comes out further up or down it.
- The whole jet behaves in the same way as the element on its centreline.
- The splitter ridge and the edge of the bucket are both on the same radial line from the runner centre.
- In the side elevation, the jet is effectively *reflected* in the bucket, so the angle at which it enters, α_1, is the same as the exit angle, α_2; $\alpha_1 = \alpha_2 = \alpha$. This is probably reasonable near the centre of the bucket, but less so towards the ends.

The force on the bucket from the jet entering it is equal to the mass flow rate into the bucket multiplied by the inlet velocity relative to the bucket, w_1. The component of this force normal to the bucket is:

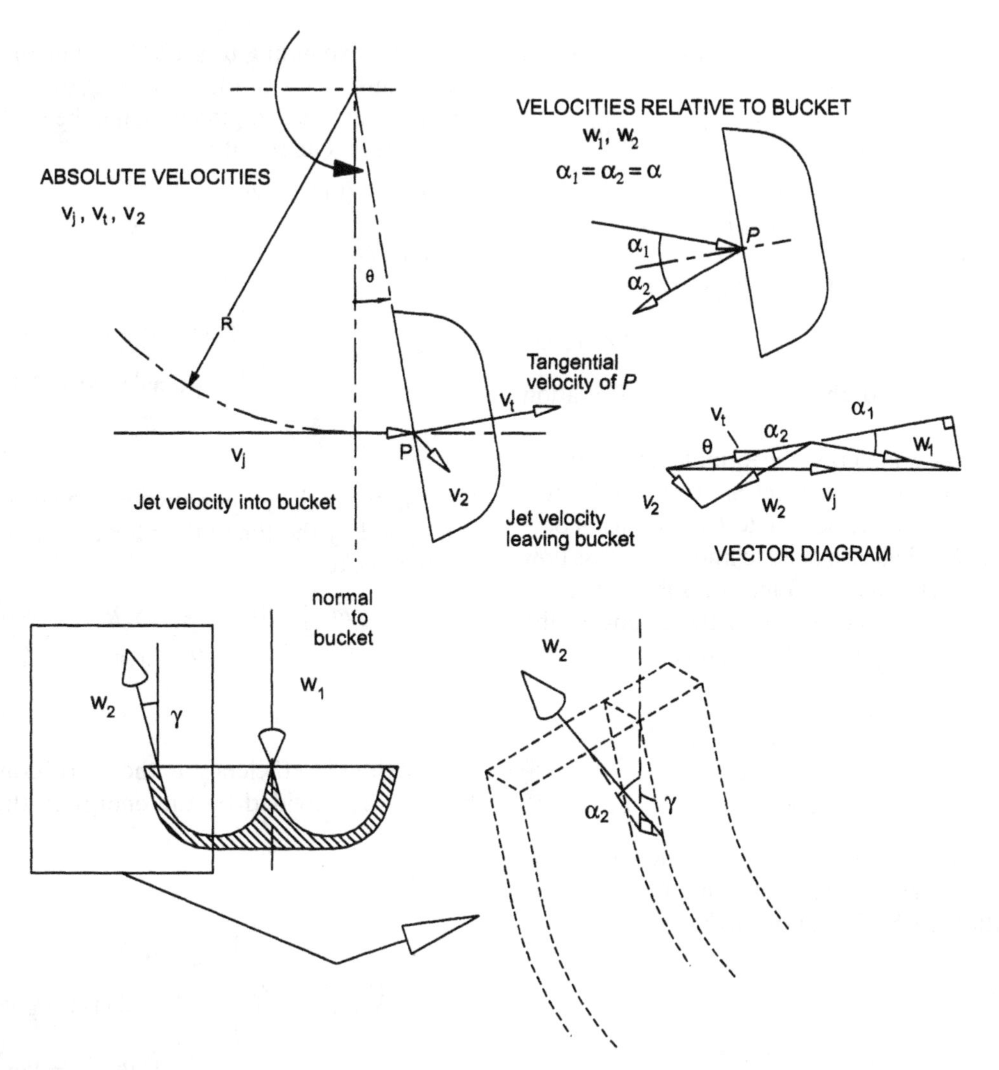

Fig. 10–8: Water velocities

$$\dot{m}_b.w_1.\cos\alpha \qquad \textit{Eq. 10–49}$$

Similarly for the water leaving the bucket. The only complication is that the velocity has an axial component, as shown in the detail in Figure 10–8. The component of the force from the water leaving the bucket, normal to the bucket, is:

$$\dot{m}_b.w_2.\cos\alpha.\cos\gamma \qquad \textit{Eq. 10–50}$$

The instantaneous rate of work done by both these forces is equal to the force multiplied by the speed, which is the tangential velocity of point P, v_t. Since both the force of the jet pushing on the bucket and the reaction force of it leaving add together, these two rates of work add to give the power output from the bucket, P_b.

$$\begin{aligned} P_b &= \dot{m}_b.w_1.\cos\alpha.v_t + \dot{m}.w_2.\cos\alpha.\cos\gamma.v_t \\ &= \dot{m}_b.v_t.\cos\alpha.(w_1 + w_2.\cos\gamma) \end{aligned}$$

Eq. 10–51

Allowing for friction in the bucket again, as in Equation 10–23:

$$w_2 = \varsigma.w_1 \qquad \textit{Eq. 10–52}$$

The velocity v_t can be derived as:

$$\begin{aligned} v_t &= R_p.\omega \\ &= \frac{R}{\cos\theta}.\frac{x.v_j}{R} \\ &= x.v_j.\sec\theta \end{aligned} \qquad \textit{Eq. 10–53}$$

Substituting from Equation 10–52 and Equation 10–53 into Equation 10–51:

$$P_b = \dot{m}_b.x.v_j.\sec\theta.w_1.\cos\alpha_1.(1 + \varsigma.\cos\gamma)$$

Eq. 10–54

From the vector diagram in Figure 10–8, and using Equation 10–53:

$$\begin{aligned} &v_j.\cos\theta = w_1.\cos\alpha + v_t = w_1.\cos\alpha + x.v_j.\sec\theta \\ &\therefore\ w_1.\cos\alpha = v_j\,(\cos\theta - x.\sec\theta) \end{aligned}$$

Eq. 10–55

Substituting this result into Equation 10–54:

$$\begin{aligned} P_b &= \dot{m}_b.v_j.x.\sec\theta.[v_j.(\cos\theta - x.\sec\theta)].(1 + \varsigma.\cos\gamma) \\ &= \dot{m}_b.v^2_j.x.(1 - x.\sec^2\theta).(1 + \varsigma.\cos\gamma) \end{aligned}$$

Eq. 10–56

Dividing this by the kinetic energy per second

going into the bucket (from Equation 10–24) gives the instantaneous efficiency at θ:

$$\eta_{inst} = \frac{\dot{m}_{b}.v_j^2.x.(1 - x.\sec^2\theta).(1+\varsigma.\cos\gamma)}{\left(\frac{\dot{m}_b.v_j^2}{2}\right)}$$

$$= 2x.(1 - x.\sec^2\theta).(1+\varsigma.\cos\gamma)$$

Eq. 10–57

Note that, for $\theta = 0$, this is the same as Equation 10–28.

We can now integrate the formula for P_b to find the work done on the bucket as it moves from a position θ_A before the bottom to θ_B after, as in Section 10.3.2. We shall also need to know the mass flow into the bucket, which is found by subtracting the bucket speed v_b (from Equation 10–42, and in the same direction as v_j) from the jet speed:

$$\begin{aligned} \dot{m}_b &= \rho.A_{jet}(v_j - v_b) \\ &= \rho.A_{jet}.(v_j - x.\ v_j.\sec^2\theta) \\ &= \rho.A_{jet}.v_j.(1 - x.\sec^2\theta) \\ &= \dot{m}(1 - x.\sec^2\theta) \end{aligned}$$

Eq. 10–58

The work done on the bucket as it moves can be found by integrating the power, using Equation 10–33, Equation 10–58, Equation 10–56:

$$W = \int_{t_{\theta_A}}^{t_{\theta_B}} P_b.dt$$

$$= \int_{-\theta_A}^{\theta_B} P_b.d\theta.\frac{dt}{d\theta}$$

$$= \int_{-\theta_A}^{\theta_B} P_b.\frac{d\theta}{\omega}$$

$$= \int_{-\theta_A}^{\theta_B} P_b.\frac{R}{x.v_j}.d\theta$$

$$= \int_{-\theta_A}^{\theta_B} \dot{m}_b.v_j^2.x.(1 - x.\sec^2\theta).(1+\varsigma.\cos\gamma).\frac{R}{x.v_j}.d\theta$$

$$= \int_{-\theta_A}^{\theta_B} \left[\begin{array}{l} \dot{m}.(1 - x.\sec^2\theta).v_j. \\ \qquad (1 - x.\sec^2\theta).(1+\varsigma.\cos\gamma).R.d\theta \end{array}\right]$$

$$= \dot{m}.v_j.R.(1+\varsigma.\cos\gamma)\int_{-\theta_A}^{\theta_B}(1 - x.\sec^2\theta)^2.d\theta$$

$$= \dot{m}.v_j.R.(1+\varsigma.\cos\gamma).$$

$$\int_{-\theta_A}^{\theta_B}(1 - 2x.\sec^2\theta + x^2.\sec^4\theta).d\theta$$

Eq. 10–59

Now the integral $\sec^2\theta$ is a standard case that can be looked up in tables. The integral of $\sec^4\theta$ can be found by expanding to $\sec^2\theta.(1 + \tan^2\theta) = \sec^2\theta + (\sec\theta.\tan\theta).(\sec\theta.\tan\theta)$, integrating the second term by parts, and rearranging to make the $\sec^4\theta$ integral the subject. The result is:

$$W_b = \dot{m}.v_j.R.(1+\varsigma.\cos\gamma).$$

$$\left[\theta + x.(x-2).\tan\theta + \frac{x^2}{3}.\tan^3\theta\right]_{-\theta_A}^{\theta_B}$$

$$= \dot{m}.v_j.R.(1+\varsigma.\cos\gamma).\left[\begin{array}{l} (\theta_B+\theta_A) + x.(x-2).(\tan\theta_B + \tan\theta_A) \\ + \frac{x^2}{3}.(\tan^3\theta_B + \tan^3\theta_A) \end{array}\right]$$

Eq. 10–60

The energy in the jet per bucket is the nozzle power multiplied by the time taken for the runner to turn by one bucket:

$$W_j = \frac{\dot{m}.v_j^2}{2}.\frac{\delta}{\omega} = \frac{\dot{m}.v_j^2}{2}.\frac{\delta.R}{x.v_j} = \frac{\dot{m}.v_j.\delta.R}{2x}$$

Eq. 10–61

The hydraulic efficiency is the work done on the bucket, W_b, divided by the energy in the jet, W_j.

$$\eta = \frac{W_b}{W_j}$$

$$= \frac{2x.(1+\varsigma.\cos\gamma)}{\delta}.\left[\begin{array}{l} (\theta_{B'} + \theta_{A'}) \\ + x.(x-2).(\tan\theta_{B'} + \tan\theta_{A'}) \\ + \frac{x^2}{3}.(\tan^3\theta_{B'} + \tan^3\theta_{A'}) \end{array}\right]$$

Eq. 10–62

The formula is completed by allowing for the water missing the runner as derived in the previous section.

Where $\theta_{A'} = \theta_A$ for $\theta_A \leq \theta_{lim}$
$\theta_{A'} = \theta_{lim}$ for $\theta_A > \theta_{lim}$

and $\theta_{B'} = \theta_B$ for $\theta_B \leq {}_{lim}$
$\theta_{B'} = \theta_{lim}$ for $\theta_B > \theta_{lim}$

The efficiency curves produced by this theory are shown in Figure 2–11, and they bear a good likeness to a real runner. A full discussion on efficiency is included in the same section.

10.3.4 Windage losses

The standard formula for drag resistance for an object moving in a fluid is:

$$D = K_D.A_f.q = K_D.A_f.\frac{\rho.u^2}{2}$$

Eq. 10–63

D – drag force (N)
K_D – drag coefficient

A_f	– frontal area of the object presented in the direction of movement (m^2)
q	– dynamic pressure (N/m^2)
ρ	– density of the fluid (kg/m^3)
u	– velocity of the object (m/s)

To adapt this to the case of a Pelton runner, we assume that the velocity u is the bucket speed at the PCD. Multiplying the drag force by the speed to obtain the power lost to the drag, and using Equation 10–24.

$$P_{windage} = D.u = K_D . A_f . \frac{\rho_c . v_b^2}{2} . v_b$$

$$= \tfrac{1}{2} K_D . A_f . \rho_c . v_b^3$$

$$= \tfrac{1}{2} K_D . A_f . \rho_c . (x.v_j)^3$$

Eq. 10–64

$P_{windage}$	– power lost to windage (W)
v_b	– tangential bucket velocity at PCD (m/s)
ρ_c	– density of air/spray within casing (kg/m^3)
v_j	– jet velocity (m/s)

Now the power in the jet is:

$$P_{jet} = \frac{\dot{m}.v_j^2}{2} = \frac{(\rho_w . v_j . A_{jet}) v_j^2}{2}$$

$$= \tfrac{1}{2} \rho_w . A_{jet} . v_j^3 \qquad \textit{Eq. 10–65}$$

$\dot{m}$	– mass flow in jet (kg/s)
ρ_w	– density of water (kg/m^3)
A_{jet}	– cross-sectional area of the jet (m)

So the efficiency loss due to the windage is:

$$\eta_w = \frac{P_{windage}}{P_{jet}}$$

$$= \frac{\tfrac{1}{2} K_D . A_f . \rho_c . x^3 . v_j^3}{\tfrac{1}{2} A_{jet} . \rho_w . v_j^3}$$

$$= K_D . \left(\frac{A_f}{A_{jet}}\right) . \left(\frac{\rho_c}{\rho_w}\right) . x^3 \qquad \textit{Eq. 10–66}$$

Note the dependence on x^3, and the use of the two ratios.

A runner is quite a complex shape, and K_D is difficult to establish. As an approximation, an individual bucket can be likened to a flat circular disc. K_D varies with the Reynolds number of the flow past the bucket, but for the conditions within a micro-hydro turbine (with Reynolds number in the range 10^3–10^6) a disc has a K_D of about 1.5. However, K_D for the runner cannot be found by multiplying 1.5 by the number of buckets. This would grossly over-estimate the drag, as the buckets follows each other round and 'clear a path' for each other. K_D of about 4 seems to give a reasonable representation of experimental results.

For the bucket pattern described in this book, A_f is approximately $0.093 \times PCD^2$, where PCD is measured in metres. For the design jet diameter of 11% PCD, the ratio $A_f/A_{jet} = 9.8$.

The fluid density in the casing is basically the density of air. Atmospheric density is about 1.23 kg/m^3 at sea level and 15°C, giving $\rho_{air}/\rho_{water} = 0.00123$. This may need to be increased to allow for water spray in certain cases. At low head there is less spray in the casing, but at high heads the casing is full of spray, and a higher ρ_{casing} is appropriate.

So for an 11% PCD jet, the windage efficiency loss measured as a fraction is approximately:

$$\eta_w = 0.05 \times x^3 \qquad \textit{Eq. 10–67}$$

This gives a loss of 0.5% at $x = 0.46$, which is realistic as windage losses are not high at operating speeds. Windage and friction together typically give 1% loss.

There are some other minor components of loss, such as the drag of the runner hub and seal discs moving in the casing fluid, but these are insignificant.

10.3.5 Mechanical friction losses

The frictional torque of the bearings and seals is basically independent of speed. Representing this as T_f the efficiency loss for friction can be calculated in the same manner as the windage loss:

$$\eta_f = \frac{P_{friction}}{P_{jet}} = \frac{T_f . \omega}{P_{jet}} = \frac{T_f}{R} . \frac{x.v_j}{P_{jet}} = \frac{F_f . v_j}{P_{jet}} . x$$

Eq. 10–68

F_f is the force at the PCD required to turn the runner, measured in Newtons.

This is the only efficiency equation derived so far that does not scale. The other efficiencies are theoretically (though not in practice) all independent of the actual head or flow or size of the turbine. Because the friction loss for a given turbine is constant, its effect on the overall efficiency is dependent on the turbine power. If the turbine is operating at maximum flow under a high head, the friction loss is relatively insignificant. If the turbine is operated with a small flow under a low head, the friction can be relatively large.

10.4 Bucket stem stress calculations

This section shows how to check the stress in the stem of the bucket. There are two load cases that can cause the bucket to break off. The first is runaway, when the external load is removed from the turbine and the runner accelerates to a high speed. This produces a large centrifugal force in the bucket, which can snap the stem. The second case is the fatigue load caused by bending stress on the stem due to the water hitting the bucket every time it passes a nozzle. Though the first case needs to be checked, it is usually the fatigue load that causes failure.

10.4.1 Runaway load

The calculation for the runaway case is quite straightforward. The centrifugal force can be calculated from the mass of the bucket and its speed, and the stress is simply the area of the stem divided by this force. The volume of the bucket design given here (without any stem) is approximately 0.0038 × D³. The mass of the stem depends on its design. With the basic stem shown in Figure 4–3, the bucket and stem have a volume of 0.0069 × D³.

Consider the specific case of the single-piece casting shown in Figure 10–9. The buckets are integral with the hub, and the stress is calculated across section A-A. The volume of the bucket outside the section is 0.0040 × D³, and the centre of this mass is at a radius 0.48 × D. The centrifugal force is given by:

$$F_{runaway} = m_{bucket} \times R_g \times \omega^2_{runaway}$$

$$= \rho_{bucket} \cdot V_{bucket} \cdot R_g \cdot \left(\frac{2\pi . N_{runaway}}{60} \right)^2 \qquad \textit{Eq. 10–69}$$

$F_{runaway}$ – centrifugal force on the bucket at runaway (N)
m_{bucket} – bucket mass (kg)
R_g – radius of bucket centre of mass from runner centre (m)
$\omega_{runaway}$ – angular speed of runner at runaway (rad/s)
ρ_{bucket} – density of bucket material (kg/m³)
V_{bucket} – volume of bucket and stem (m³)
$N_{runaway}$ – runner speed at runaway (rpm)

If the turbine is being operated at its optimum point, the runaway speed is about 1.8 times the optimum speed. The area of the section A-A has an area of 0.16 × 0.122 = 0.0195 × D². The stress in the stem is therefore:

$$\sigma_t = \frac{F_{runaway}}{A_{stem}} = \frac{\rho_{bucket} \cdot V_{bucket} \cdot R_g \cdot \left(\frac{\pi . N_{runaway}}{30} \right)^2}{A_{stem}}$$

$$= \frac{\rho_{bucket} \times 0 \cdot 0040 \times D^3 \times 0 \cdot 48 \times D . \pi^2 . N^2_{runaway}}{30^2 \times 0 \cdot 0195 . D^2}$$

$$= 0 \cdot 0011 \times \rho_{bucket} \cdot D^2 . N^2_{runaway} \qquad \textit{Eq. 10–70}$$

σ_t – tensile stress at runaway in bucket stem (N/m²)
A_{stem} – area of stem at weakest section (m²)
D – runner PCD (m)

This can be multiplied by 10⁶ to give a result in N/mm². This stress should be less than the yield point of the material, with a suitable safety factor.

The method is similar for clamped buckets, but the weakest section is at a different place. For the clamping arrangement shown in Figure 10–10(a), the fracture will occur at the section through the outer bolt hole, B-B. The part of the bucket outside

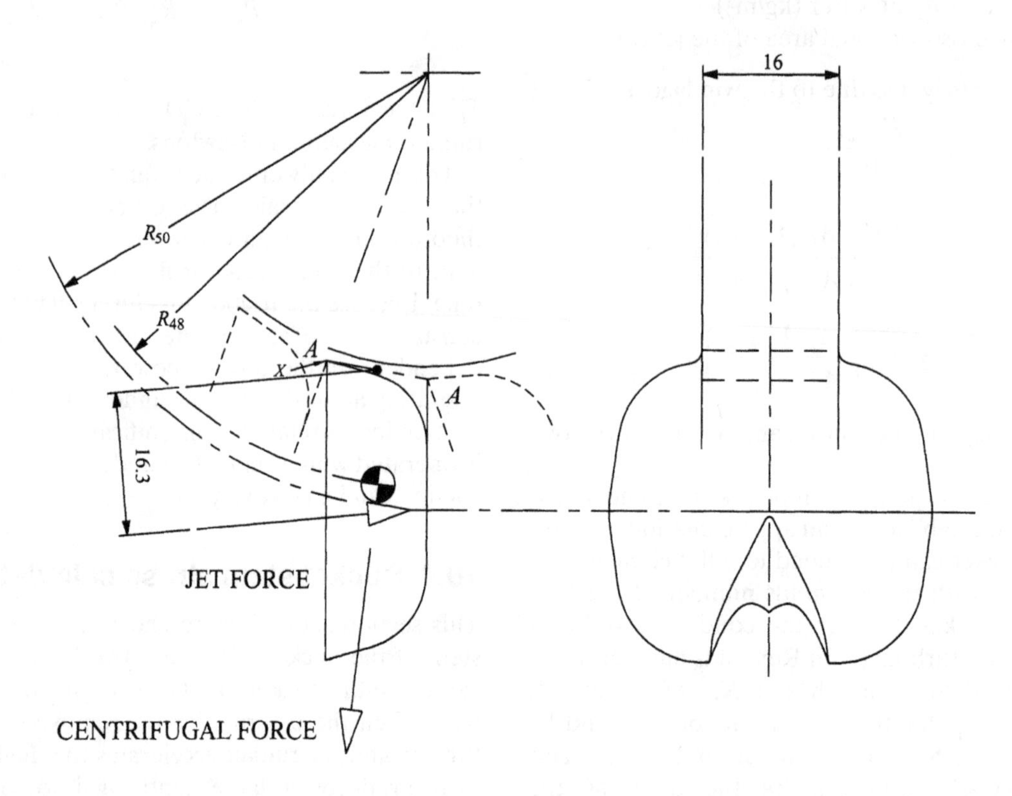

Fig. 10–9: Single-piece casting, buckets integral with the hub. Dimensions are in % PCD

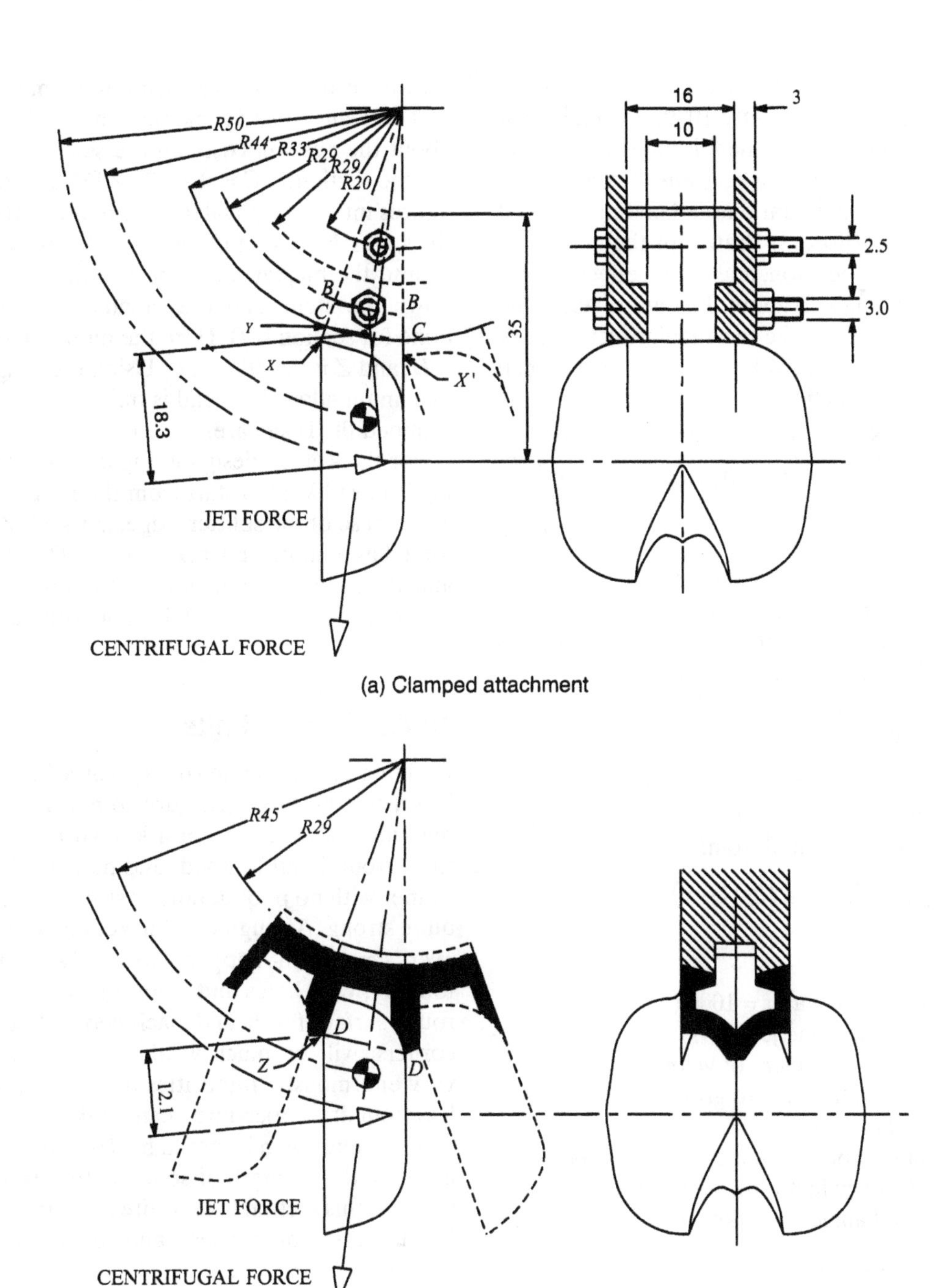

(a) Clamped attachment

(b) Welded attachment

Fig. 10–10: Clamped and welded bucket attachment designs for medium heads

B-B has a volume of $0.0052 \times D^3$, the centre of mass is at a radius of $0.44 \times D$, and the area of the section is $0.0071 \times D^2$. Equation 10–70 can be used to calculate the stress. Note that the holes shown in Figure 10–10(a) are 2.5% and 3% PCD, but this may need to be varied for a given PCD to fit actual bolt sizes.

The welded construction in Figure 10–10(b) is based on Figure 4–11. The volume of the bucket (including half the weld fillets) is approximately $0.0049 \times D^3$, and the centre of gravity is at a radius of $0.45 \times D$. The force is carried by the weld connecting the bucket to the hub, and the weld area per bucket is $0.0108 \times D^2$. The welds between the buckets are neglected. This is because all the buckets are being flung out at the same time, so they all pull on the hub welds together. In practice, there is a 'hoop stress' contribution from between the bucket welds, and the real stress will be somewhat lower than calculated here.

10.4.2 Fatigue load

The main fatigue load on a bucket comes from the jet force. This creates a bending stress in the stem every time a bucket passes a jet. Most turbines will exceed one million bending cycles on the buckets within a few weeks, or even days, of operation. The point at which the worst fatigue stress occurs depends on how the buckets are fixed. For the single-piece casting shown in Figure 10–9, it is plain that the maximum bending moment is going to occur at the section where each bucket joins the disk, A-A. This will give a maximum tensile bending stress at point 'X'. The jet force is drawn for the maximum bending moment position, and has an effective moment arm of 0.163 × D about the neutral axis. The section modulus Z for the rectangular section A-A is $0.16 \times 0.122^2/6 = 0.00040 \times D^3$. The stress is then found from the equation:

$$\sigma_f = \frac{M}{Z} = \frac{F_{jet} \times a_{jet}}{Z} = \frac{F_{jet} \times 0 \cdot 163 \times D}{0 \cdot 00040 \times D^3} = \frac{408 \times F_{jet}}{D^2}$$

Eq. 10–71

σ_f – fatigue stress in stem (N/mm²)
M – bending moment (N.mm)
Z – stem section modulus = I/y_{max} (mm)
F_{jet} – jet force on bucket (N)
a_{jet} – moment arm from jet to section (m)
D – runner PCD (mm)

Note that by putting D into the equation in (mm), instead of (m), the stress comes out in N/mm². The jet force can be calculated from:

$$F_{jet} = \rho_w . Q_{jet} . C_v . \sqrt{2g.H_n} . (1-x)^2 . (1+\varsigma.\cos\gamma)$$

Eq. 10–72

ρ_w – density of water = 1000 (kg/m³)
Q_{jet} – flow from one jet (m³/s)
C_v – nozzle coefficient of velocity
g – acceleration due to gravity (m/s²)
H_n – net head (m)
x – ratio of bucket velocity to jet velocity
ζ – efficiency factor for flow in bucket
γ – outlet angle of bucket sides

This formula is derived from Equation 10–31. At optimum operating speed, $x = 0.46$. γ is 15° for the bucket here. For most calculations, C_v can be taken as 0.97, and ζ as 0.95.

Now consider the clamped attachment in Figure 10–10(a). At first sight, it looks as if each bucket rests on the bucket behind it, touching at point 'X'. This would make a solid disk in the centre, equivalent to the single-piece casting above. In practice, this is not the case. Variations in the machining of each bucket will mean that there will be gaps between some of the buckets. If there is no gap when the runner is made, corrosion and movement will create one. So at what point do we consider the bucket to be fixed? The trick is to make the notch on the sides of the stem a press fit on the hub side plates. By doing this we know that the buckets will be held at section C-C, and the maximum bending moment will be at point 'Y'. Since this is a thin point in the stem, Y will also have the maximum tensile bending stress. The jet force is drawn for the maximum bending moment position and has an effective moment arm of 0.183 × D about the neutral axis, and the section modulus Z for the section C-C is $0.00022 \times D^3$. Feeding in the figures into the equivalent to Equation 10–71 gives the stress. Note that if the hub sides are not a press fit into the bucket sides, the bending occurs lower down the stem, and the calculations need to be done for section B-B. Here the moment arm is $0.22 \times D$, and Z is $0.00015 \times D^3$. Using these figures, the maximum allowable head is only 56% of that for a clamped fit. Take care!

For the welded design in Figure 10–10(b), cracking is most likely to start from the root of the weld at the base of the splitter ridge, marked 'Z'. The jet force has a moment arm of $0.121 \times D$ to this point, and the section modulus Z of the bucket through the section marked 'D-D' is approximately $0.00025 \times 10^{-4} \times D^3$.

10.4.3 Stress limits

Defining the allowable stresses for a Pelton runner is difficult, because there are so many variables. A high-quality casting from a known material, carefully ground and finished, and machined to fit the runner with no poor details or stress raisers, will be quite strong. A rough-cast bucket made from whatever scrap iron happens to be lying around a foundry, with holes and inclusions in it, left with a rough surface finish, and machined with some sharp corners, will be much weaker. Materials too can vary enormously. Impurities in the mix, and careless foundry procedure, can make the material much weaker than it should be. Heat treatment can improve the strength of some materials. For these reasons, this section takes quite conservative values for material properties, and uses large safety factors.

Table 10–1 gives the properties of possible bucket materials described in Section 4.1.3. Note that the fatigue design stress in the table is the maximum recommended stress for a *non-reversing* bending load. Where possible, these figures are based on published data. Safety factors for tensile load are between 2.5 and 4, depending on the material. Where the endurance limit, σ_e, is not known for a particular steel, it can be estimated at 40% of the ultimate tensile strength. This is the figure for a fatigue test piece subjected to a fully reversing load that gives a stress $\pm\sigma_e$. The fatigue case we are considering here has the stress going from zero to a maximum value of σ_f, as given by Equation 10–71. A Goodman diagram can be drawn to show that the maximum stress is approximately $1.4 \times \sigma_e$ when the endurance limit is 40% of the ultimate tensile strength. However, the allowable stress is much lower than this, because the bucket will have stress concentrations and a much poorer surface finish

Table 10–1: Properties of materials for Pelton bucket castings. Figures in italics are estimated

Material	*Ultimate tensile stress*	*Proof or yield stress*	*Endurance limit*	*Density*	*Tensile design stress*	*Fatigue design stress range*
	σ_{ult}	σ_p or σ_y	σ_e	ρ	σ_t	σ_f
	(N/mm²)	(N/mm²)	(N/mm²)		(N/mm²)	(N/mm²)
Brass, 60/40	*280*	*90*		*8300*	30	20
Brass, Grade SCB4	250	80		8300	30	20
Silicon Bronze, C87200	380	172		*8600*	60	30
Cast Iron, Grade 220	220	140	100	7150	54	25
Cast Iron, Grade 260	260	170	120	7200	65	30
Cast Steel, Grade A1	430	230	*172*	7830	58	35
Cast Steel, Grade A3	540	295	*216*	7830	74	45
Stainless Steel, 304C12	430	215	*172*	7970	54	35
Welded joints					as material	20

than the test piece. It is also subject to water corrosion. Putting these factors all together, the working stress is lower than the endurance limit by a factor of 4 to 5, and lower than the ultimate tensile strength by a factor of between 10 and 12.5.

The brasses and bronze listed in Table 10–1 do not have defined fatigue limits, but the fatigue life does increase substantially as the stress is reduced to low levels. The design stresses here are set low to give adequate fatigue life. Many stainless steels also have no fatigue limit, and again the stresses need to be kept low to give an appropriate life. Welded joints do not, in theory, have a fatigue limit, but the prediction of life at very high cycles is uncertain (for further information see BS 7608:1993). The figures below derive from experience of welded micro-hydro runners surviving.

Table 10–2 summarizes the bucket properties for the bucket designs given above, for various attachment methods.

If these figures are substituted into the formula in Sections 10.4.1 and 10.4.2, the attachment stresses can be calculated, and compared with the figures in Table 10–1.

For a given geometry of turbine, both the tensile stress due to runaway and the fatigue stress due to the jet force are independent of the PCD. This can readily be shown by substituting equations for runaway speed and jet force into the equations in the preceding sections. The bucket shape, including the stem and attachment system, and the ratio of the jet diameter to the PCD, must be kept the same for this to hold true. This means that one can estimate the maximum heads allowable for each material and attachment system. The results of all the above calculations for the standard buckets and attachment methods used here and a 12% PCD jet (the largest recommended) are tabulated in terms of the limiting heads in Table 4-1 in Section 4.1.4.

All these calculations err on the safe side. If a jet diameter less than 12% PCD is being used, the head can be increased, but it may also be possible to increase the allowable head by reducing the safety factors. This should only be done if one can control the quality of the buckets. If sample pieces of the material are tested and found to meet or exceed the specification, if the castings are of good quality, with no inclusions or defects, and if the finishing and machining are done to a high standard, then the safety factors of 3 (for tensile load) and 4 (for fatigue load) can be reduced. Only increase the allowable weld stress if the weld is carefully controlled, crack-detection is done, and the weld itself is ground or peened to a smooth finish. It is also possible to increase the allowable head by modifying the stem to lower the stresses.

Table 10–2: Summary of bucket dimensions for stress calculations for different attachment methods

Attachment method	*Bucket volume*	*Radius to centre of gravity*	*Stem area for tensile failure*	*Moment arm of jet about A_{stem}*	*Section modulus for fatigue*
	V_b	R_g	A_{stem}	a_{jet}	Z
	($\times D^3$)	($\times D$)	($\times D^2$)	($\times D$)	($\times D^3$)
No stem	0.0038	0.48	–	–	–
Stem to Figure 4–3	0.0069	0.39	–	–	–
Single-piece casting	0.0040	0.48	0.0195	0.163	0.00040
Clamped	0.0052	0.44	0.0071	0.183	0.00022
Welded	0.0049	0.45	0.0108	0.121	0.00025

10.5 Bolted buckets

This section gives the background for the calculation of the bolted clamping friction in Section 4.2.2. The bolt forces are calculated for a simple two-bolt clamped case, but the general theory is presented too, so that the reader can do calculations for more complex arrangements.

10.5.1 Standard analysis

Consider the group of n bolts shown in Figure 10–11. The force F applies both a direct shear force to the bolts, and a bending moment. A standard text-book analysis, for example Shigley & Mischke (1989), considers these two components separately. If it is assumed that the structure around the bolts is completely rigid, then the direct, or primary, shear force is distributed among the bolts in proportion to their area.

$$S_i = F.\frac{A_i}{(A_1 + A_2 + \ldots + A_n)} \quad Eq.\ 10\text{–}73$$

S_i – direct shear force component on bolt i
A_i – cross-sectional area of bolt i

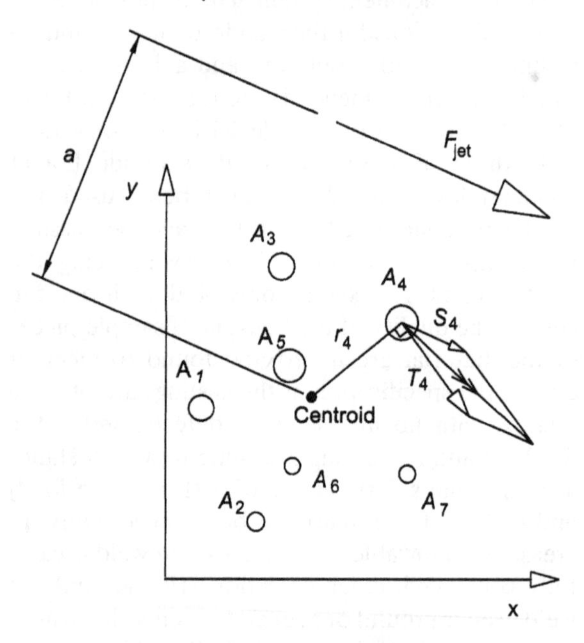

Fig. 10–11: A group of n bolts of varying cross-sectional area, with a force F acting on the joint they are clamping

To find the effect of the moment, the centroid of the bolts has to be found. In the x-direction, this is given by:

$$x_c = \frac{(A_1.x_1 + A_2.x_2 + \ldots + A_n.x_n)}{(A_1 + A_2 + \ldots + A_n)} \quad Eq.\ 10\text{–}74$$

Similarly for y_c. If the shear force on bolt 'i' from the moment is T_i, then for equilibrium:

$$F.a = T_1.r_1 + T_2.r_2 + \ldots + T_n.r_n \quad Eq.\ 10\text{–}75$$

Assuming infinite rigidity again, the amount of the moment or *secondary shear* that a bolt takes is proportional to its area multiplied by its moment arm from the centroid:

$$\frac{T_1}{A_1.r_1} = \frac{T_2}{A_2.r_2} = \ldots = \frac{T_n}{A_n.r_n} \quad Eq.\ 10\text{–}76$$

Combining Equation 10–75 and Equation 10–76:

$$T_i = F.a.\frac{(A_1.r_1^2 + A_2.r_2^2 + \ldots + A_n.r_n^2)}{A_i.r_i} \quad Eq.\ 10\text{–}77$$

Both S_i and T_i can be drawn as vectors, and the actual force on bolt 'i' is the vector sum of these two, as shown in Figure 10–11. This is a standard method for calculating bolt forces. The assumption that the parts being bolted are completely rigid is never actually true, but the simplicity of the method has lead to it being widely adopted. It is also assumed that the bolt centres and the hole centres match each other exactly, so that the loads are distributed into the bolts evenly. Therefore this calculation does not account for internal loads generates by misalignment as the bolts are fitted and tightened – which can be considerable. Nevertheless, it usually gives results that are a reasonable approximation to the real loads, and is certainly better than doing no calculations at all.

10.5.2 Analysis of a bolted bucket

Now consider the bolted bucket in Figure 10–12. The jet force is drawn in the position that gives the maximum bending moment as discussed in Section 10.4.2. From Equation 10–73, the direct shear force in each bolt is given by:

$$S_1 = F_{jet}.\frac{A_1}{(A_1 + A_2)}$$

$$S_2 = F_{jet}.\frac{A_2}{(A_1 + A_2)} \quad Eq.\ 10\text{–}78$$

By symmetry, the centroid C must be on the centreline of the two bolts, so Equation 10–74 can be simplified to:

$$R_c = \frac{(A_1.R_1 + A_2.R_2)}{(A_1 + A_2)} \quad Eq.\ 10\text{–}79$$

Equation 10–77 becomes:

$$T_1 = F_{jet}.a.\frac{A_1.(R_c - R_1)}{\left[A_1.(R_c - R_1)^2 + A_2.(R_2 - R_c)^2\right]}$$

$$T_2 = F_{jet}.a.\frac{A_2.(R_2 - R_c)}{\left[A_1.(R_c - R_1)^2 + A_2.(R_2 - R_c)^2\right]} \quad Eq.\ 10\text{–}80$$

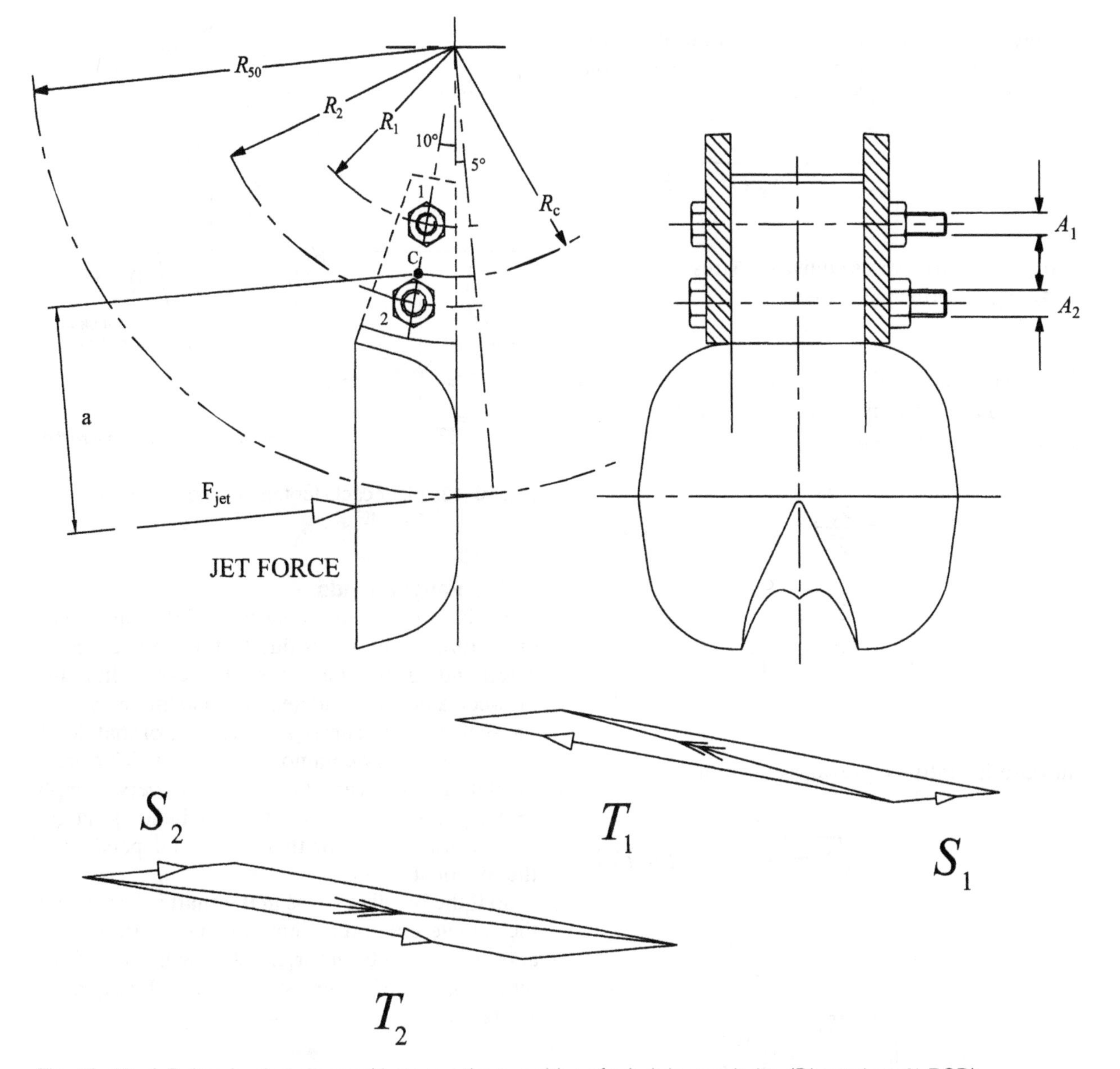

Fig. 10–12: A Pelton bucket clamped between the two sides of a hub by two bolts. (Dimensions % PCD)

However, for the two-bolt case, the moment is resolved as a simple couple, and the above formulae can be much reduced to:

$$T_1 = T_2 = \frac{F_{jet} \cdot a}{(R_2 - R_1)} \qquad \textit{Eq. 10–81}$$

The primary and secondary shear forces can be added vectorially as shown in Figure 10–12. Section 4.2.2 in the main text gives an example of the calculation procedure.

10.6 Shaft design

This section derives and explains the background of the formulae used in Section 4.8.2. It is a calculation procedure for finding the material diameter required for a rotating shaft only subjected to torsion and bending moments.

10.6.1 Static loads

First consider the static case. The bending and torsion stresses in a solid round shaft can be found from basic beam and torsion theory:

$$\sigma_x = \frac{32M}{\pi d^3}$$

$$\tau_{xy} = \frac{16T}{\pi d^3} \qquad \textit{Eq. 10–82}$$

σ_x – bending stress (N/m²)
τ_{xy} – torsion stress (N/m²)
M – bending moment (Nm)
T – torsion (Nm)
d – shaft diameter (m)

These need to be added and compared to the failure stress of the material. For ductile materials such as steel, the von Mises-Hencky distortion-energy theory gives the best prediction of failure. For the two-dimensional stress case being considered here, the von Mises stress is:

$$\sigma_{vm} = \sqrt{\sigma_A^2 - \sigma_A . \sigma_B + \sigma_B^2} \quad \textit{Eq. 10–83}$$

σ_{vm} – von Mises equivalent stress
σ_A, σ_B – principal stresses

For torsion and bending only, the Mohr's circle can be drawn as shown in Figure 10–13, and the principal non-zero stresses are:

$$\sigma_A = \frac{\sigma_x}{2} + \tau_{max}$$

$$\sigma_B = \frac{\sigma_x}{2} - \tau_{max}$$

$$\tau_{max}^2 = \left(\frac{\sigma_x}{2}\right)^2 + \tau_{xy}^2 \quad \textit{Eq. 10–84}$$

Substituting into the previous equation:

$$\sigma_{vm} = \sqrt{\sigma_x^2 + 3\tau_{xy}^2} \quad \textit{Eq. 10–85}$$

Then substituting from Equation 10–82:

$$\sigma_{vm} = \sqrt{\left(\frac{32M}{\pi d^3}\right)^2 + 3\left(\frac{16T}{\pi d^3}\right)^2}$$

$$= \frac{32}{\pi d^3}\sqrt{M^2 + \frac{3T^2}{4}} \quad \textit{Eq. 10–86}$$

This stress needs to be less than the yield stress of the material, with an appropriate safety factor:

$$\frac{\sigma_{yield}}{SF_y} = \frac{32}{\pi d_y^3}\sqrt{M^2 + \frac{3T^2}{4}}$$

$$\therefore \; d_y = \left[\frac{32\, SF_y}{\pi \sigma_{yield}}\left(M^2 + \frac{3T^2}{4}\right)^{\frac{1}{2}}\right]^{\frac{1}{3}} \quad \textit{Eq. 10–87}$$

d_y – minimum shaft diameter (m)
σ_{yield} – yield stress for shaft material (N/m2)
SF_y – safety factor

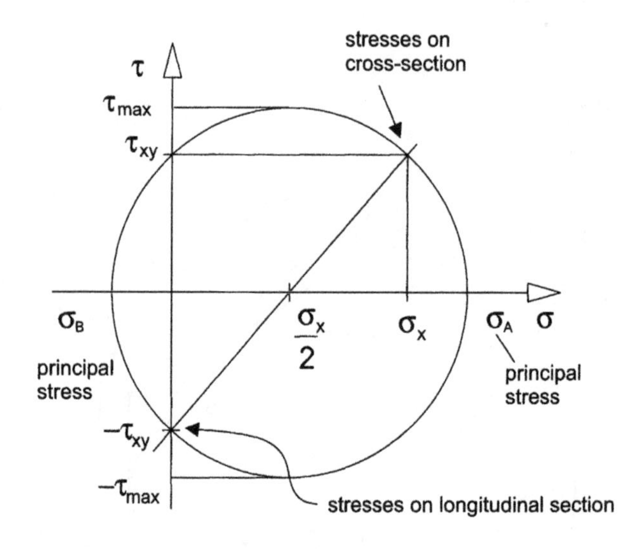

Fig. 10–13: Mohr circle for torsion and bending only in a shaft

10.6.2 Fatigue loads

Consider a point on the surface of the shaft. At a given torque, the stress due to torsion is constant. The bending moment stress, however, alternates between a positive and negative maximum value as the shaft rotates. This is a special case of shaft loading – albeit a very common one – for which experimental results can be analysed very simply (Shigley, 1986). It is found that the bending-fatigue strength of a shaft for this case is independent of the torsional stress.

So if the bending stress in the shaft is kept below the fatigue endurance limit for the shaft, with an appropriate safety factor, the shaft will have infinite fatigue life. The bending stress is given by Equation 10–82, so:

$$\frac{\sigma_{ec}}{SF_f} = \frac{32M}{\pi d^3}$$

$$\therefore \; d_f = \left(\frac{32M.SF_f}{\pi \sigma_{ec}}\right)^{\frac{1}{3}} \quad \textit{Eq. 10–88}$$

d_f – -minimum shaft diameter
σ_{ec} – corrected endurance limit for the shaft material
SF_f – safety factor

The endurance limit found from a standard test has to be corrected for a real shaft. This is because the rotating-beam specimens used to find endurance limits have cross-sections equal to ϕ75 mm, have a shape that is free of stress concentrations, and have polished surfaces. Most turbine shafts are larger, have a variety of shapes, and have machined surfaces, all of which reduce the fatigue life. The endurance limit from a rotating beam test has to be reduced by approximately 0.7 for an average machined micro-hydro shaft (0.8 if the shaft is ground). If the endurance limit is not known, it can be estimated as half the ultimate tensile strength. Correcting this for a real shaft:

$$\sigma_{ec} = 0.7 \times \sigma_{ult}/2 = 0.35 \times \sigma_{ult}$$

A word of caution, though. The estimates for σ_e above are approximations only. In most cases they will lead to reasonable designs, but they cannot be guaranteed to be reliable in all cases. Fatigue is a complicated subject, and a full treatment is beyond the scope of this manual. If the fatigue life is critical, you are advised to consult a specialist text. Shigley (1986) gives a good introduction to the subject.

Note that the method used here to find the diameter for a shaft is based on recent advances in the understanding of fatigue. Many references still use the 1927 American Society of Mechanical Engineers (ASME) code, even though this has been obsolete for many years. It uses some rules-of-thumb for calculating the allowable shear stress within a material which are not accepted nowadays. Nevertheless, it can be used to design shafts provided an ample safety allowance is included.

10.7 Manifold loss calculations

The following sections give data for calculating the head losses in a manifold. As was stated in the discussion on manifolds in Section 4.10, it is not usually necessary to do all these calculations – the losses are generally small. If small pipes and high water velocities are being used, or a very low head is being considered where the manifold loss could be quite high proportion of it, it is worth doing design calculations. It is instructive to do the full calculations at least once just to get an idea of where the losses come from, and to check what design features to avoid.

The losses in any particular route through a manifold are found by adding the pipe friction losses for the straight lengths of pipe to the losses due to the bends, bifurcations, valves and diameter changes. This calculation actually slightly overestimates the losses if the features are within a few pipe diameters of each other, as the turbulence caused by the first one will not have fully dissipated by the second, and so on. Nevertheless, the result gives an indication of the overall effect. For more detailed information on loss calculation, the reader is referred to Miller (1990).

10.7.1 Pipe losses

The head losses in the following sections refer to the 'velocity head', $V^2/2g$, which is the head that would appear if all the momentum of water travelling at velocity V were turned into pressure. The various pipework components are assigned coefficients 'K', and the head loss across them is equal to the velocity head multiplied by this coefficient. By using equivalent values of K for all the components, the relative size of each loss in each can easily be seen. So, for a straight section of manifold pipework:

$$H_f = K_f \frac{V^2}{2g} \qquad \textit{Eq. 10–89}$$

H_f – head loss in a given section of pipe (m)
K_f – pipe friction head loss coefficient
V – water velocity in the pipe (m/s)
g – gravity (9.8 m/s²)

The coefficient K_f for a pipe depends on a number of factors: the diameter, the length, the velocity, and various water properties. In order to show the link with conventional formulae, the equation below shows K_f in terms of a friction coefficient f. The value of f can be calculated from Equation 10–90, or looked up in a Moody diagram in a hydraulics book, or in Kempe (1989).

$$K_f = f\frac{L}{d}$$

$$f = \frac{1}{4.\left[\log\left(\frac{k}{3\cdot 7d} + \frac{5\cdot 74}{\mathrm{Re}^{0.9}}\right)\right]^2}$$

$$\mathrm{Re} = \frac{V.d}{\nu} \qquad \textit{Eq. 10–90}$$

f – friction coefficient
L – length of the pipe (m)
d – pipe bore (internal diameter) (m)
k – friction coefficient (m) – note units for calculation!
Re – Reynolds number for flow
V – water velocity (m)
ν – kinematic viscosity of water (1.57×10^{-6} m²/s at 5°C)

Typical values for k are given in Table 10–3. They are given in millimetres, as universally found in the literature. Remember to convert to metres to put in the formula, or the results will be wildly wrong! The value of k is roughly the average value for the roughness on the surface of the material. A pipe with heavy deposits on it can have a k of a few millimetres.

A word of warning about using values of k or f from other literature. There are many different friction loss formulae for pipes, and they use different roughness coefficients, which are not interchangeable. Another common pipe loss equation, the Manning equation, which is used in Inversin (1986), uses a roughness coefficient normally labelled n, but n is not the same as k, and cannot be

Table 10–3: Roughness coefficients *k* for manifold pipes

Type of pipe	*Roughness coefficient* *k* (mm)
Welded steel, bare or light rust	0.02
Welded steel, rough bitumen paint or brush enamel	0.10
HDPE, PVC with smooth joints	0.003
HDPE with internal beads at the joints	0.2

used in Equation 10–90. For reference, the equation used in this text, Equation 10–90 is a variant of the Colebrook-White equation, with this version taken from Miller (1990). The Moody diagrams found in books are a graphical form of the Colebrook-White equation, and normally do use the same definition of *k* as here. Be aware too that some books use a different definition of the friction coefficient *f*, which is still normally labelled *f*, but is a quarter of the value used here. The definition of *f* used here is easier to use, but the other definition arises more naturally in the derivation of the equation. It works best to calculate the value of *f* for each section of pipe with different diameter or flow, and then calculate a value of K_f for each length of pipe.

10.7.2 Bend losses

Manifold bends as shown in Figure 10–14 are usually made as mitred joints, which are much easier to make than radiused bends, certainly in steel. The bend has (n–1) intermediate sections of pipe covering $\phi°$ each. Note the pipes on either side extend into the bend area by r.sin (ϕ/2). The head losses due to bends are calculated from the loss coefficient, K_b, in a similar way to the pipe losses above.

$$H_b = K_b \frac{V^2}{2g} \qquad \textit{Eq. 10–91}$$

H_b – head loss in bend (m)

The graphs shown in Figure 10–15 and Figure 10–16 give K_b for common micro-hydro bends. The graphs are drawn for typical conditions (equivalent to a 2.5 m/s flow in a 100mm bore welded mild steel pipe). In theory, the loss coefficient varies with Reynold's Number and roughness, but the graphs give values of K_b accurate enough for most micro-hydro. For more detailed information on loss coefficients, see Miller (1990).

Figure 10–15 shows the loss coefficient against the bend angle for mitred joints and for curved

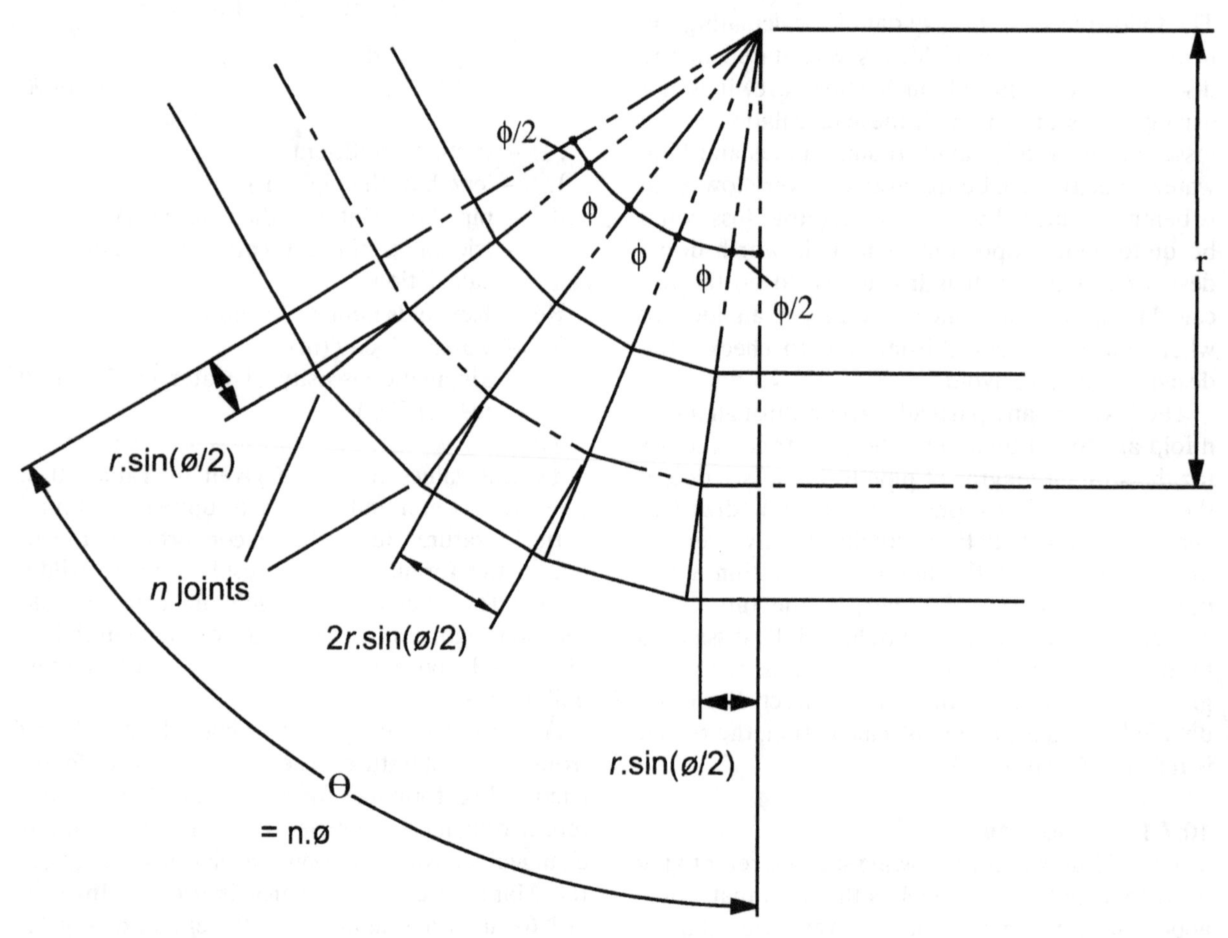

Fig. 10–14: Compound mitred θ° bend with n joints

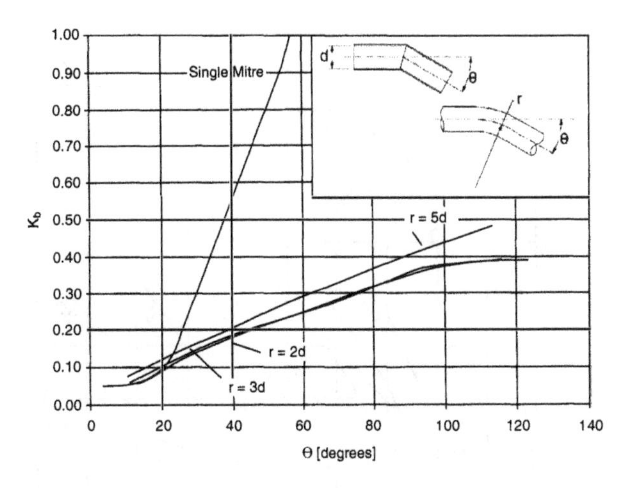

Fig. 10–15: Loss coefficients for various bends (Miller, 1990)

bends with various radii. The figure is drawn for Re = 1.5 × 10⁵ and roughness of k = 0.25mm. (Note that most literature has these curves in idealised form – for smooth pipes and Re = 10⁶ – which give rather lower values of K_b.) As would be expected, radiused bends are better than mitred joints. Although it looks as if $r = 5d$ bends have worse losses than the smaller bends, which is not what you would expect, this is misleading because K_b includes the pipe losses, and larger bends are longer than short ones; if lengths of straight pipe were added to the $2d$ bend to make the same length as the $5d$ bend, the overall loss would be higher.

Figure 10–16, shows the loss coefficients for 90° bends made out of multiple mitred joints. The diagram shows a 3 × 30° compound mitre bend, with a total angle of 90°. It is drawn for the same flow conditions as Figure 10–15. The line for a smooth, radiused 90° bend is shown for comparison. Having more segments in the bend lowers K_b, but there is no need to have too many. If $r \geq 3d$, mitre angles of 22½° are nearly as good as a radiused bend. (This is one reason for recommending that bend radii should generally be kept above $3d$ in Section 4.10.1).

Though the two graphs do not cover every bend, it is possible to estimate K_b for most bends by extrapolating from values for similar types.

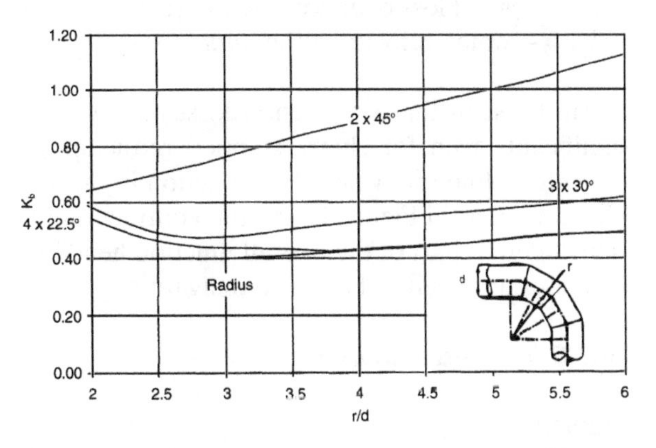

Fig. 10–16: Loss coefficients for composite mitre bends (Miller, 1990)

10.7.3 Bifurcation losses

Bifurcations are where the flow is divided in two, the junctions in the manifold to take the flow off to the jets. The losses at bifurcations can be much higher than for valves or bends, and careful design is essential. The following section allows the losses to be calculated – or at least, estimated. More importantly, studying the graphs shows which sorts of bifurcations are reasonable and which ones are likely to give large losses.

The losses for geometry shown in Figure 10–17 are again calculated from coefficients, as follows:

$$H_{31} = K_{31}\frac{V_3^2}{2g}$$

$$H_{32} = K_{32}\frac{V_3^2}{2g} \qquad \textit{Eq. 10–92}$$

H_{31}	– head loss between inlet 3 and branch 1 (m)
H_{32}	– head loss between inlet 3 and branch 2 (m)
K_{31}, K_{32}	– loss coefficients
V_3	– water velocity at inlet 3 (m/s)

There is one coefficient for each branch, 1 and 2, and the inlet is given the label 3. All coefficients are referenced to the inlet velocity, V_3. Note that for some branch conditions, the coefficients can be negative, corresponding to a head gain. The coefficients K are affected both by the geometry of the bifurcation, and by the relative flow in each branch.

There are so many variables for bifurcations that it is not practicable to present all possibilities. The K coefficients for joints not covered can be estimated by finding similar junctions in this text. If

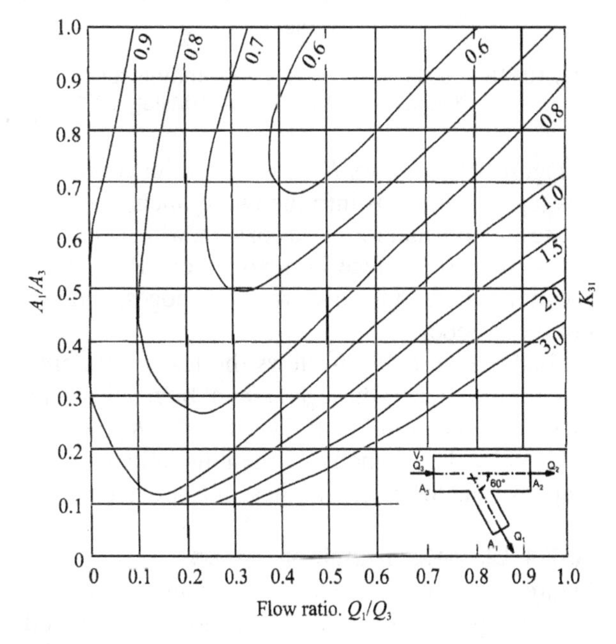

Fig. 10–17: Loss coefficient K_{31} for sharp-edged 60° branch junctions. $A_3 = A_2$. (Miller, 1990)

more detailed information is required, the reader is referred to Miller (1990).

Figure 10–17 shows the side branch coefficient, K_{31}, for a rather poor, sharp-edged junction. This is produced by simply welding a branch pipe on to the main pipe at the required angle. Q_1/Q_3 and A_1/A_3 are the ratios of the flows and areas of the branch and inlet respectively. In certain areas of this graph, the losses are not too bad, but if the design operating conditions are near the bottom of the graph, the coefficients can be high.

Better are Y-junctions, shown in Figure 10–18. The graph is based on test pieces with rounded internal corners, and standard joints will be somewhat worse. Even so, Y-junctions are good for most flow conditions. Note that the graph can also be used to determine the loss in both branches, because the junction is symmetrical. Comparing the relevant parts of Figure 10–18 and Figure 10–17 gives an indication of the difference sharp edges make.

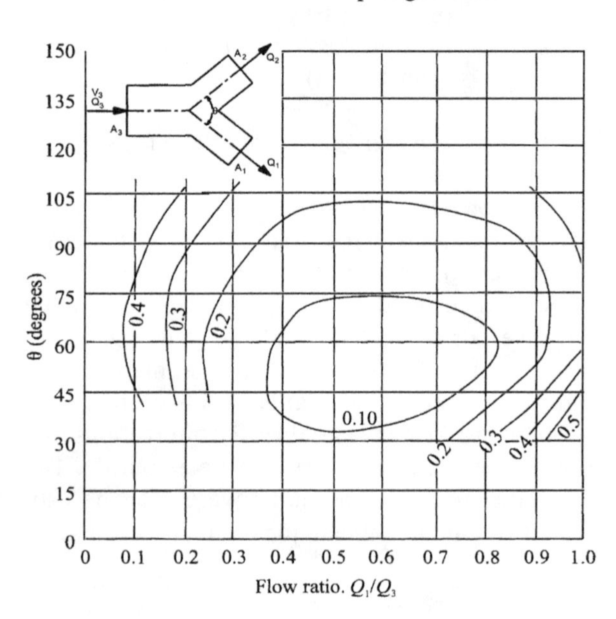

Fig. 10–18: Loss coefficient K_{31} for symmetrical Y-junctions. $A_3 = A_1 + A_2$. (Miller, 1990)

Figure 10–19 shows K_{31} for a good junction, that has taper reductions into the two branches. Again, the test pieces had radiused corners, which are time consuming to produce in a workshop. Real joints are more likely to have sharp inside edges, and will not be so good.

Finally, Figure 10–20 shows the loss coefficients, K_{32}, for the straight-through paths of bifurcations like those in Figure 10–17 and Figure 10–19. K_{32} is little affected by the geometry or the flow ratio, and this graph is valid for most branches off a straight pipe.

10.7.4 Diameter transition losses

Figure 10–21 shows the loss coefficient for contractions in a pipe. Note that the head loss is calculated from the water velocity in the smaller pipe, the outlet of the contraction:

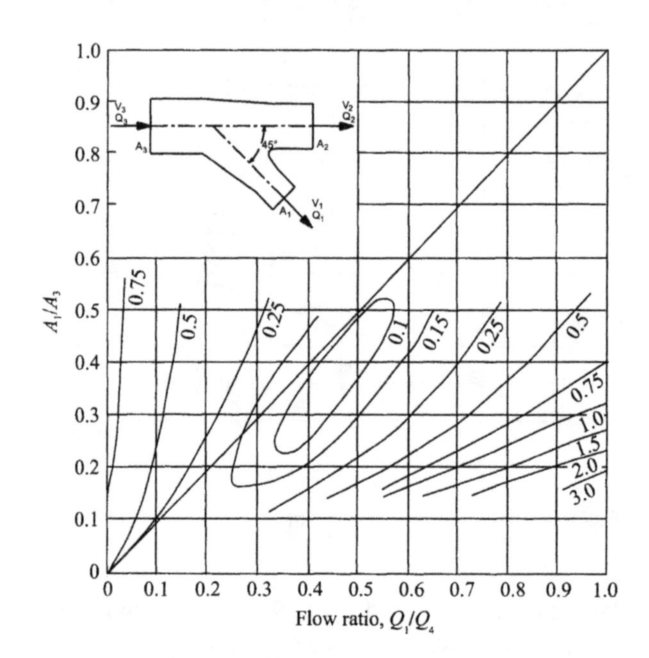

Fig. 10–19: Loss coefficient K_{31} for tapered 45° junctions. $A_3 = A_1 + A_2$. (Miller, 1990)

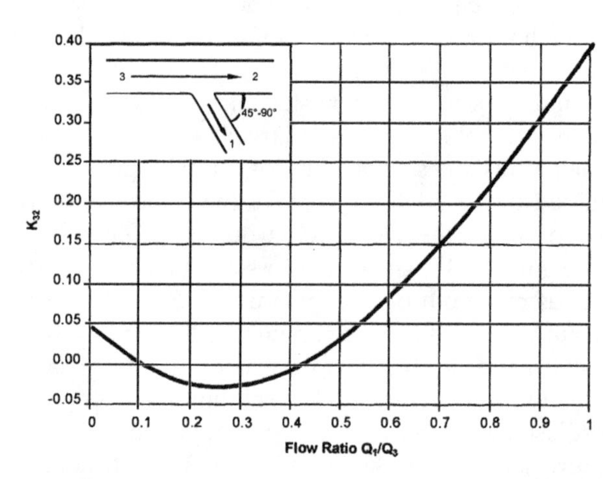

Fig. 10–20: Loss coefficient K_{32} for the straight-through branch of an angled junction. (Miller, 1990)

$$H_c = K_c \frac{V_2^2}{2g} \qquad \textit{Eq. 10-93}$$

H_c – head loss over contraction (m)
V_2 – water velocity at the *outlet* (m/s)

It can be seen that contraction losses are not too significant, even for sharp, stepped reductions in diameter. Note that when the reduction is part of a bifurcation, as in Figure 10–20, the contraction loss is included as part of the overall junction head loss, and does not need to be added in again.

10.7.5 Worked example

Problem

Consider the welded steel manifold in Figure 10–22. The water has come down a Ø100mm bore

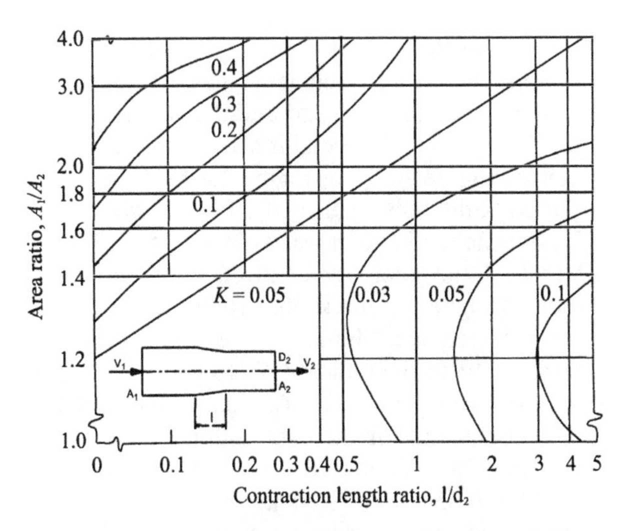

Fig. 10–21: Loss coefficient K_c for contractions. (Miller, 1990)

penstock, and enters the powerhouse through a 35° mitre bend. A ∅70mm pipe branches off to the lower nozzle, and the main pipe contracts to ∅70mm for the upper nozzle. Each jet takes 10l/s of water. What are the head losses from the end of the penstock to the nozzles when both jets are operating?

Solution

Each branch can take 10 litres/s (0.01m/s), so the velocity in the ∅70mm (0.07m) branches is

$$V = \frac{Q}{A} = \frac{0 \cdot 010}{\left(\frac{\pi}{4} 0 \cdot 07^2\right)} = 2 \cdot 60\,\text{m/s}$$

Similarly, when both jets are open, the velocity in the penstock is 2.55m/s, and after the bifurcation, the velocity of 10 l/s in the ∅100mm pipe is 1.27m/s.

The losses in the pipes comes from Table 10–3 and Equation 10–90. For the ∅70mm pipe, using k = 0.02 mm:

$$\text{Re} = \frac{V.d}{\nu} = \frac{2 \cdot 60 \times 0 \cdot 07}{1 \cdot 57 \times 10^{-6}} = 115{,}900$$

$$f = \frac{1}{4.\left[\log\left(\frac{k}{3 \cdot 7d} + \frac{5 \cdot 74}{\text{Re}^{0 \cdot 9}}\right)\right]^2}$$

$$= \frac{1}{4.\left[\log\left(\frac{0 \cdot 00002}{3 \cdot 7 \times 0 \cdot 07} + \frac{5 \cdot 74}{[115{,}900]^{0 \cdot 9}}\right)\right]^2}$$

$$= 0.0190$$

Similarly, for the ∅100mm pipe, Re = 162 400, f = 0.0176 before the junction, Re = 80 890, f = 0.0197 after.

These values can be used to calculate K_f values for each length of pipe. So, for example, in the 795mm long section of the ∅100mm pipe:

$$K_f = f\frac{L}{d} = 0.0176\,\frac{0 \cdot 795}{0 \cdot 1} = 0.14$$

Other values are given in Table 10–4.

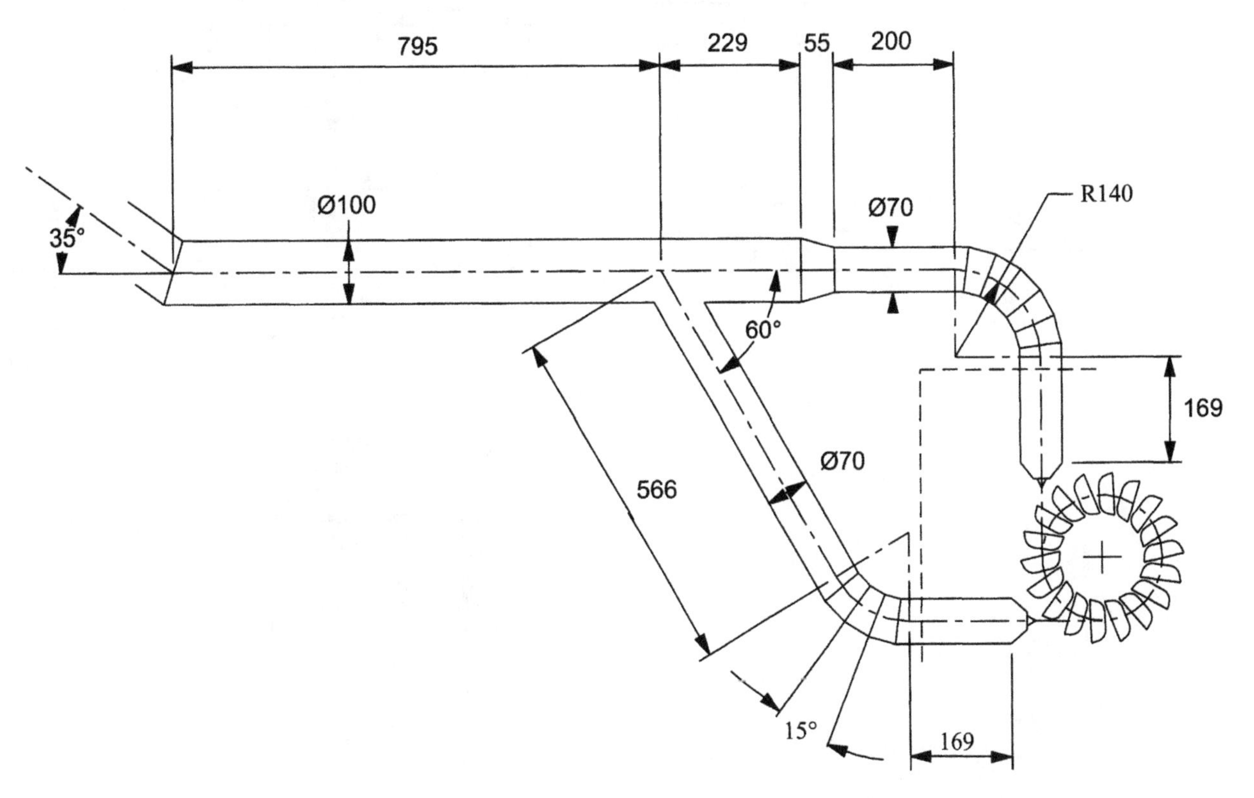

Fig. 10–22: Manifold system

Table 10–4: Worked example results

Component	*Loss coefficient K*	*Reference velocity* (m/s)	*Head loss* (m)
MANIFOLD ROUTE TO TOP BRANCH:			
Mitred bend, 35°	$K_b = 0.37$	2.55	0.12
Straight, Ø100 × 0.795m	$K_f = 0.14$	2.55	0.05
Bifurcation	$K_{32} = 0.03$	2.55	0.01
Straight Ø100 × 0.229m	$K_f = 0.05$	1.27	0.00
Contraction	$K_c = 0.05$	2.60	0.02
Straight, Ø70 × 0.200m	$K_f = 0.05$	2.60	0.02
Compound mitred bend, 90°	$K_b = 0.50$	2.60	0.17
Straight, Ø70 × 0.169m	$K_f = 0.05$	2.60	0.02
Total			0.41m
MANIFOLD ROUTE TO BOTTOM BRANCH:			
Mitred bend, 35°	$K_b = 0.37$	2.55	0.12
Straight, Ø100 × 0.795m	$K_f = 0.14$	2.55	0.05
Bifurcation	$K_{31} = 0.77$	2.55	0.26
Straight Ø70 × 0.566m	$K_f = 0.15$	2.60	0.05
Compound mitred bend, 60°	$K_b = 0.30$	2.60	0.10
Straight, Ø70 × 0.169m	$K_f = 0.05$	2.60	0.02
Total			0.60m

Next, consider the bends. To find the loss in the 35° bend, read K_b from Figure 10–15; $K_b = 0.37$. The bends on the branches are compound mitres. The upper branch bend is 90°, and Figure 10–16 shows K_b for this type of bend. The bend has $r = 2d$, and is made of 6 × 15° segments, so it will be somewhere between the 4 × 22.5° and the radiused bend line, $K_b = 0.5$. The 60° bend is not covered properly by either Figure 10–16 or Figure 10–15. If it were a radiused bend, then Figure 10–15 would give K_b as 0.25. For the top bend, it was estimated from Figure 10–16 that the compound mitred joint had a K_b about 0.1 higher than a similar radiused bend. The losses incurred by 60° bends will be less, so we might add, say, 0.05 to the radiused K_b to get the effect of the mitres, giving a total of $K_b = 0.25 + 0.05 = 0.3$.

The coefficient for the contraction on the upper line comes from Figure 10–21. $A_1/A_2 = 100^2/70^2 = 2$, and $l/d_2 = 55/70 = 0.79$, giving $K_b = 0.05$.

Lastly, consider the bifurcation. For the upper branch, look at the straight-through loss graph, Figure 10–17. For $Q_1/Q_3 = 10/20 = 0.5$, $K_{32} = 0.03$. For the upper branch, consult Figure 10–17, $Q_1/Q_3 = 10/20 = 0.5$, $A_1/A_3 = 70^2/100^2 = 0.5$, giving $K_{31} = 0.77$.

All these results shown in Table 10–4, together with the reference velocities used to calculate the velocity head for the coefficient. To illustrate this, the equation below shows the calculation for the 35° mitre bend.

$$H_b = K_b \frac{V^2}{2g} = 0 \cdot 37 \frac{2 \cdot 55^2}{2 \times 9 \cdot 8} = 0 \cdot 12$$

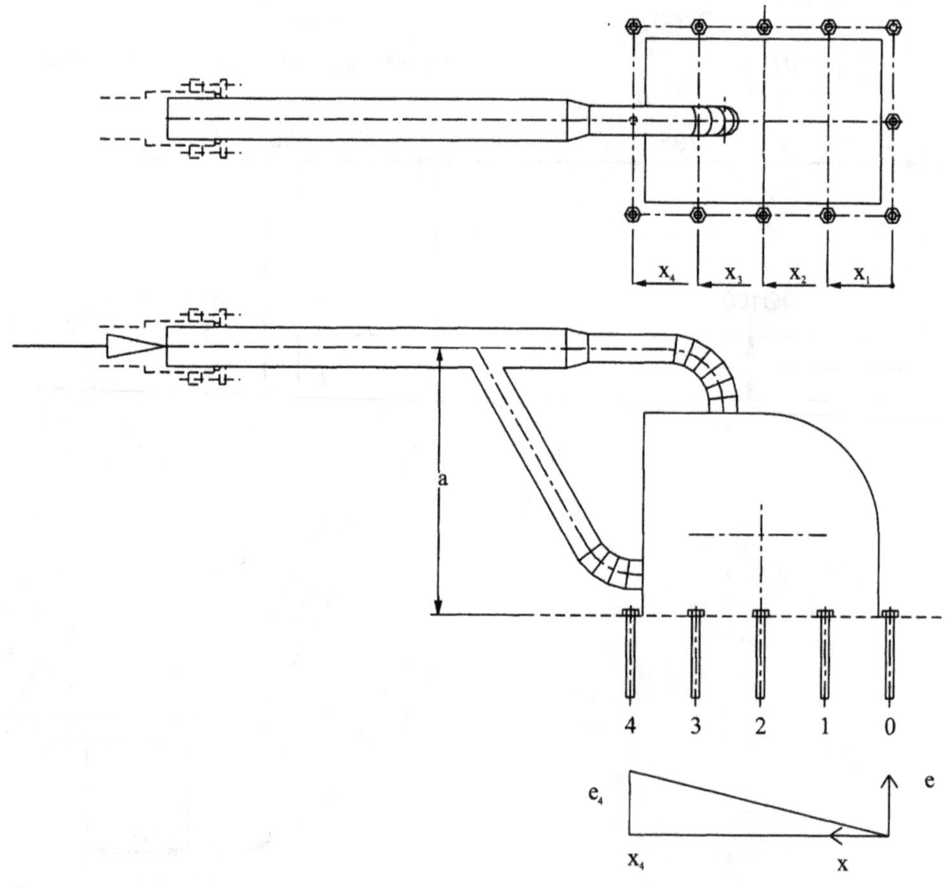

Fig. 10–23: Forces acting on a turbine housing

10.8 Baseframe anchor bolt loads

This section derives and explains the background of the formulae used in Section 4.13.2.

In Figure 10–23, a turbine and manifold is considered as a free body. The manifold is joined to the penstock by an expansion joint. If the friction in the joint is ignored, the only force acting along the penstock is the pressure force, *P*, equal to the pressure multiplied by the pipe cross-sectional area. The turbine is bolted to the floor of the powerhouse with anchor bolts around the baseframe.

The force *P* is taken by the bolts in shear. Assuming that the housing and frame are completely rigid, if there are *N* bolts, each bolt will take a shear force of *P*/*N*.

Working out the tensile force in the bolts is no easy matter. The force *P* induces a moment on the turbine, and this will tend to pivot it around its right hand edge, pulling upwards on the anchor bolts. How the load is distributed in the bolts depends on the relative stiffness of different parts of the casing. Also, when an external load is applied to a pretensioned bolt, not all of the force goes into the bolt. The amount that the bolt carries depends on the relative stiffnesses of the bolt and the assembly it is clamping. The analysis in this text assumes a completely rigid turbine frame, that all the load goes into the bolts, and that the bolts are all the same. Because these assumptions are never true, the results calculated are indicative only. With a bit of experience, the figures can be modified to estimate the real loads, and allow the size of the anchor bolts to be calculated. A more detailed analysis is given in Shigley & Mischke (1989).

Assume that the turbine pivots around the line of the centres of the bolts on the right side. The bolts will stretch slightly, and because of the assumption that the turbine is rigid (i.e. there is no deflection or bending in it), the amount if stretch e_i at a point x_i from the pivot line will be proportional to the distance x_i. The extra force induced in the bolts is proportional to the stretch, and can be written with a constant '*k*' as:

$$F_i = k.e_i = k.e_4.\frac{x_i}{x_4} \qquad \textit{Eq. 10–94}$$

The force *P* has a moment about the pivot line of $M = P \times a$. This must be equal to the sum of the moments due to the bolt forces, and this allows us to find the value of each bolt force.

$$P.a = \sum n_i.F_i.x_i = \sum n_i.k.e_4.\frac{x_i}{x_4}.x_i = k\frac{e_4}{x_4}\sum n_i.x_i^2$$

$$\therefore\ F_i = k.e_4.\frac{x_i}{x_4} = \frac{P.a.x_4}{\sum n_i.x_i^2}.\frac{x_i}{x_4} = \frac{P.a.x_i}{\sum n_i.x_i^2} \qquad \textit{Eq. 10–95}$$

x_i – horizontal distance from pivot
n_i – number of bolts at position x_i
F_i – force in each bolt at position x_i

An example of this calculation is given in Section 4.13.2, which also contains a discussion on how to adjust the result for a real turbine situation.

Note that this formula works if the penstock, and force *P*, is at an angle instead of horizontal. Only the moment induced by *P* needs to be considered. A downwards vertical components of *P* goes directly into the floor, not through the bolts. An upwards vertical component only affects the calculations if a gap appears between the floor and the turbine – which should never happen (if it did, the bolting arrangement would be shown to be totally inadequate).

11

APPENDIX: FULL SYSTEM CALCULATION SPREADSHEET

The calculations required to choose a Pelton turbine are quite laborious if done manually, but are easily handled on a personal computer. Once a spreadsheet has been set up, numerous options can be tested very quickly. This section shows a spreadsheet set up for Microsoft Excel, though it can easily be adapted for other spreadsheet software.

Consider the process of starting a Pelton turbine. Initially, when the nozzle valves are shut, the pressure at the turbine is the full static head. When a valve is first opened, this head accelerates the water in the penstock, and pushes it out of the nozzle as a jet. As the speed of the water in the penstock increases, so do the losses. This has the effect of reducing the head at the nozzle, and consequently reducing the flow in the jet. There is an equilibrium flow at which the losses in the penstock give a head at the nozzle which gives exactly that flow in the jet. This is the operating flow, and it is this flow that the spreadsheet is calculating.

The spreadsheet (Table 11–1) is divided into inputs – on the left hand side, and outputs – on the right. The outputs are calculated from the inputs, and the formulae used are shown in Table 11–3. All the input and output value cells in the spreadsheet have been given names (shown in the 'Symbol' columns) so that these range names can be used in the formulae. It would be possible to use the cell addresses instead, but using the symbol names makes the formulae much easier to read and understand. The formulae are correct for the units given in Tables 11–1 and 11–2. Remember that if the units for any value are changed, all the formulae that refer to it must be adjusted.

When first setting up this spreadsheet on a computer, it is recommended that the values given in the sample spreadsheets are tried out to check that everything is correct. Two different examples are given because the penstock thickness calculations have an 'IF' statement, and two different sets of equations are used depending on whether the closure time is greater than, or less than, the critical time. The full spreadsheet shown in Table 11–1 has values which test for instantaneous flow blockage, the part spreadsheet in Table 11–2 tests for slow valve closure. Table 11.2 uses the same variables as Table 11–1 except for the values in the penstock thickness calculations. Test the spreadsheet with both values of `z` and `t_close` (`z` = 65%, `t_close` = 0 or `z` = 100%, `t_close` = 0.8s) to check it.

The formulae used are the same as those used in other places in this book, with the exception of the slow valve closure surge formula; this latter is taken from Inversin (1986), though it is easily derived and can be found in most standard hydraulics textbooks. (Note that there is a slight error in Inversin, in that the formula omits the term for the gross head.) The basic turbine equations are given in Section 3.1, the penstock loss is calculated using the same equations as for manifold pipe losses given in Section 10.7.1, and the thickness comes from the equations in Section 4.10.3.

11.1 Using the spreadsheet

The procedure for using the spreadsheet is to fill in the inputs on the left hand side, and then check the outputs on the right hand side. The spreadsheet has three parts which can be solved separately. It is easiest to sort out Part 1, 'Penstock losses, Turbine jet size and speed calculations' first, then move on to Part 2, 'Power calculations', and finally do Part 3, 'Penstock thickness calculations'. The bottom part, 'Constants', needs to be filled in for all calculations.

11.1.1 Part 1, Penstock losses, turbine jet size and speed calculations

The 'Constants' `g`, `K_w`, `ro` and `nu` can be filled in with the values given in the example. Provided fresh water is being used, and the scheme is located on earth, these values should suffice. In Part 1, the values for flow `Q`, gross head, `H_g`, and penstock length `L`, come from site survey data, and can be filled in straight away. Values for the penstock material roughness coefficient `k`, depend on the type of penstock. Values of `k` are given in in Table 10–3 in Section 10.7.1. A figure of 0.1 is used here as a general value for rough-painted fabricated mild steel pipes.

The inlet and penstock minor loss coefficients `K_m`, are the sum of the loss coefficients for all the bends, valves, and the inlet. If the penstock has an

Table 11–1: Spreadsheet values for instantaneous valve closure

	A	B	C	D	E	F	G	H
1	**PELTON TURBINE FULL SYSTEM CALCULATION SPREADSHEET**							
2	**INPUT DATA**				**CALCULATED OUTPUTS**			
3	**SITE NAME**	**Sample**						
4	**1. Penstock Losses, Turbine Jet Size & Speed Calculations**							
5	***Quantity***	***Symbol***		***Unit***	***Quantity***	***Symbol***		***Unit***
6	Flow	Q	32	l/s	Velocity in penstock	V	1.811	m/s
7	Gross head	H_g	60	m	Friction head loss in penstock	H_f	2.676	m
8	Penstock length	L	120	m	Minor head losses in penstock	H_c	0.502	m
9	Penstock diameter	d	150	mm	Net head at end of penstock	H_n	56.82	m
10	Penstock roughness coefficient	k	0.1	mm	Penstock efficiency	e_pen	94.7	%
11	Inlet & penstock minor coefficient	K_m	3		Nett head at turbine	H_tur	55.69	m
12	Manifold efficiency	e_man	98	%	Jet velocity	v_jet	32.05	m/s
13	Pelton PCD	PCD	250	mm	Jet size (water)	d_jet	25.21	mm
14	Number of jets	n_jet	2		Max. allowable jet size	d_max	27.5	mm
15	Max. jet dia./PCD	beta	11	%	Turbine optimum speed	N	1126	rpm
16	Nozzle velocity coefficient	c_v	0.97		Turbine runaway speed	N_r	2027	rpm
17								
18	**2. Power Calculations**							
19	Drive efficiency	e_drv	95	%	Total system efficiency	e_tot	56.41	%
20	Generator efficiency	e_gen	85	%	Turbine mechanical power	P_tur	13.14	kW
21	Turbine efficiency	e_tur	80	%	Electrical power	P_elec	10.61	kW
22								
23	**3. Penstock Thickness Calculations**							
24	Penstock thickness	t	3	mm	Wave velocity in penstock	V_wave	1183	m/s
25	% of flow stopped	z	63	%	Penstock critical time	T_crit	0.203	s
26	Valve closure time	T_close	0	s	Surge head	H_sur	137.7	m
27	Corrosion allowance	t_cor	1.5	mm	Total head at surge	H_tot	197.7	m
28	Penstock UTS	s_ult	410	N/mm2				
29	Penstock Young's Modulus	E_p	210	kN/mm2	Required penstock thickness	t_req	2.904	mm
30	Overall safety factor	SF_tot	3.96					
31								
32	**Constants**							
33	Gravity	g	9.8	m/s				
34	Bulk density of water	K_w	2.1	kN/mm2				
35	Density of water	ro	1000	kg/m3				
36	Kinematic viscosity of water (5°C)	nu	1.57	cSt				

Table 11–2: Part Spreadsheet for slow valve closure

	A	B	C	D	E	F	G	H
	3. Penstock Thickness Calculations			23				
24	Penstock thickness	t	2.2	mm	Wave velocity in penstock	V_wave	1117	m/s
25	% of flow stopped	z	100	%	Penstock critical time	T_crit	0.215	s
26	Valve closure time	T_close	0.8	s	Surge head	H_sur	34.85	m
27	Corrosion allowance	t_cor	1.5	mm	Total head at surge	H_tot	94.85	m
28	Penstock UTS	s_ult	410	N/mm2				
29	Penstock Young's modulus	E_p	210	kN/mm2	Required penstock thickness	t_req	2.173	mm
30	Overall safety factor	SF_tot	3.96					
31								

Table 11–3: Formulae used in Full System Calculation Spreadsheet

Cell	*Cell name*	*Formula*
G6	V	=4*(Q/1000)/(PI()*(d/1000)^2)
G7	H_f	=V^2/(2*g)*L/(d/1000)*1/(4*(LOG10((k/1000)/(3.7*d/1000) +5.74/(V*(d/1000)/(nu/1000000)^0.9))^2)
G8	H_c	=K_m*V^2/(2*g)
G9	H_n	=H_g-H_f-H_c
G10	e_pen	=H_n/H_g*100
G11	H_tur	=H_n*(e_man/100)
G12	v_jet	=c_v*SQRT(2*g*H_tur)
G13	d_jet	=SQRT(4*(Q/1000)/(n_jet*PI()*v_jet))*1000
G14	d_max	=PCD*beta/100
G15	N	=0.46*v_jet*60/(PI()*PCD/1000)
G16	N_r	=N*1.8
G19	e_tot	=e_pen/100*e_man/100*c_v^2*e_tur/100*e_drv/100*e_gen/100*100
G20	P_tur	=H_tur*ro*g*Q/1000*c_v^2*e_tur/100*1/1000
G21	P_elec	=H_g*ro*g*Q/1000*e_tot/100*1/1000
G24	V_wave	=1/SQRT(ro*(1/(K_w*1000000000) +(d/1000)/(E_p*1000000000*t/1000)))
G25	T_crit	=2*L/V_wave
G26	H_sur	=IF(T_close<T_crit,V_wave*z/100*Q/1000/(PI()4*(d/1000)^2*g), +H_g*((L*z/100*V/(g*H_g*T_close))^2/2 +SQRT((L*z/100*V/(g*H_g*T_close))^2 +(L*z/100*V/(g*H_g*T_close))^4/4)))
G27	H_tot	=H_g+H_sur
G29	t_req	=H_tot*ro*g*(d/1000)/(2*s_ult*1000000/SF_tot)*1000+t_cor

inlet loss coefficient of K_i, and three bends with coefficients of K_{b1}, K_{b2} and K_{b3}, then

$$K\_m = K_i + K_{b1} + K_{b2} + K_{b3}.$$

As a rule-of-thumb for initial calculations, a value of $K_b = 0.5$ can be used for each bend. More detailed values of K_b can be looked up in Figure 10–15 or Figure 10–16. Inlet losses are shown in Figure 11–1. Generally speaking, the minor losses do not greatly affect the performance, and a rough guess for K_m will be sufficient for most calculations.

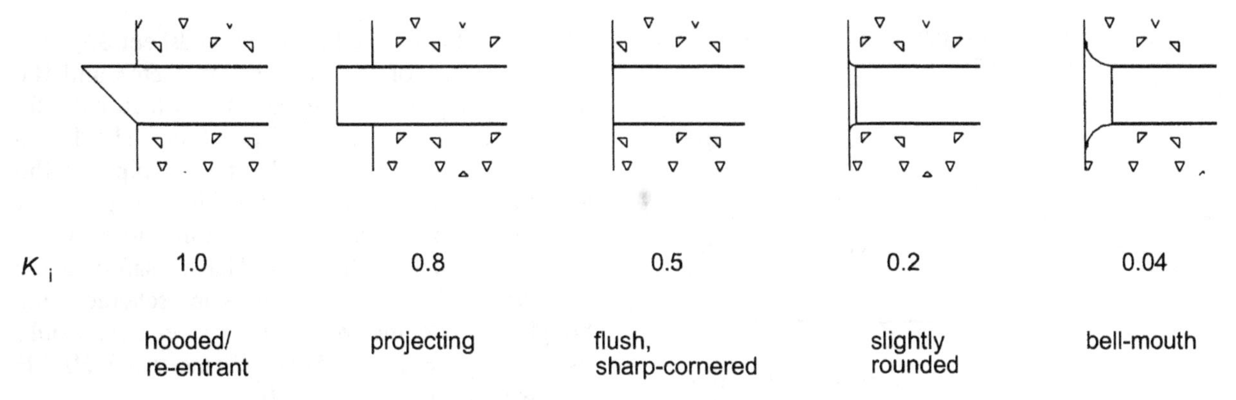

Fig. 11–1: Loss coefficients for various penstock inlets (after Inversin, 1986; and Harvey et al, 1993)

The manifold efficiency will typically be 93–97%. A straight forward, smooth reduction for a single jet will have high efficiency, a complicated manifold with many branches will be somewhat lower.

The nozzle coefficient c_v, depends on the type of nozzle being used. A value of 0.97 is acceptable for preliminary calculations. Better values can be obtained from Section 2.2.1 or Section 2.2.2.

Beta is the ratio of the maximum jet diameter to the PCD, and is determined by the bucket design. The bucket used in this manual is designed for a beta of 11% nominally, and can be pushed to 12%. See the discussion in Section 3.1.4 for more detail.

Having put in the basic data, it is now time to make some guesses for the penstock diameter d, the PCD, and the number of jets n_jet. For those not familiar with Pelton calculations, it is advisable to go through the selection procedure in Chapter 3, and use the values obtained from there in this section. The alternative approach is to start with a large value of d (say 20–30% more than would be expected, and the adjust the PCD until approximately the required speed, N, is achieved. Next, adjust the values of n_jet until d_jet is less than the maximum allowable jet size d_max. If this gives too many jets, the PCD will have to be increased, and the speed will have to be accepted. Next, reduce d until the penstock efficiency e_pen, comes to about 95%. Finally, refine all the values until a reasonable looking design is achieved.

The spreadsheet is set to determine the value of d_jet for a given value of Q. Once a turbine has been designed, it is often useful to calculate the other way round. For example, a 2-jet turbine may be designed for a maximum flow, but one may wish to calculate the flow and efficiency with only one jet. This can be done using the 'goal-seek' option in Excel, giving a value for jet diameter and changing Q to achieve this. Goal-seeking can obviously be used to try other options too, though one has to choose variables that are related through the equations.

Note that the spreadsheet assumes a single penstock pipe of the same diameter for its whole length. It can be modified for multiple pipes or for changes in diameter, though the latter requires a good understanding of the equations. Similarly, it is possible to include the manifold as a series of pipe pieces plus bend and junction loss coefficients, but this is quite complicated to do, and changes for each manifold layout. Losses in the manifold are usually not too significant, and making allowances for them as an efficiency is a reasonable approximation.

11.1.2 Part 2, Power calculations

This part of the spreadsheet processes the information from Part 1. The input values are the drive efficiency e_drv, the generator efficiency e_gen, and the turbine efficiency e_tur.

The drive efficiency will be 100% for direct-coupled arrangements, and perhaps 94–97% for belt drives. The generator efficiency should be available from generator manufacturers. Figures of 80–85% are reasonable for small induction motors, up to 90% for larger (50kW+) synchronous generators.

The turbine efficiency depends on the size and the accuracy of manufacture. A general figure of 75–80% is reasonable for micro-hydro, though better efficiencies may be obtained for large units (50kW+), worse for very small ones (<5kW).

Once these efficiency figures are put in, the spreadsheet returns the power outputs, which can be checked against the expected values.

11.1.3 Part 3, Penstock thickness calculations

Part three calculates the required thickness of the penstock for surge induced by the turbine. It is not necessary to use this part of the spreadsheet unless one is choosing a penstock thickness.

The Young's Modulus E_p, and Ultimate Tensile Strength s_ult, for the penstock material can be put in as soon as the material is chosen. Typical values for the most common materials are given in Table 11–4.

The corrosion allowance t_cor, is the amount of the penstock wall that can be lost to corrosion over the design life of the penstock – which might be 15 or 20 years for micro-hydro plants (large schemes work on 50 years for the civil works). The corrosion

Table 11–4: Typical properties of common penstock materials

	Young's Modulus E_p (kN/mm²)	*Ultimate Tensile Strength* s_ult (N/mm²)	*Density* (kg/m³)
Mild steel	210	350	7850
HDPE	0.2–0.8	6–9	900
uPVC	2.8	28	1400

allowance depends on the corrosion protection applied to the penstock. Typical values are given in Table 4–7. For plastic penstocks, no corrosion allowance is necessary.

The valve closure time `T_close`, and the percentage of the flow stopped `z`, are discussed in Section 4.3.5. If the flow is blocked instantaneously, say by the spear valve end falling into the nozzle, `T_close` is zero. If it can be guaranteed that such a failure can never occur, then the most rapid possible valve closure time can be put in. The amount of flow stopped depends primarily on how many jets there are: for a 4-jet Pelton, the value of `z` is likely to be 25%, but see Section 4.3.5 for more detail.

The spreadsheet calculates the stress in the penstock from the expected surge or water hammer head. The allowable head depends on the safety factor used. Choosing a safety factor is something of an art, relying mainly on experience. Safety factors for penstocks tend to be large, as there is usually quite a degree of uncertainty about the loads. Suggested safety factors are given in Table 11–5. The safety factor used in the spreadsheet `SF_tot`, is the product of the basic safety factor and the hand-welding factor. The basic safety factor should be applied to any type of penstock. The hand-welding factor is only used for steel pipes if the longitudinal welds are hand-welded. A penstock made from rolled mild steel plate in a local workshop for a 15kW scheme would have a safety factor `SF_tot` = 3.5 × 1.2 = 4.2. The same scheme using bought-in, machine welded mild steel tube would use `SF_tot` = 3.5. A 5kW scheme using HDPE pipe would use `SF_tot` = 3.0.

Table 11–5: Penstock thickness safety factors (Waltham, 1994)

Size of scheme	*Very small* (<8kW)	*Small* (8–20kW)	*Medium* (20–50kW)	*Large* (>50kW)
Basic safety factor	3.0	3.5	4.0	4.0
Hand-welding factor	1.2	1.2	1.1	1.1

Once the values of `z`, `T_close`, `t_cor`, `s_ult`, `E_p` and `SF_tot` have been put in, a trial value of thickness, `t`, can be inserted. The spreadsheet will calculate the required thickness `t_req`, and `t` should be adjusted until it is a standard thickness that is greater than `t_req`. Note that the penstock thickness calculations are done for the bottom of the penstock only. Thickness can be reduced because of the lower static head higher up the penstock.

12
APPENDIX: O-RING SEALS

O-rings are cheap, reliable seals that are very useful for micro-hydro. They form good, leak-free pipe flange seals, and can also be used to seal spear valves. A standard rubber O-ring will readily seal up to a 1000m head. To perform well and give good life, O-rings need to be installed correctly. This means choosing the correct size of O-ring and groove, and making sure the machining is accurate.

12.1 Sealing arrangements

The most common ways of using O-rings are shown in Figure 12–1. The first diagram shows a flange seal, and the other two show internal and external rod seals. First decide which type of sealing method is to be used. Static sealing is where the sealing surfaces do not move relative to the O-ring. Dynamic sealing is where one of the surfaces moves relative to the O-ring. A flange-type seal is assumed to be static, but a rod seal can be static or dynamic. The seal on the shaft of a spear valve is dynamic, because the shaft moves, and the O-ring rubs along the sealing surface. If a governor is used to control the spear, the application is truly dynamic, because the spear valve can be moving continuously. If the spear is controlled manually, the application is semi-dynamic, and slightly more squeeze can be used if problems are experienced with leakage.

12.2 O-ring section and groove details

Check which O-ring cross-sections (dimension '*a*') are available, and see which can be fitted. Within reason, use the largest section possible. In order to maintain a pressure inside the rubber to achieve sealing, the O-ring has to be squeezed in the

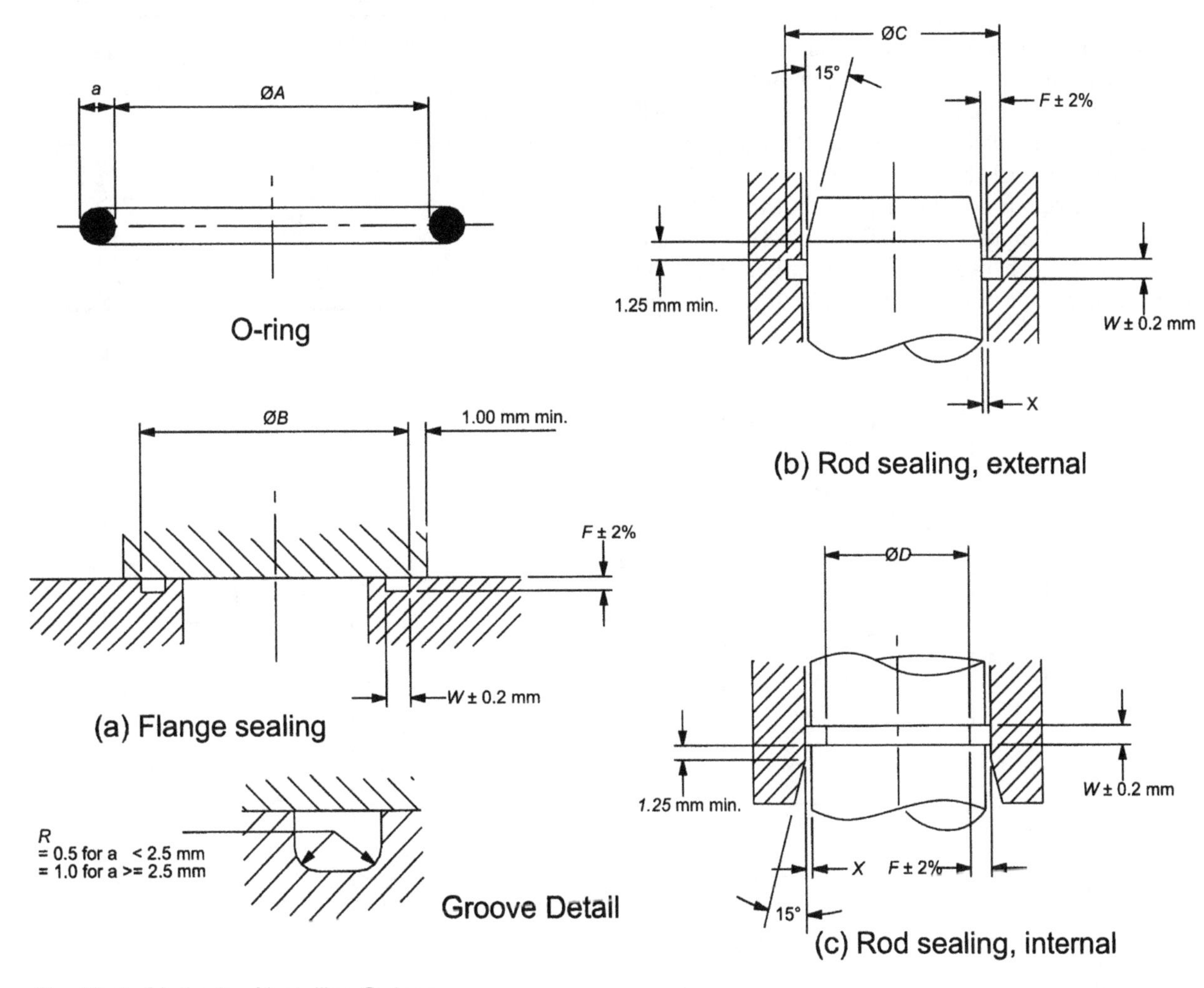

Fig. 12–1: Methods of installing O-rings

groove. This means that the dimension F must be less than the O-ring section, a. The amount of squeeze depends on the O-ring section, and the application. Small O-rings need more squeeze to make sure they seal. Static seals can tolerate a high amount of squeeze, but dynamic applications use less squeeze to reduce the friction and wear on the O-ring. If possible, followed the suppliers recommendations for selecting the size. If no recommended groove sizes are available, then determine the amount of squeeze from Figure 12–2. This graph is plotted from manufacturers recommended groove sizes for a number of different sizes (Dowty, 1978).

$$F = a.\left(1 - \frac{\%\ squeeze}{100}\right) \qquad Eq.\ 12\text{–}1$$

The next stage is to determine the width of groove required, W. Surprisingly, perhaps, rubber is incompressible. If it is squeezed in one direction, it needs space to expand in another. Rubber also has a relatively high coefficient of volumetric expansion with temperature. The groove needs to have space both for the O-ring to be squeezed sideways and for thermal expansion. The maximum width for static applications is not too critical, but for dynamic applications the width needs to be kept to a minimum, to stop the O-ring shuffling about in the groove and wearing. The recommended width of the groove can be read off from Figure 12–3 – if better information is not available. The graph gives values of W/a for various values of a.

The groove has now been sized, but the detail of the groove is important too. The corners must not have too large a radius, R, or it will reduce the volume available for the O-ring. R should be no larger than 0.5mm for O-rings up to 2.5mm section, and 1.0mm for larger sizes.

The surface finish depends on the application. Published recommended values are shown in Table 12–1. Machining needs to be done carefully to achieve these values. A dynamic rod surface should ideally be ground, though for a manually adjusted spear valve, a turned finish is acceptable. Also, there is no point having a perfect surface finish if the surface is going to corrode in service. If the surface is exposed to water, the material needs to be corrosion resistant, ideally stainless steel.

Table 12–1: O-ring housing surface finish for different applications

	Static	*Dynamic*
	Ra (μm)	
Housing	1.6	0.8
Mating surface (rod)	0.8	0.4

For rod seals, the gap between the rod and the housing must be checked. If this gap is too big the O-ring may be extruded into it. Table 12–2 gives recommendations for the maximum values of X.

Table 12–2: Clearance between shaft and housing

O-ring section (mm)		X_{max} (mm)
Over	Incl.	
	2.50	0.12
2.50	3.00	0.15
3.00	4.00	0.16
4.00	5.00	0.17
5.00	8.00	0.18
8.00	10.00	0.20

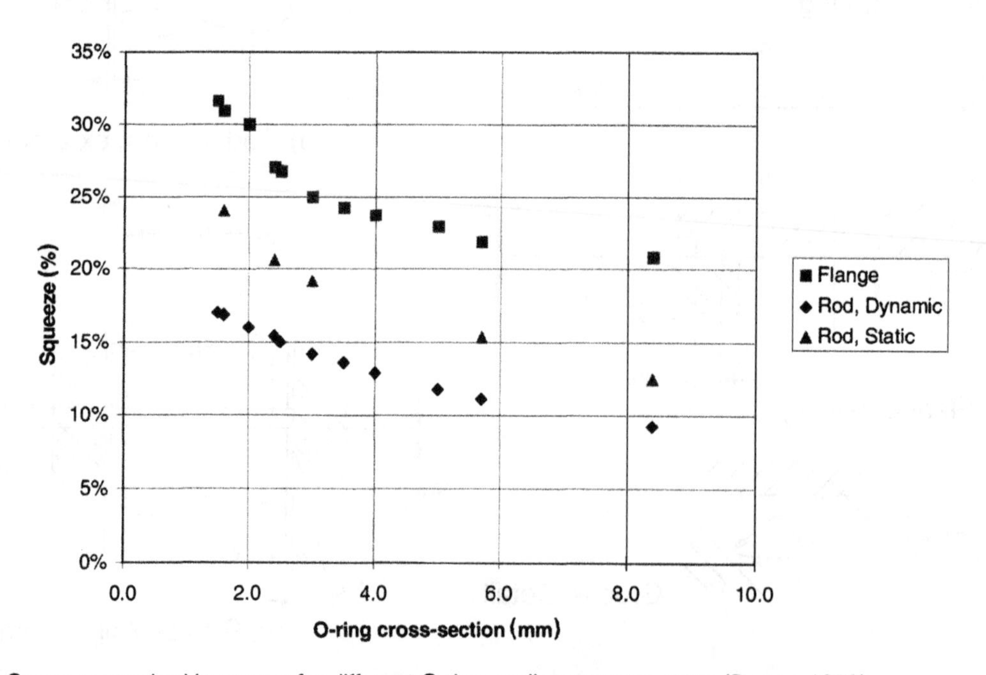

Fig. 12–2: Squeeze required in groove for different O-ring sealing arrangements (Dowty, 1978)

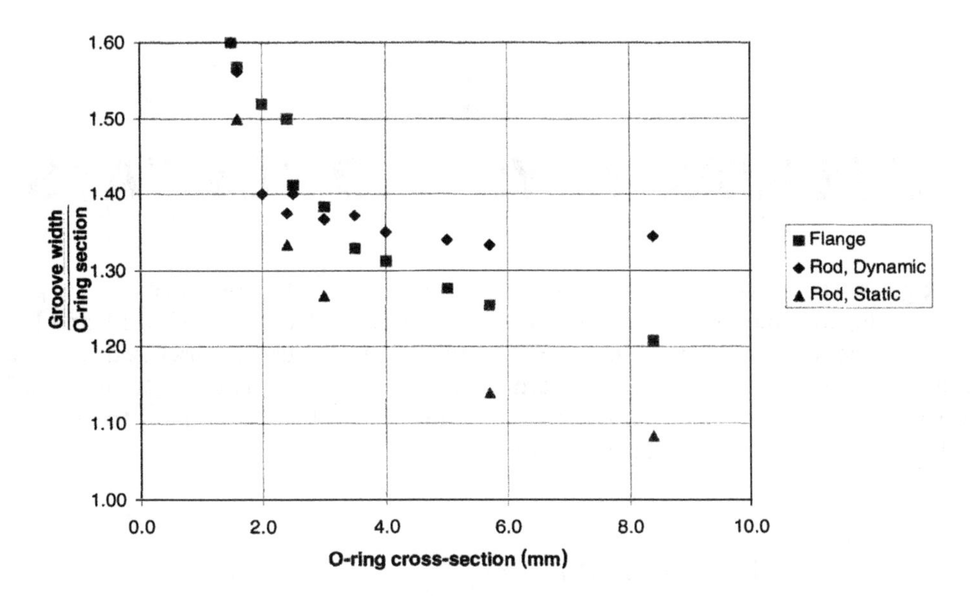

Fig. 12–3: Groove width/section for different O-ring sealing arrangements

12.3 O-ring diameter

The final parameter to choose is the diameter of the O-ring itself. For a flange seal (with internal pressure, as assumed here) the O-ring outside diameter needs to be the same size or slightly larger than the housing diameter B. For external rod sealing, the O-ring outside diameter should be the same size or slightly larger than the groove outside diameter. For internal rod sealing, the O-ring internal diameter should be the same size or slightly smaller than the groove internal diameter. These relationships are expressed in Equation 12–2, and W should be chosen so that it satisfies the appropriate equation.

Flange Seal:

$$\frac{W+2a}{B} = 1 \cdot 00 \text{ to } 1 \cdot 01$$

External Rod Seal:

$$\frac{W+2a}{C} = 1 \cdot 00 \text{ to } 0 \cdot 97$$

(1·00 to 0·92 in special circumstances - see below.)

Internal Rod Seal:

$$\frac{D}{A} = 1 \cdot 00 \text{ to } 1 \cdot 05$$

(1·00 to 1·08 in special circumstances - see below.)

Eq. 12–2

When designing O-ring housings, the effect of tolerancing on the O-ring should be considered. The need to have reasonable machining tolerances on the various components may make it impossible to keep the O-ring within the above limits, especially for small O-rings. If this is the case, the maximum compression or stretch can be increased to 8% to ensure that some compression or stretch remains in the worst tolerance condition.

12.4 Material

For use in water, oil or grease, most standard stock, nitrile-rubber O-rings will work. The recommended rubber for use with water has a nominal hardness of 70 IHRD (International Rubber Hardness test Shore type -D), and this should be requested for critical applications.

Large-diameter O-rings for pipe flanges can be made from continuous lengths of material. Calculate the length of rubber required as per the equations above, cut the ends at an angle, and glue together with cyanoacrylate adhesive ('superglue'). Smaller diameter O-rings should be bought moulded as one piece, without a join.

12.5 Installation

Note in Figure 12–1 that 15° chamfers are included on the rod or housing for the rod-type seals. This is to ensure that the O-rings are not damaged during assembly. Care should be taken not to catch the O-rings on sharp corners or threads during assembly, and not to use sharp tools to put them in place.

13
APPENDIX: KEYS AND KEYWAYS

This appendix gives recommended dimensions for both parallel and tapered metric keyways. Figure 13–1 shows the general layout, and Table 13–1 gives the dimensions and tolerances for standard keyways for various shaft diameters from Ø12mm–110mm, which should cover most micro-hydro applications. The tolerance classes given for the key width in Table 13–1 (N9, J_s9 and D10) are ISO tolerances, which are discussed more fully in Appendix 14 in the section on 'Limits and Fits'. For a more general discussion on the use of keys, see Section 4.8.3.

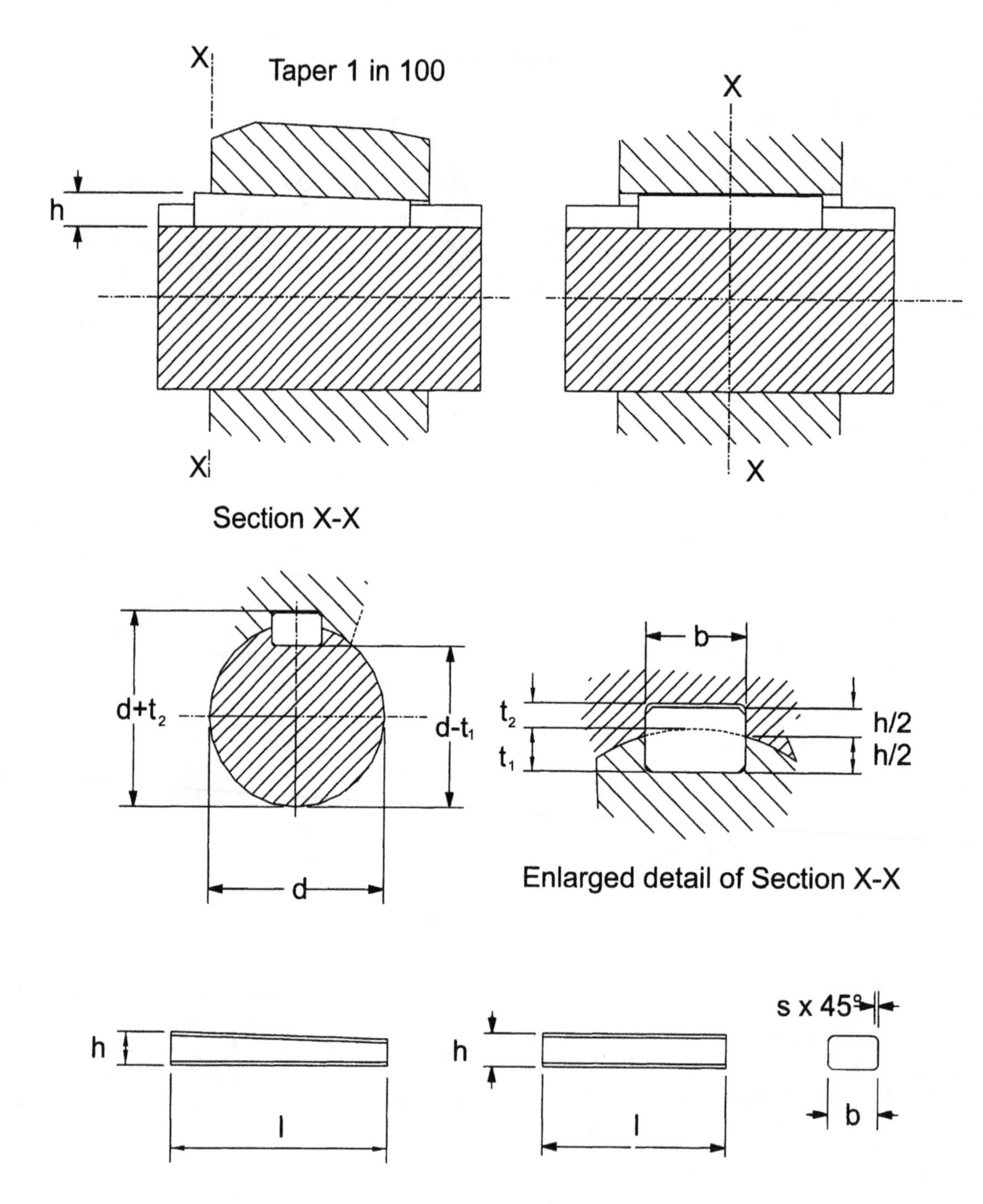

Fig. 13–1: Parallel and taper metric keys and keyways to BS 4235

Table 13–1: Selected parallel and taper metric key and keyway dimensions to BS 42345: Part 1: 1972. Reproduced with permission

SHAFT		*KEY*								*KEYWAY*											
Nom. dia.		*Size*								*Width, b*				*Depth*						*Radius*	
d		*Width b*		*Height h*		*Chamfer s*		*Length l*			*Tol. on parallel shaft*	*Tol. on parallel hub*	*Tol. on taper shaft & hub*	*Shaft, t_1*		*Parallel key hub t_2 (parallel)*		*Taper key hub t_2 (taper)*			
Over	*Incl.*	*Nom*	*Tol*	*Nom*	*Tol*	*Min*	*Max*	*Min*	*Max*	*Nom*	*(N9)*	*(J_s9)*	*(D10)*	*Nom*	*Tol*	*Nom*	*Tol*	*Nom*	*Tol*	*Max*	*Min*
(mm)	(mm)	(mm)	(µm)	(mm)	(µm)	(mm)	(mm)	(mm)	(mm)	(mm)	(µm)	(µm)	(µm)	(mm)	(mm)	(mm)	(mm)	(mm)	(mm)	(mm)	(mm)
12	17	5	0	5	0	0.25	0.40	10	56	5	0	+0.015	+0.078	3	+1.0	2.3	+0.1	1.7	+0.1/0	0.25	0.16
17	22	6	–0.030	6	–0.030	0.25	0.40	14	70	6	–0.030	–0.015	+0.030	3.5	0	2.8	*0	2.2	+0.2	0.25	0.16
22	30	8	0	7	0	0.25	0.40	18	90	8	0	+0.018	+0.098	4	+0.2	3.3	+0.2	2.4	0	0.25	0.16
30	38	10	–0.036	8	–0.090	0.40	0.60	22	110	10	–0.036	–0.018	+0.040	5	0	3.3	0	2.4		0.40	0.25
38	44	12	0	8		0.40	0.60	28	140	12	0	+0.021	+0.120	5		3.3		2.4		0.40	0.25
44	50	14	–0.043	9		0.40	0.60	36	160	14	–0.043	–0.021	+0.050	5.5		3.8		2.9		0.40	0.25
50	58	16		10		0.40	0.60	45	180	16				6		4.3		3.4		0.40	0.25
58	65	18		11	0	0.40	0.60	50	200	18	0	+0.026		7		4.4		3.4		0.60	0.25
65	75	20	0	12	–0.110	0.60	0.80	56	220	20	–0.052	–0.026	+0.149	7.5		4.9		3.9		0.60	0.40
75	85	22	–0.052	14		0.60	0.80	63	250	22			+0.065	9		5.4		4.4		0.60	0.40
85	95	25		14		0.60	0.80	70	280	25				9		5.4		4.4		0.60	0.40
95	110	28		16		0.60	0.80	80	320	28				10		6.4		5.4		0.60	0.40

(* +0.2/0 for taper keys)

14
APPENDIX: LIMITS AND FITS

When a bush has to be fitted in a housing, the size and tolerance of both components affects how they go together. A heavy press fit can be produced, where the bush has to hammered or pressed into the housing, and it is almost impossible to get it out again. Alternatively, a loose clearance fit means that the bush can be easily slipped into the housing, and can fall out again. By putting the correct tolerances on component drawings, a designer can determine exactly what sort of fit results. All too often, lack of knowledge of how to do tolerancing stops people from attempting to design machinery. However, it is not magic, and by following a few simple guidelines components can be made to fit together as required.

Metric tolerances are usually bases on the ISO system of tolerances, deviations, limits and fits (see, for example, BS 4500: 1969). Each component has a tolerance band, which is the difference between the largest and smallest allowable component sizes. The ISO system assigns different *tolerance grades*, which control the size of the tolerance band. A *fundamental deviation* is then assigned, which is the distance from the nominal size to the start of the tolerance band. The fundamental deviations are given capital letters for holes, A to ZC, and lower-case letters for shafts, a to zc. The tolerance grades are given numbers from 1 to 16. A hole might be H7, and a shaft k6. Giving a shaft as ∅30 mm, h6, completely specifies its size for manufacture.

Suppose shaft needs to be put into a hole. The nominal size is chosen, and the hole is made to this size with a tolerance of H7, the shaft with a tolerance of g6. The shaft will fit into the hole with a good, close sliding fit. This will be the case whatever the nominal size, whether is it 30mm or 300mm. The tolerances H7 and g6 can be looked up in tables for the nominal size chosen.

Tables 14–2 and 14–3 give the tolerances for *hole-basis* and *shaft-basis* fits. For a hole-basis fit, the minimum size of the hole is always equal to the nominal diameter (i.e. the fundamental deviation is zero). For a shaft-basis fit, the maximum size of the shaft is always equal to the nominal diameter (again, the fundamental deviation is zero). In the ISO system, hole-basis fits are preferred, but shaft-basis fits may be needed if a piece of equipment comes with an h-deviation shaft on it.

14.1 Selected fits

Table 14–1 contains a list of recommended fits, with a brief description of the type of fit and the appropriate places that they may be used for Pelton turbines or micro-hydro generally.

Having chosen a fit, the tolerances for hole-basis fits can be read from Table 14–1, and for shaft-basis fits from Table 14–3.

Example

A phosphor bronze bush is to be used for the front bearing of a spear valve in a Pelton turbine. The bush is to be fitted into a mild steel housing, and will be held in place with a circlip. The outside diameter of the bush is a nominal 50mm.

Solution

There is no need to use shaft-basis fits, so use the hole-basis scheme. The bearing could be put in place with an interference press fit (H7-p6), but since it is retained by a circlip this is not necessary. The forces on it are not high, and there has to be a certain amount of clearance between the bearing and the shaft anyway, so any location fit will be good enough. Use a H7-k6 fit, which is a transition fit which will normally require pushing in.

From Table 14–2, for a 50mm nominal diameter, the tolerances for the hole, H7, are +0.025/–0.000, so the housing should be made 50.025mm to 50.000mm. The k6 bush is +0.018/–0.002, so should be made 50.018mm to 49.998mm. The loosest fit that can result is 0.027mm clearance. The tightest fit is 0.018mm interference.

Table 14–1:

Hole basis	*Shaft basis*	
Hole – Shaft	Shaft – Hole	
H9 – d10	h9 – D10	*Loose running fit.*
H9 – e9	h9 – E9	*Easy running fit.* Widely space bearings. Spear valve front bearing.
H8 – f7	h7 – F8	*Close running fit.* Easily produced running fit for bearings.
H7 – g6	h6 – G7	*Sliding fit.* Not intended to run free, but a small clearance fit which can be moved and turned, yet which locates accurately. Precision fit for spigots and pins. Taper locking sleeves on shafts.
H7 – h6	h6 – H7	*Clearance location fit.* Minimum clearance is zero, but usually gives a small clearance. For accurate location. A compromise between ease of assembly and accurate location. Flange spigots.
H7 – k6	h6 – K7	*Transition location fit.* A transition fit that usually gives a small clearance. Not always easy to assemble or disassemble. Hubs and pulleys on keyed shafts. Spear valve plain bearings into housing.
H7 – n6	h6 – N7	*Transition location fit.* May give clearance but is usually a tight fit.
H7 – p6	h6 – P7	*Interference location fit.* A true interference fit. Can be pressed into place, and should be possible to pull apart. Bushes and hubs that need to be well located. Collars on spear shaft for bearings.

Table 14–2: Selected ISO fits – hole basis. Reproduced with permission from BS Data Sheet 4500A, 1970

		Clearance										*Transition*				*Interference*			
		Loose running fit		*Easy running fit*		*Close running fit*		*Sliding fit*		*Clearance location fit*		*Transition location fit*		*Transition location fit*		*Interference location fit*			
Nominal size		*Tolerance*		*Tolerance*		*Tolerance*		*Tolerance*		*Tolerance*		*Tolerance*		*Tolerance*		*Tolerance*		*Nominal size*	
Over	*Incl.*	*H9*	*d10*	*H9*	*e9*	*H8*	*f7*	*H7*	*g6*	*H7*	*h6*	*H7*	*k6*	*H7*	*n6*	*H7*	*p6*	*Over*	*Incl.*
(mm)	(mm)	(mm)	(mm)	(mm)	(mm)	(mm)	(mm)	(mm)	(mm)	(mm)	(mm)	(mm)	(mm)	(mm)	(mm)	(mm)	(mm)	(mm)	(mm)
–	3	+0.025 0.000	–0.020 –0.060	+0.025 0.000	–0.014 –0.039	+0.014 0.000	–0.006 –0.016	+0.010 0.000	–0.002 –0.008	+0.010 0.000	0.000 –0.006	+0.010 0.000	+0.006 0.000	+0.010 0.000	+0.010 +0.004	+0.010 0.000	+0.012 +0.006	–	3
3	6	+0.030 0.000	–0.030 –0.078	+0.030 0.000	–0.020 –0.050	+0.018 0.000	–0.010 –0.022	+0.012 0.000	–0.004 –0.012	+0.012 0.000	0.000 –0.008	+0.012 0.000	+0.009 +0.001	+0.012 0.000	+0.016 +0.008	+0.012 0.000	+0.020 +0.012	3	6
6	10	+0.036 0.000	–0.040 –0.098	+0.036 0.000	–0.025 –0.061	+0.022 0.000	–0.013 –0.028	+0.015 0.000	–0.005 –0.014	+0.015 0.000	0.000 –0.009	+0.015 0.000	+0.010 +0.001	+0.015 0.000	+0.019 +0.010	+0.015 0.000	+0.024 +0.015	6	10
10	18	+0.043 0.000	–0.050 –0.120	+0.043 0.000	–0.032 –0.075	+0.027 0.000	–0.016 –0.034	+0.018 0.000	–0.006 –0.017	+0.018 0.000	0.000 –0.011	+0.018 0.000	+0.012 +0.001	+0.018 0.000	+0.023 +0.012	+0.018 0.000	+0.029 +0.018	10	18
18	30	+0.052 0.000	–0.065 –0.149	+0.052 0.000	–0.040 –0.092	+0.033 0.000	–0.020 –0.041	+0.021 0.000	–0.007 –0.020	+0.021 0.000	0.000 –0.013	+0.021 0.000	+0.015 +0.002	+0.021 0.000	+0.028 +0.015	+0.021 0.000	+0.035 +0.022	18	30
30	50	+0.062 0.000	–0.080 –0.180	+0.062 0.000	–0.050 –0.112	+0.039 0.000	–0.025 –0.050	+0.025 0.000	–0.009 –0.025	+0.025 0.000	0.000 –0.016	+0.025 0.000	+0.018 +0.002	+0.025 0.000	+0.033 +0.017	+0.025 0.000	+0.042 +0.026	30	50
50	80	+0.074 0.000	–0.100 –0.220	+0.074 0.000	–0.060 –0.134	+0.046 0.000	–0.030 –0.060	+0.030 0.000	–0.010 –0.029	+0.030 0.000	0.000 –0.019	+0.030 0.000	+0.021 +0.002	+0.030 0.000	+0.039 +0.020	+0.030 0.000	+0.051 +0.032	50	80
80	120	+0.087 0.000	–0.120 –0.260	+0.087 0.000	–0.072 –0.159	+0.054 0.000	–0.036 –0.071	+0.035 0.000	–0.012 –0.034	+0.035 0.000	0.000 –0.022	+0.035 0.000	+0.025 +0.003	+0.035 0.000	+0.045 +0.023	+0.035 0.000	+0.059 +0.037	80	120
120	180	+0.100 0.000	–0.145 –0.305	+0.100 0.000	–0.084 –0.185	+0.063 0.000	–0.043 –0.083	+0.040 0.000	–0.014 –0.039	+0.040 0.000	0.000 –0.025	+0.040 0.000	+0.028 +0.003	+0.040 0.000	+0.052 +0.027	+0.040 0.000	+0.068 +0.043	120	180
180	250	+0.115 0.000	–0.170 –0.355	+0.115 0.000	–0.100 –0.215	+0.072 0.000	–0.050 –0.096	+0.046 0.000	–0.015 –0.044	+0.046 0.000	0.000 –0.029	+0.046 0.000	+0.033 +0.004	+0.046 0.000	+0.060 +0.031	+0.046 0.000	+0.079 +0.050	180	250
250	315	+0.130 0.000	–0.190 –0.400	+0.130 0.000	–0.110 –0.240	+0.081 0.000	–0.056 –0.108	+0.052 0.000	–0.017 –0.049	+0.052 0.000	0.000 –0.032	+0.052 0.000	+0.036 +0.004	+0.052 0.000	+0.066 +0.034	+0.052 0.000	+0.088 +0.056	250	315
315	400	+0.140 0.000	–0.210 –0.440	+0.140 0.000	–0.125 –0.265	+0.089 0.000	–0.062 –0.119	+0.057 0.000	–0.018 –0.054	+0.057 0.000	0.000 –0.036	+0.057 0.000	+0.040 +0.004	+0.057 0.000	+0.073 +0.037	+0.057 0.000	+0.098 +0.062	315	400
400	500	+0.155 0.000	–0.230 –0.480	+0.155 0.000	–0.135 –0.290	+0.097 0.000	–0.068 –0.131	+0.063 0.000	–0.020 –0.060	+0.063 0.000	0.000 –0.040	+0.063 0.000	+0.045 +0.005	+0.063 0.000	+0.080 +0.040	+0.063 0.000	+0.108 +0.068	400	500

Table 14–3: Selected ISO fits – shaft basis. Reproduced with permission from BS Data Sheet 4500B, 1970

		Clearance										*Transition*				*Interference*			
		Loose running fit		*Easy running fit*		*Close running fit*		*Sliding fit*		*Clearance location fit*		*Transition location fit*		*Transition location fit*		*Interference location fit*			
Nominal size		*Tolerance*		*Tolerance*		*Tolerance*		*Tolerance*		*Tolerance*		*Tolerance*		*Tolerance*		*Tolerance*		*Nominal size*	
Over	*Incl.*	*h9*	*D10*	*h9*	*E9*	*h7*	*F8*	*h6*	*G7*	*h6*	*H7*	*h6*	*K7*	*h6*	*N7*	*h6*	*P7*	*Over*	*Incl.*
(mm)	(mm)	(mm)	(mm)	(mm)	(mm)	(mm)	(mm)	(mm)	(mm)	(mm)	(mm)	(mm)	(mm)	(mm)	(mm)	(mm)	(mm)	(mm)	(mm)
–	3	0.000 –0.025	+0.060 +0.020	0.000 –0.025	+0.039 +0.014	0.000 –0.010	+0.020 +0.006	0.000 –0.006	+0.012 +0.002	0.000 –0.006	+0.010 0.000	0.000 –0.006	0.000 –0.010	0.000 –0.006	–0.004 –0.014	0.000 –0.006	–0.006 –0.016	–	3
3	6	0.000 –0.030	+0.078 +0.030	0.000 –0.030	+0.050 +0.020	0.000 –0.012	+0.028 +0.010	0.000 –0.008	+0.016 +0.004	0.000 –0.008	+0.012 0.000	0.000 –0.008	+0.003 –0.009	0.000 –0.008	–0.004 –0.016	0.000 –0.008	–0.008 –0.020	3	6
6	10	0.000 –0.036	+0.098 +0.040	0.000 –0.036	+0.061 +0.025	0.000 –0.015	+0.035 +0.013	0.000 –0.009	+0.020 +0.005	0.000 –0.009	+0.015 0.000	0.000 –0.009	+0.005 –0.010	0.000 –0.009	–0.004 –0.019	0.000 –0.009	–0.009 –0.024	6	10
10	18	0.000 –0.043	+0.120 +0.050	0.000 –0.043	+0.075 +0.032	0.000 –0.018	+0.043 +0.016	0.000 –0.011	+0.024 +0.006	0.000 –0.011	+0.018 0.000	0.000 –0.011	+0.006 –0.012	0.000 –0.011	–0.005 –0.023	0.000 –0.011	–0.011 –0.029	10	18
18	30	0.000 –0.052	+0.149 +0.065	0.000 –0.052	+0.092 +0.040	0.000 –0.021	+0.053 +0.020	0.000 –0.013	+0.028 +0.007	0.000 –0.013	+0.021 0.000	0.000 –0.013	+0.006 –0.015	0.000 –0.013	–0.007 –0.028	0.000 –0.013	–0.014 –0.035	18	30
30	50	0.000 –0.062	+0.180 +0.080	0.000 –0.062	+0.112 +0.050	0.000 –0.025	+0.064 +0.025	0.000 –0.016	+0.034 +0.009	0.000 –0.016	+0.025 0.000	0.000 –0.016	+0.007 –0.018	0.000 –0.016	–0.008 –0.033	0.000 –0.016	–0.017 –0.042	30	50
50	80	0.000 –0.074	+0.220 +0.100	0.000 –0.074	+0.134 +0.060	0.000 –0.030	+0.076 +0.030	0.000 –0.019	+0.040 +0.010	0.000 –0.019	+0.030 0.000	0.000 –0.019	+0.009 –0.021	0.000 –0.019	–0.009 –0.039	0.000 –0.019	–0.021 –0.051	50	80
80	120	0.000 –0.087	+0.260 +0.120	0.000 –0.087	+0.159 +0.072	0.000 –0.035	+0.090 +0.036	0.000 –0.022	+0.047 +0.012	0.000 –0.022	+0.035 0.000	0.000 –0.022	+0.010 –0.025	0.000 –0.022	–0.010 –0.045	0.000 –0.022	–0.024 –0.059	80	120
120	180	0.000 –0.100	+0.305 +0.145	0.000 –0.100	+0.185 +0.085	0.000 –0.040	+0.106 +0.043	0.000 –0.025	+0.054 +0.014	0.000 –0.025	+0.040 0.000	0.000 –0.025	+0.012 –0.028	0.000 –0.025	–0.012 –0.052	0.000 –0.025	–0.028 –0.068	120	180
180	250	0.000 –0.115	+0.355 +0.170	0.000 –0.115	+0.215 +0.100	0.000 –0.046	+0.122 +0.050	0.000 –0.029	+0.061 +0.015	0.000 –0.029	+0.046 0.000	0.000 –0.029	+0.013 –0.033	0.000 –0.029	–0.014 –0.060	0.000 –0.029	–0.033 –0.079	180	250
250	315	0.000 –0.130	+0.400 +0.190	0.000 –0.130	+0.240 +0.110	0.000 –0.052	+0.137 +0.056	0.000 –0.032	+0.062 +0.017	0.000 –0.032	+0.052 0.000	0.000 –0.032	+0.016 –0.036	0.000 –0.032	–0.014 –0.066	0.000 –0.032	–0.036 –0.088	250	315
315	400	0.000 –0.140	+0.440 +0.210	0.000 –0.140	+0.265 +0.125	0.000 –0.057	+0.151 +0.062	0.000 –0.036	+0.075 +0.018	0.000 –0.036	+0.057 0.000	0.000 –0.036	+0.017 –0.040	0.000 –0.036	–0.016 –0.073	0.000 –0.036	–0.041 –0.098	315	400
400	500	0.000 –0.155	+0.480 +0.230	0.000 –0.155	+0.290 +0.135	0.000 –0.063	+0.165 +0.068	0.000 –0.040	+0.083 +0.020	0.000 –0.040	+0.063 0.000	0.000 –0.040	+0.018 –0.045	0.000 –0.040	–0.017 –0.080	0.000 –0.040	–0.045 –0.108	400	500

15

APPENDIX: SPECIFIC SPEEDS

15.1 Definition of specific speed

It is quite possible to select, design and install turbines without ever considering *specific speed.* It is a slightly awkward concept to grasp, and is mainly of use for large hydropower stations where it is important to know exactly which type of turbine is best. However, books on turbines often start with a chapter on specific speed, so for completeness, this brief summary is included.

For a given head, a turbine will have one speed at which it gives peak efficiency. This is the speed at which the water flows most smoothly around the turbine. If the head is increased, there will be a new optimum speed, higher than the first. At this speed, the water will be flowing faster than before, but the paths followed by the flow will be the same. It is possible to imagine the turbine run under a head of only one metre, at a speed which gives the same flow pattern.

When operating under a 1m head, the turbine would produce a certain amount of power. By scaling the size of the turbine, making it bigger or smaller as required, a turbine could be made that produced 1kW of power at 1m head. It would be operating at optimum speed, and the flow patterns would be the same as the original turbine at optimum speed. This optimum speed for a 1kW machine operating under a 1m head is the specific speed of that turbine design.

For historical reasons, it is more common to use one metric horsepower as the unit of power, instead of one kilowatt. The specific speed for a particular design of turbine is then defined as the speed, in rpm, at which the turbine, with its valves fully open, would give best efficiency with a 1m head and with its size scaled down to give an output of 1 metric horsepower. In terms of an equation, this can be expressed as:

$$N_{s\text{-metric}} = \frac{N.\sqrt{P}}{H_n^{\frac{5}{4}}} \qquad \textit{Eq. 15–1}$$

$N_{s\text{-metric}}$ – metric specific speed (rpm.mhp$^{0.75}$/m$^{1.25}$)
H_n – net head at turbine (m)
P – turbine output power (metric hp)
N – turbine speed (rpm)

This is the conventional form. Note carefully the strange units for power. One metric horsepower is defined as 75kgf.m/s = 735.5W. The following form of the equation is required to use kW for power:

$$N_{s\text{-metric}} = \frac{1{\cdot}166N\sqrt{P}}{H_n^{\frac{5}{4}}} \qquad \textit{Eq. 15–2}$$

H_n – (m)
P – (kW)
N – (rpm)

Note also that specific speed is not actually a speed, because its units are not simply [rpm]. This means that when the units are changed, the value of N_s changes. In the literature N_s is often quoted in imperial, or British, units. The formula is the same as Equation 15–1, but the units used are:

$$N_{s\text{-imperial}} = \frac{N.\sqrt{P}}{H_n^{\frac{5}{4}}} \qquad \textit{Eq. 15–3}$$

$N_{s\text{-imperial}}$ – Imperial specific speed (rpm.hp$^{0.75}$/ft$^{1.25}$)
H_n – (ft)
P – (hp)
N – (rpm)

Before comparing values of N_s, note which systems of units are being used. To convert between metric and imperial specific speeds:

$$N_{s\text{-metric}} = N_{s\text{-imperial}} \times 4.446 \qquad \textit{Eq. 15–4}$$

The reason for doing this exercise is to compare different turbine designs. By mathematically scaling down turbines to a standard head and power, the different speeds at which they operate can be seen.

For Pelton turbines, the specific speed is proportional to the bucket width divided by the PCD.

15.2 Choosing a turbine

Turbines with higher specific speeds run faster, as would be expected, and are usually lighter and smaller that lower N_s turbines. Thus it makes sense to use a turbine with as high a specific speed as possible. Advances in the design and materials of turbines usually increase the specific speed of that type of turbine.

However, specific speed is by no means the only criterion for selecting a turbine. The other major factor is head, and high-N_s turbines generally only work with low heads. Figure 15–1 shows the specific speed and head ranges for a variety of turbines, and it can be seen that Pelton turbines have low specific speeds but can work under very high heads. Propeller turbines have very high N_s but can only work under limited heads.

Other factors that limit the use of turbines are:

- Efficiency at partial flow: if the turbine is required to run for substantial periods at partial flow, its efficiency away from its optimum point is important.
- Strength and hydraulic stability: the turbine may not actually be able to function over the full range of powers and speeds deduced from its N_s and head limitations.
- Site-specific features: it may not be possible to excavate deep enough to fit a particular type of turbine, or the generator design may limit the allowable turbine speed.

In micro-hydro, where there are often only one or two types of turbine available, and then in only a few sizes, the choice of turbine is often quite straightforward, and can safely be made without reference to specific speed. But it is a good general principle to use faster-running turbines to keep the size and cost down.

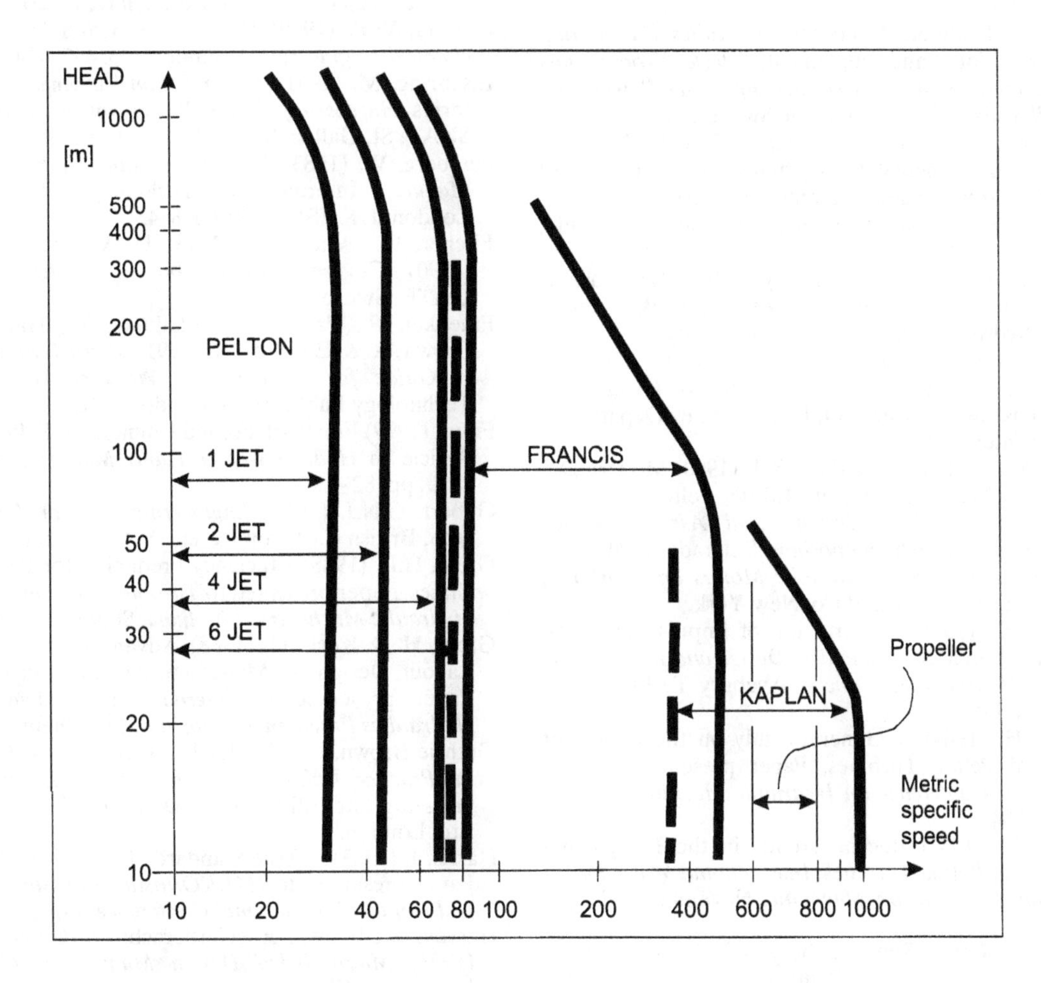

Fig. 15–1: Specific speed/head ranges for various types of turbines

REFERENCES

British and International Standards referred to in the text (e.g. BS 1400:1985 or ISO 1191 Part 1) can be obtained by post from BSI Customer Services, 389, Chiswick High Road, London W4 4AL, U.K.

ASCE, (1993) American Society of Civil Engineers, 'Steel Penstocks', *ASCE Manuals and Reports on Engineering Practice* No. 79, ASCE, New York. ISBN 0 87262 951 1.

Arter, A. & Meier, A. (1990) *Hydraulics Engineering Manual*, Vol. 2 and supplement, *Selected Nomograms and Diagrams* of series *Harnessing Water Power on a Small Scale*, SKAT, St. Gallen, Switzerland.

Bachmann, P., Schärer, Ch., Staubli, T. & Vullioud, G. (1990) Experimental Flow Studies on a 1-jet Model Pelton Turbine. Paper presented to *International Association for Hydraulic Research Symposium*, Belgrade, September 1990.

Baines, J. & Williams. A. (2000) Website and various downloadable documents: eee/ntu.ac.uk/research/microhydro. Nottingham Trent University.

Bier, P.J. (1986) *Welded Steel Penstocks, Water Resources Publication, Engineering Monograph No. 3*: Bureau of Reclamation, Denver, CO, United States Department of the Interior, USA.

Bjerke, M., Brekke, H. & Berg, K.I. (1990) Modern Design and Manufacturing of Multijet Pelton Turbines. Paper presented to *International Association for Hydraulic Research Symposium*, Belgrade, 1990.

Bodmer, G. R. (1900) *Hydraulic Motors and Turbines,* 3rd ed. D. van Nostrand Co. New York.

Borciani, G. (1991) Construction of Impulse Turbines, (ed. Zu-Yan Mei) *Mechanical Design and Manufacturing of Hydraulic Machinery*, Avebury Technical, Aldershot, UK.

Brekke, H. (1984) A General Study on the Design of Vertical Pelton Turbines, Paper presented to *Turboinstitut Conference on Hydraulic Machinery*, Ljubljana, 1984.

Brekke, H. (1987) Recent Trends in the Design and Layout of Pelton Turbines, *International Water Power & Dam Construction Magazine*, November 1987, pp. 13–17.

Brekke, H. (1994) State of the Art in Pelton Turbine Design, *Hydropower & Dams Magazine*, March 1994, pp. 21–28.

Buckner, B. (1989) Pelton Wheel Using Lost Wax Casting, Unpublished notes from course run by ITDG and the Agricultural Development Bank of Nepal; 23 July 1989.

Castrol (1984) *Ball and Roller Bearing Lubrication*, Booklet IND/60c/2/84, Industrial Lubricants Division, Burmah-Castrol Industrial Ltd, UK.

Chapuis, L. & Fröschl, K. (1998) Optimized Fabrication of Pelton Turbine Runners, *Hydropower & Dams Magazine*, Volume 5, Issue 2, pp. 30–32.

Cunningham, P. (2000) Energy Systems and Design, Canada, *Pico Hydro Magazine*, April 2000.

Dansie, J. & Bonifay, J. (1996) High Head, Record Output, Publicity leaflet on the Cleuson-Dixense scheme, Sulzer Hydro.

Daugherty, R.L. (1920) *Hydraulic Turbines*, McGraw-Hill.

Doble, W.A. (1899) Paper presented to the *ASME California Meeting*, September.

Douglas, J.F. (1986) *Solving Problems in Fluid Mechanics*, Volume 1, Longman. ISBN 0 582 28641 7.

Dowty (1978): *'O'-Ring Catalogue, Part 1 – Technical Data*; also *Guide for 'O'-Ring Selection*, Dowty Seals Limited, Ashchurch, Gloucestershire, GL20 8JS, UK.

Durand, W.F. (1939) 'The Pelton Water Wheel', *Mechanical Engineering*, Volume 6, pp. 447–454.

Eisenring, M. (1991) *Micro Pelton Turbines*, Vol. 9 of series *Harnessing Water Power on a Small Scale*, SKAT, St. Gallen, Switzerland. ISBN 3–908001-34 X.

Feinberg, W. (1983) *Lost-wax Casting, A Practitioner's Manual*, Intermediate Technology Publications, London, UK. ISBN 0903031884.

Fischer, G., Arter, A., Meier, U. & Chapallaz, J.M. (1990) *Governor Production Information*, SKAT-GATE, Switzerland.

Fraenkel, P., Paish, O., Bokalders, V., Harvey, A., Brown, A. & Edwards, R. (1991) *Micro-Hydro Power, A Guide for Development Workers*, Intermediate Technology Publications, London, UK.

Fravit (1999) Fravit srl, Forged Runners for Pelton Units, Article in *Hydropower & Dams Magazine*, Issue 3, 1999, pp. 82–83.

Gilbert, G.N.J. (1968) *Engineering Data on Grey Cast Iron*, British Cast Iron Research Association.

Grein, H.L. (1988) Efficiency Prediction for Pelton Machines. Paper presented to *Turboinstitut Conference on Hydraulic Machinery, Ljubljana*, September 13–15.

Grein, H. & Keck, H. (1988) Advanced Technology in Layout, Design and Manufacturing of Pelton Turbines. Paper presented to *International Association for Hydraulics Research Symposium*, Trondhein.

Guthrie Brown, J. (ed.) (1984) *Hydro-Electric Engineering Practice, Volume II, Mechanical and Electrical Engineering*, 2nd Edition, CBS, Delhi, and Blackie & Son, Ltd. London.

Harris, J. (1983) A True Standard of Gravity for Malawi. Paper presented to *IMEKO/Institute of Measurement and Control International Conference*, London.

Harvey, A., Brown, A., Hettiarachi, P. & Inversin, A. (1993) *Micro-Hydro Design Manual, A Guide to Small-Scale Water Power Schemes*, London, UK. ISBN 1 85339 103 4.

Hill, D. (1984) *A History of Engineering in Classical and Medieval Times*, Routledge. ISBN 0 415 15291 7.

Hodge, T. (1990) A Roman factory, *Scientific American*, November.

Hulscher, W. & Fraenkel, P. (1994) *The Power Guide, An International Catalogue of Small-Scale Energy Equipment*, 2nd Edition, Intermediate Technology Publications, London, UK.

Hurst, S. (1996) *Metal Casting: Appropriate Technology in the Small Foundry*, Intermediate Technology Publications, London, UK. ISBN 1 85339 197 2.

Inversin, A.R. (1986) *Micro-Hydropower Sourcebook, A Practical Guide to Design and Implementation in Developing Countries*, NRECA International Foundation, Washington DC.

Jordan Jnr., T.D. (1996) *A Handbook of Gravity-Flow Water Systems For Small Communities*, Intermediate Technology Publications, London, UK. ISBN 0 946688 50 8.

Kempe (1989) *Kempe's Engineers Handbook*, Edited by Carill Sharpe, 94th Edition, Morgan-Grampian.

Kisioka, E. & Osawa, K. (1972) Investigation into the Problem of Losses of the Pelton Wheel. Paper presented to *Second International JSME Symposium on Fluid Machinery and Fluidics*, Tokyo, September.

Lazzaro, B. & Rossi, G. (1990) Pelton Turbine: Model Investigation on Some Peculiar Geometrical Features. Paper presented to *International Association for Hydraulics Research Symposium*, Belgrade, Yugoslavia.

Macken, P.J. (1970) *Copper Alloy Casting Design, Reference Data and Foundry Design*, Copper Development Association, London.

Maguire, D. (1998) *Engineering Drawing from First Principles*, Arnold. ISBN 0 340 69198 0.

Miller, D.S. (1990) *Internal Flow Systems*, 2nd Edition, BHRA (Information Services).

Moritz, L.A. (1958) *Grain-Mills and Flour in Classical Antiquity*, Clarendon Press, Oxford. ISBN for set of volumes 0–405-12345–0.

Nechleba, M. (1957) *Hydraulic Turbines, Their Design and Equipment*, Artia, Prague.

Nixon, N. & Hill, J. (1987) *Water Power*, Quarry Bank Mill Publishing. ISBN 0 946414 06 8.

Oberg, E., Jones, F.D., Holbrook L., Horton, H.L., Henry, H. and Ryffel, H.H. (1992) Robert E. Green (ed.) *Machinery's Handbook*, 24th Edition, Industrial Press Inc., New York. ISBN 0 8311 2492 X.

Perera, L. (1995) 'Making Micro-hydro Turbines in Sri Lanka', *Hydronet* magazine, 1/95, p. 10.

Raabe, I.J. (1985) *Hydro Power, The Design, Use, Function of Hydromechanical, Hydraulic, and Electrical Equipment*, VDI-Verlag Gmbh, Dusseldorf.

Roark, W.C. (1989) *Roark's Formulas for Stress & Strain*, 6th International Edition, McGraw Hill. ISBN 0 07 100373 8.

Saubolle, B.R. & Bachmann, A. (1978) *Mini Technology I*, Sahayogi Press, Kathmandu.

Schneebeli, F., Baltis, E. & Keck, H. (1996) 'New Technology Earns Acceptance', Sulzer Technical Review, 1/96, Sulzer Hydro, Zürich.

Shaeffer, J. (1991): *Alternative Energy Sourcebook*, 9th Edition, Real Goods Trading Corporation, Ukiah, CA 95482, USA.

Shigley, J.F. (1986) *Mechanical Engineering Design*, First Metric Edition, McGraw-Hill, Singapore. ISBN 0 07 100292 8.

Shigley, J.E. & Mischke, C.R. (1989) *Mechanical Engineering Design*, International Fifth Edition, McGraw-Hill. ISBN 0 07 100607 9.

Siervo, F. de & Lugaresi, A. (1978) 'Modern Trends in Selecting and Designing Pelton Turbines', *Water Power & Dam Construction Magazine*, December, pp. 40–48.

Smith, N.A.F. (1984) The origins of water power: a problem of evidence and expectations, *Transactions of the Newcomen Society*, Volume 55, 1983–4.

SKF (1994) SKF General Catalogue 4000/IV E.

SKF (1991) SKF Bearing Maintenance Handbook 4100E.

Smith, N. (1994) *Motors as Generators for Micro-Hydro Power*, Intermediate Technology Publications, London. ISBN 1 85339 286 3.

Vivier, L. (1966) *Turbines Hydrauliques et Leur Régulation*, Editions Albin Michel, Paris, France.

Waltham, M. (1991) The Manufacture of Pelton Wheel Runners, unpublished booklet for Agricultural Development Bank of Nepal/ITDG, July.

Waltham, M. (1994) Mechanical Guidelines for Micro-Hydro-Electric Installations, unpublished booklet, ITDG Nepal, January.

17
GLOSSARY

γ, gamma
bucket exit angle

ζ, zeta
efficiency of flow in bucket

bore
internal diameter of a pipe

cavitation
air/vapour bubbles which can appear in low pressure fluid, often causing damage as they collapse

direct-coupled turbine
running the turbine at the same speed as the driven machinery

head
pressure, measured as height of column of water

induction generator
generator where magnetic field in rotor induced by rotating currents in stator; speed may vary even at constant frequency

manifold
pipework connecting penstock to the turbine

monobloc runners
runners made as a single-piece casting

surge pressure
pressure rise in a penstock due to a decrease in the flow

synchronous generator
generator rotor has coils or permanent magnets to create a magnetic field; speed constant for a given frequency

tangential turbine
turbine in which a jet or jets of water act tangentially on a runner; a Pelton is one form of a tangential turbine

turbine setting/set
the vertical position of the turbine, used for large installations where the powerhouse is underground; a deep set turbine is lower, giving higher head, but giving a lower drop for the tailrace

water hammer
extreme surge pressure due to sudden blocking of flow

18
INDEX